Hormones and Animal
Social Behavior

MONOGRAPHS IN BEHAVIOR AND ECOLOGY

Edited by John R. Krebs and
Tim Clutton-Brock

Hormones and Animal Social Behavior

ELIZABETH
ADKINS-REGAN

Princeton University Press
Princeton and Oxford

Published by Princeton University Press,
41 William Street, Princeton, New Jersey 08540
In the United Kingdom: Princeton University Press,
3 Market Place, Woodstock, Oxfordshire OX20 1SY

LIBRARY OF CONGRESS CATALOGING-IN-PUBLICATION DATA

Adkins-Regan, Elizabeth, 1945–
Hormones and animal social behavior /
Elizabeth Adkins-Regan
p.   cm.   — (Monographs in behavior and ecology)
Includes bibliographical references (p.).
ISBN 0-691-09246-X (cloth : alk. paper) —
ISBN 0-691-09247-8 (pbk. : alk. paper)
1. Social behavior in animals. 2. Animal behavior—
Endocrine aspects. I. Title. II. Series
QL775.A4 2005
591.56—dc22   2004054935

British Library Cataloging-in-Publication Data is available

This book has been composed in Times Roman and
Univers Light 45

Printed on acid-free paper. ∞

pup.princeton.edu

Printed in the United States of America

10 9 8 7 6 5 4 3 2 1

Contents

Illustrations and Tables

Figures

Tables

Preface

These are exciting times for the science of animal social behavior. Researchers have succeeded in eavesdropping on the social lives of marked individuals in free-living populations of an impressive array of species. New methods for determining genetic relationships and parentage have revealed previously hidden forms of social organization. Sophisticated yet testable theories and models have accounted for some of the unity and the diversity that have been discovered. Important advances have been made in understanding why the social behavior of wild species is adaptive for the individuals engaging in it, however odd their practices may seem to us. With the successful sequencing of several animal genomes and the development of new methods for studying the involvement of multiple genes in complex processes, scientists are on the threshold of bridging the gap between genes and adaptive ecologically relevant behavior.

This book about the role of steroid hormones in the social life of animals is motivated by the belief that further progress in understanding social behavior requires considering these hormones and their actions. Why? One simple reason is that no behavior is completely understood without knowledge of its underlying physiological mechanisms. This point was made long ago by the ethologist Niko Tinbergen (1963), who provided a simple and clear agenda for a science of animal behavior in the form of four basic aims or problems that must be solved: causation, survival value, ontogeny, and evolution. Today these are often phrased as proximate mechanisms, function or adaptive value, development, and phylogeny. As a major category of physiological mechanisms for the development and expression of social behavior, hormones are an essential ingredient for two of the four aims.

A second reason is the potential of hormonal studies to contribute to integration among the four Tinbergian categories. Such integration is vitally important because none of the four problems can be solved in isolation. Some social behaviors with hormonal causes or consequences are closely linked to fitness. Hormones give the behavior some of the properties that make it adaptive. Some difficult questions (for example, about the costs of social behavior) might be easier to answer if hormonal and other mechanisms were more deeply understood. The mechanistic assumptions and predictions of theories of behavioral evolution need to be clarified and tested. As new behavioral strategies are discovered, asking how the animals are able to produce them may lead to the discovery of new hormonal and neurochemical mechanisms. "Why" questions need to be asked about the mechanisms themselves, in a phylogenetic context that respects their history. Hormonal mechanisms include gene products that affect brain function. They are one of the links between behavior, brain, and the genome that allow scientists to open Darwin's black box to understand how genetic change results in the developmental change in the nervous system that is manifest as behavioral evolution. The four aims inform each other, inspiring new and different hypotheses and constraining the set of possible explanations. This kind of integration requires a dialog between people studying mechanisms and development and people studying function and phylogeny.

The goal of the book is to encourage this dialog by providing an accessible bridge between animal behavior and behavioral ecology, on one shore, and behavioral endocrinology and neuroendocrinology, on the other shore, allowing two-way traffic to flow more easily, inspiring more people to visit the other side, and giving armchair travelers a look at an exciting scientific landscape. There has long been such a bridge but it has not always seen enough use. It needs some repairs based on evolutionary theory grounded in individual selection and a gene-centered perspective, on current phylogenetic concepts and methods, on knowledge of the strategies and tactics in which animals are engaged in the real world, on recent discoveries about how and where hormones act, and on a modern understanding of how developmental change leads to evolutionary change. This book synthesizes the current state of knowledge to provide those repairs.

The book is aimed especially at graduate students and researchers at any level interested in ecologically relevant social behavior who would like to know about hormones but need a map designed for their interests. Hormones and behavior is a century-old multidisciplinary enterprise. The literature is voluminous and scattered over a huge array of journals and books. Some of the key concepts lie in publications that cannot be accessed over a computer and some of the older ideas are still essential to the field. Others have been eliminated or replaced but the word hasn't always reached other fields. New discoveries about hormone action with big implications for animal behavior appear in neuroscience or endocrinology journals, not animal behavior journals. There are two excellent undergraduate textbooks on hormones and behavior (Nelson 2005; Becker et al. 2002) and two excellent textbooks of comparative endocrinology (Norris 1996; Bentley 1998). There is an authoritative multivolume series on behavioral neuroendocrinology providing exhaustive in-depth coverage (Pfaff et al. 2002). What is missing is something in between—a single volume emphasizing concepts, theories, and hypothesis testing focused on the naturally occurring social behavior of animals whenever possible and grounded in evolutionary thinking. Such a volume should find a welcome home not only in a hormones and behavior course, but also as a textbook supplement in courses in animal behavior, behavioral ecology, comparative psychology, comparative endocrinology, and comparative reproductive biology. In keeping with the intended audience, there will be discussion of what isn't known as well as what is known. Promising but untested ideas will appear and questions will continually be raised that don't yet have answers. Critical thinking will be encouraged along with constructive critiques of current ideas. The emphasis will be on experimental and comparative approaches to getting answers.

An endocrinologist might find it odd that the behavior is limited to social behavior but an animal behaviorist won't. Social behavior has properties that make it a special domain of behavioral evolution. For example, whether the behavior is fitness enhancing or not depends on what other individuals are doing, and success may require coordination between individuals with different genes and interests. The hormones will be limited mainly, but not entirely, to steroids, because these gene transcription regulators have close links to social behavior. Understanding how they work will require the occasional foray into neuropeptides and into peptide and protein hormones. The animals will be mainly but not entirely vertebrates. The book is organized so that it moves to progressively higher levels of biological organization both spatially and temporally. It begins with internal hormonal mechanisms, then moves to the specific behaviors of individual animals, to relationships between individuals, to the development of individuals, to evolutionary change over relatively short timescales, to life histories, and finally to broad-scale evolution (vertebrate phylogeny).

To do this in a single volume requires an idiosyncratic selection of the literature. Much excellent work by significant contributors to the field is not included. Citations are provided for all of the material directly related to hormones and behavior. Fewer citations are provided for basic background material about endocrinology, behavior, and evolution, and some familiarity with contemporary animal behavior and behavioral ecology is assumed. For avian brain structures, the nomenclature of Reiner et al. (2004) is used.

I owe an enormous debt of gratitude to the National Science Foundation, which has supported me in one way or another throughout my research life, beginning in my freshman year at the University of Maryland. NSF is a national treasure, steadfast in its adherence to scientific and academic values in the face of criticism from ignorant quarters. The research from my lab referred to in this book was supported by NSF grants BNS76-24308, BNS-8204462, BNS-8412083, INT-8603645, BNS-8809441, IBN-9514088, IBN-0104907, and IBN-0130986, with additional support from NIMH grants MH21435, MH39295-01, and MH46457-01, and from Hatch NY(C)-191406.

I would also like to thank Norman Adler, who inspired me to study hormones and behavior and encouraged my interest in a comparative approach, Alan Leshner and Doug Candland, who provided me with a welcoming academic home and lab space at a critical career stage, and my Cornell University colleagues in the many departments and disciplines related to hormones and behavior, for their support and contributions to my continuing education.

I am deeply grateful to all of my co-researchers whose insights and efforts contributed to my published papers cited in this book. These include collaborators (Timothy DeVoogd, Pierre Orgeur, Mary Ann Ottinger, Michael Romero, J.-P. Signoret, Juli Wade), graduate students (James Goodson, Christine Lauay, Viveka Mansukhani, Paul Mason, Kevin McGraw, Bruce Nock, Kevin Pilz, Luke Remage-Healey, Emilie Rissman, Michael Ruscio, Cynthia Seiwert, Richmond Thompson, James Watson), undergraduate students (Michelle Friedman, Jeanne Park, Ellen Pniewski, Larry Schlesinger, Sharlene Yang), and research assistants (Mary Ascenzi, Véronique Connolly, Gretchen Gilbert, Emiko MacKillop, Andrew Mong, Laura Moore, Maggie Smith, Joyce Swartzman, Timothy Van Deusen). Thanks are also due to the participants in my weekly lab meetings for penetrating comments about the recent literature during the period when I was writing this book: Stephanie Correa, Christine Lauay, Jon Lee, Jill Mateo, Kevin McGraw, Kevin Pilz, Luke Remage-Healey, Dustin Rubenstein, Michael Ruscio, Maria Sandell, and Michelle Tomaszycki. Cary Leung, Emiko MacKillop, and Timothy Van Deusen ensured that the lab ran smoothly while I disappeared to write.

Ellen Ketterson has provided important support for my goal of integrating behavioral endocrinology and behavioral ecology. The book benefitted mightily from the comments of those who generously reviewed the manuscript, including David Crews, Wendy Hill, Kevin Pilz, Luke Remage-Healey, Steven Schoech, and Katherine Wynne-Edwards. A number of colleagues around the country responded quickly and positively to my request for animal images, including John Buntin, Robert Denver, John Godwin, Karen Hollis, Robert Johnston, Ellen Ketterson, David Lank, Catherine Marler, Mary Mendonça, Sally Mendoza, Frank Moore, Michael Moore, and Steven Schoech.

Sam Elworthy and John Krebs provided critical encouragement for the writing of this book, and Rosemary Knapp and David Winkler's enthusiasm for the project spurred me on to the finish line. The production team and staff at Princeton University Press (Dimitri Karetnikov, Sara Lerner, Alycia Somers, Sarah Stengle, and Hanne Winarsky), the copyeditor (Virginia Dunn), and the indexer (Dorothy Hoffman) shepherded the book through the publication stages with skill and patience.

I would like to thank Philip Lewis, former Dean of the College of Arts and Sciences at Cornell University, for allowing the sabbatic leave in the middle of my term as a department chair that enabled completion of this book, and (together with Katherine Porter) for making available a lovely Paris apartment, the ideal writing environment.

Finally, a thousand thanks to Dennis Regan, for making sure that I played as well as worked.

Hormones and Animal
Social Behavior

1 Hormonal Mechanisms

Social behavior has been linked to many hormones. The best established connections are to steroid hormones, pituitary prolactin, and a few peptide neurohormones (table 1.1). What are these hormones that have been shown to be important for social behavior? How do they work? How are they regulated by the physical environment and the social world? This chapter presents some essential facts and concepts that will inform the rest of the book. It also shares exciting discoveries and new ideas with major implications for behavior.

Hormonal mechanisms, like everything else about organisms, are a fascinating mixture of old and new. Modern biology has revealed astounding similarity in genomes and biochemical signaling pathways across long-separated lineages of the animal kingdom differing in ecology, morphology, and behavior. Similarly, some hormones and their actions are highly conserved and taxonomically widespread. Such mechanisms are very old and could reflect the limited number of workable solutions to basic problems of living given the physicochemical realities of the earth. Along with these are hormonal mechanisms that are more recent, more derived, more taxon specific, and serve different functions in different kinds of animals. This chapter highlights the relatively conserved hormonal mechanisms, occasionally hinting at some of the diversity that is seen. Subsequent chapters will draw on both types. Chapter 7 will return to a focus on the hormonal mechanisms themselves to look at broad-scale evolutionary patterns and differences between major lineages.

Like most aspects of the living world, hormonal mechanisms resist easy categorization. Even the concept of a "hormone" has undergone significant change. Hormones used to be defined as the secretions of the ductless (endocrine) glands. Hormones were thought of as relatively long-distance internal messengers related to regulatory functions such as growth, metabolism, or reproduction. They were contrasted, for example, with neurotransmitters such as dopamine, serotonin (5-HT), or acetylcholine, chemicals for rapid communication across the tiny gaps between neurons and between neurons and muscles. A prototypical vertebrate example would be testosterone, which is carried by the circulation from the testes to target tissues such as the rooster's comb and the neurons and muscles that produce crowing. The discovery that hormones can be made in the brain itself upset this neat dichotomy. Some vertebrate neural tissue can release neurohormones into the circulation and most insect hormones are neurohormones (Nijhout 1994). If hormones define an endocrine organ, then the vertebrate brain is the biggest endocrine organ in the body! The boundaries between hormones, neurohormones, neuromodulators,

TABLE 1.1
Steroids with known links to social behavior, along with neuromodulators and neurohormones
that are important for understanding hormone action in relationship to social behavior

Chemical	Acronym	Type of chemical	Comments
Estradiol (1,3,5[10]-estratrien-3,17β-diol)	E_2	Steroid	An estrogen
Progesterone (4-pregnen-3,20-dione)	P	Steroid	A progestogen
Testosterone (4-androsten-17β-ol-3-one)	T	Steroid	An androgen
Androstenedione (4-androsten-3,17-dione)	A4 (AE)	Steroid	An androgen
Dihydrotestosterone (5α-androstan-17β-ol-3-one)	DHT	Steroid	An androgen
11-Ketotestosterone (4-androsten-17β-ol-3,11-dione)	11-ketoT	Steroid	An androgen; teleost fishes only
Corticosterone (4-pregnen-11β,21-diol-3,20-dione)	CORT (B)	Steroid	A glucocorticoid
Cortisol (4-pregnen-11β,17,21-triol-3,20-dione)		Steroid	A glucocorticoid
20-Hydroxyecdysone		Ecdysteroid	Arthropods
Gonadotropin-releasing hormone	GnRH	Decapeptide	See also table 7.2
Corticotropin-releasing hormone	CRH (CRF)	Peptide	
Adrenocorticotrophic hormone	ACTH	Peptide	
Prolactin	PRL	Protein	
Vasopressin (arginine vasopressin)	AVP (VP)	Nonapeptide	See also table 7.1
Vasotocin (arginine vasotocin)	AVT (VT)	Nonapeptide	See also table 7.1
Oxytocin	OT (OXY)	Nonapeptide	See also table 7.1
Melatonin	MT (MEL)	Indoleamine	
Thyroxine (3,5,3′,5′-tetraiodothyronine)	T_4		
Triiodothyronine (3,5,3′-triiodothyronine)	T_3		
Prostaglandins	PGs	Eicosanoids	
Juvenile hormones	JHs	Terpenoids	Arthropods

Source. Sources include Nijhout (1994) and Norris (1996).

and neurotransmitters are now rather fuzzy in both vertebrates and invertebrates, and the requirement of long-distance communication to be a hormone no longer seems necessary or desirable. Both nervous and endocrine systems originally evolved from a system of cell–cell chemical messengers, which is why chemically they overlap so much and why table 1.1 includes a number of substances not considered to be "classic" hormones.

As with other signaling systems, we can ask about the source of the signal (where the hormone is produced), the nature of the signal (the hormone), and the receiver for the signal (the target organ, tissue, or cell), which for behavior often means the nervous system and neurons. The brain is both a source and a target. Targets can even include other individuals in the case of those pheromones derived from hormones.

Why Does Social Behavior Need Hormonal Regulation?

Before delving into hormonal nuts and bolts, it is only fair to first provide some clues as to how hormones help animals solve real-world problems and achieve fitness. In general, hormones are coordinators: of reproduction, of suites of physiological and behavioral components, of different parts of the brain, of brain with body. On both short- and long-term (life history) scales, they coordinate behavioral and physiological sequences over time, establish the duration of events and sequences by regulating onset and offset, and modify the nervous system appropriately (Truman 1994). They help adjust behavior to circumstances and contexts: physical, social, and developmental.

Mating behavior illustrates well the merits of this functional approach. Mating has significant costs, such as increased risk of predation and communicable disease. The obvious benefit of mating is the achievement of reproductive success, but this benefit is possible only if there are mature gametes (eggs and sperm) ready for fertilization. Gamete maturation is hormonally regulated, and so one reason that mating behavior is hormonally regulated is to ensure that the behavior is coordinated with the presence of fertilizable gametes. Vertebrates achieve this by having the gonads produce both gametes and hormones, with the same hormones regulating gametogenesis and mating behavior. Insects achieve this by having the hormones from elsewhere that stimulate the gonads also stimulate sexual behavior. One of the major achievements of animal behavior research has been to show how other social behaviors such as territoriality and dominance also serve reproductive interests. To the extent that they should occur only when fertility is possible (their occurrence at other times would be too costly), they also need hormonal regulation.

Social behavior is often age related, and hormones are a mechanism that can ramp up the behavior at the appropriate age. The onset of adult reproductive behavior at puberty in mammals is a widely known example because of our own personal familiarity with the phenomenon. In animals with indeterminate growth, size rather than age may be the trigger for the onset of adult social behavior, and hormones can be a messenger between size and behavior. In adulthood the behavior may need to ramp up and down seasonally, again requiring some kind of hormonal regulation.

Other reasons why hormonal regulation is needed are closely tied to the social nature of social behavior. Responding in appropriate ways to other individuals (to competitors or potential reproductive partners) is crucial for reproductive success. In many (not all) species, for mating to increase fitness, both the male and female have to be fertile at the same time. For a pair of birds to raise young together, both have to be in the parental mood at the same time. Hormones are a major mechanism ensuring coordination between different individuals. Another major achievement of animal behavior research has been to show how such "coordination" is often a mixture of conflict and cooperation. Hormones are likely to be responsive to this give and take.

Finally (to conclude this preview of coming attractions), social behavior is often different in quantity or quality between or even within the sexes. Animal behaviorists have devoted a great deal of attention to understanding why such differences have evolved. There need to be hormonal mechanisms to produce the behavioral phenotypes during development and cause the behavior to be expressed more in one sex or within-sex type in adulthood.

All these are compelling reasons for hormonal regulation of social behavior. Where none of these conditions apply, we would not expect the behavior to be hormonally regulated. If all individuals engage in the behavior regardless of age, sex, breeding condition, or social context, it wouldn't make sense to go looking for hormonal regulation.

The word "regulation" is appearing here for a reason. Unlike light switches turning light bulbs on and off, hormones don't make behavior happen in a deterministic manner. It would be a very poorly adapted male who would begin a mating sequence just because he had a lot of testosterone even if no female was present and he was surrounded by predators. Rather, hormones are one of several factors that go into the nervous system's decision. They may change the thresholds for other factors that enter into the decision (for example, thresholds for responding to stimuli from another animal) but are not normally the sole triggering agent. Thus, words like "regulate" or "prime" are often used, and "permit behavior" is usually preferred to "cause behavior." This important concept will be a recurring theme.

Steroids

The use of steroids for internal signaling is probably universal in metazoans (Wang et al. 2001). Like other chemical regulators, steroids are normally pres-

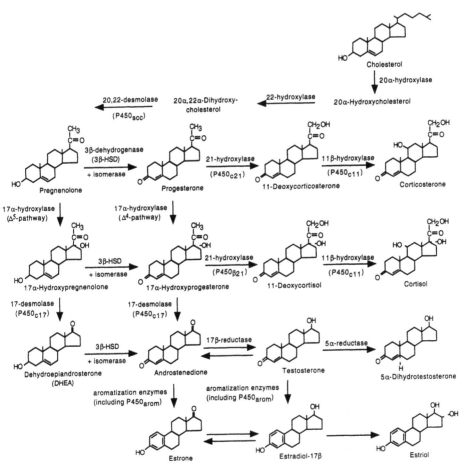

FIGURE 1.1. Pathways for the synthesis of sex steroids and glucocorticoids in gonadal and adrenal (interrenal) tissue. Note that most arrows are unidirectional (the reverse transformation does not occur) while others are bidirectional. The P450 enzymes are members of the cytochrome P oxidase family and have alternative names such as CYP19 (for P450$_{arom}$) or CYP11A (for P450$_{scc}$, where "scc" stands for "side chain cleavage"). Redrawn with minor modifications from Bentley (1998). © 1998 Cambridge University Press.

ent in minuscule amounts. Their mechanisms of action involve cascades of events that amplify these tiny signals. This property helps explain how a very dilute concentration of circulating steroid can dramatically change the behavior of a very large animal, as when estradiol induces estrus behavior in a lion.

Estradiol is a steroid. What are steroids? How and where are they made? In today's world the word "steroid" triggers images of bulked up super-athletes. Steroids are a much larger category than this, for a steroid is any molecule (and there are many) with the distinctive structure shown in fig. 1.1. Steroids are small molecules; molecular weights are usually in the 250–350 range. The structure they all have in common is the four rings in the arrangement shown: three hexagons and one pentagon. Different steroids vary in what is attached to the carbons of these rings and whether the rings contain double bonds. Rings with three double bonds, like the A rings of the estrogens, are said to be aromatized. They get this way through an enzymatic process (aromatization, see next section) that is very important for behavior and will be referred to repeatedly. Names like "testosterone" or "corticosterone" then refer to specific steroid molecules, to the different variations on the basic structural theme. These specific steroids are identical across species. For example, testosterone, a potent androgen and the predominant circulating androgen in many vertebrates, is the exact same molecule in anchovies, axolotls, adders, antbirds, aardvarks, and apes. This makes steroids very different from peptide neurohormones and amino-acid-based hormones such as prolactin, which vary structurally among species.

These steroids are then lumped into categories according to chemical structure, bodily source, or physiological functions. One category is the gonadal steroids, also called sex steroids, those manufactured by vertebrate testes and ovaries. These are predominantly androgens ("masculinizing" steroids), estrogens ("feminizing" steroids), and progestogens. Another is the adrenal steroids, those manufactured by the adrenal cortex or interrenal tissue (in fishes and amphibians), including the glucocorticoids and the mineralocorticoids, which together constitute the corticosteroids. Another category is the ecdysteroids, which function in molting or ecdysis and are limited to insects. Vertebrate sex steroids have been found in invertebrates as well, but their functions, if any, are still unclear.

As is usually the case when humans try to carve up nature, the so-called gonadal sex steroids keep slipping the boundaries of their name in ways that will become very important for understanding their relationship to social behavior. Both androgens and estrogens are produced by cells located outside the gonads, for example, by the adrenal cortex (interrenal tissue), by fat, and by the brain. This means that removal of the gonads does not always eliminate or even reduce circulating androgens or estrogens.

The association of androgens with masculine traits and estrogens with feminine traits is also a poor fit with nature's ways. No sex-specific steroids (or other hormones, for that matter) have been discovered in vertebrates. Furthermore, androgens can have feminizing actions and estrogens can have masculinizing actions. It is common to find sex differences in circulating steroid levels (including corticosteroids), especially at times when animals are breeding, but this is not always the case. Male pigs and horses have as much estradiol in the blood as estrous females. Some female fish, lizards, and amphibians produce as much or more testosterone than males do (Borg 1994; Staub and DeBeer 1997). Furthermore, even with "conventional" sex differences (higher androgens in males, higher estrogens in females) females often have more androgen than estrogen in absolute terms (numbers of molecules per unit volume).

Steroid Synthesis and Metabolism

Vertebrate gonadal and adrenocortical (interrenal) tissues do not produce the exact same steroids, although there is some overlap. Gonads usually release more androgens, estrogens, and progestogens, whereas adrenals usually release more glucocorticoids and mineralocorticoids. In addition, not all species produce the exact same steroids. For example, in primates such as humans, and in teleost fish such as salmon, cortisol is the predominant glucocorticoid, whereas birds and most rodents produce corticosterone. In many nonprimate mammals the weakly androgenic steroid dehydroepiandrosterone (DHEA) comes mainly from the gonads but in primates it comes mainly from the adrenals. What is it about the manufacture of steroids that accounts for these species and tissue variations? Steroids are produced in steps by a series of synthesizing enzymes. Which steroids are found where is a function of which enzymes are produced there. This in turn is a function of whether the animal has the gene for the enzyme, in which tissues the gene is being transcribed into mRNA for the enzyme, and in which tissues the mRNA is being translated.

The synthesizing pathway for the production of androgens and estrogens by vertebrate gonads and for the production of glucocorticoids by adrenal tissue has been well established (fig. 1.1). Both pathways begin with the abundant sterol cholesterol. Steroid synthesizing tissue contains a store (depot) of cholesterol obtained from plasma lipoproteins that is mobilized for steroid synthesis by steroidogenic acute regulatory protein (StAR). The enzyme that converts cholesterol to pregnenolone and thus is essential for all steroid production is cytochrome P450scc. Its name indicates that it is a member of the large and very ancient cytochrome P450 family of enzymes and that this particular member performs the side-chain cleavage (scc) step in steroidogenesis. Steroid secretion and concentration in circulation is more a matter of synthesis than storage (little is stored), so synthesis needs to happen fast, as does putting the brake on it. Conversion of cholesterol to pregnenolone by P450scc is the rate-limiting step, with StAR rapidly mobilizing cholesterol for synthesis.

Different enzymes then turn pregnenolone into progesterone and other progestogens and turn progestogens into either corticosteroids or androgens. The aromatizable androgens such as testosterone and androstenedione are the substrate for the production of estrogens (which have a fully aromatized A ring), including estradiol. This feat is carried out by another member of the cytochrome P450 family, P450arom, also called aromatase or estrogen synthase. Regulation of the gene for aromatase determines how much aromatizable androgen is turned into estrogen, that is, sets the ratio of those androgens to estrogens (Lephart 1997). Aromatization is a one-way street; estrogens cannot be turned back into androgens. The synthesis of estrogens from androgens will be important for understanding steroids and social behavior, and is one of several facts that flies in the face of the common assumption that androgens are strictly male hormones and estrogens are strictly female hormones. Not all androgens can be aromatized; 5α-dihydrotestosterone (5α-DHT) cannot, nor can the teleost fish androgen 11-ketotestosterone.

The gonads and adrenals are not the only parts of the body that express these enzymes and can carry out these conversions. The liver, fat tissue, and the mammalian placenta can carry out one or more of the steps. Steroid targets, including the brain, also have this capability, as when a rooster's comb converts testosterone to the more potent 5α-DHT or the hypothalamus of the brain, a region rich in aromatase, converts testosterone to estradiol. This phenomenon in which a steroid precursor or prohormone made elsewhere is turned at the target into a locally more potent steroid is important for understanding steroid action.

Steroids are small lipophilic molecules that can easily move between cells and tissues. They can easily pass from the circulation to the brain or from the brain to the circulation (Schlinger and Arnold 1991). The yolky (lipid) eggs of nonmammalian vertebrates invariably contain steroids, often in the same amounts as in the maternal circulation (Dickhoff et al. 1990; Schwabl 1993; Adkins-Regan et al. 1995).

Regardless of where the steroid originates and is transported, the eventual fate of most circulating steroid molecules is metabolism to other forms by the liver, which then pass out of the body in the urine and other excretory products. Endocrinologists describe these forms as "weak" or "inactive" on the basis of their minimal biological activity in the producer, but these molecules can be highly informative to conspecifics (chapter 2).

More recently, it has been proposed that the brain not only can convert steroids made elsewhere, but also can manufacture steroids such as estrogens de novo ("from scratch") from cholesterol (Baulieu and Robel 1990; Tsutsui and Schlinger 2001; Mellon and Griffin 2002). Evidently, neurons or glia (the "other" brain cells) express the necessary steroid-synthesizing enzymes. These "neurosteroids" are thought to differ from other steroids found in the brain (those made from gonadally or adrenally produced precursors) in three important ways. First, they can be made in parts of the brain that are not conventional targets for circulating steroids, for example, in the mammalian isocortex ("neocortex") or avian caudal nidopallium. Second, it is suspected that parts of the steroidogenic pathways might differ from those in fig. 1.1. Estrogens might be made from nonandrogenic precursors, and the different steps in the synthesis sequence might take place in different cell groups rather than all in the same cells. Third, these neurosteroids are thought to act locally, possibly on the same cells where the final synthesis step occurs. These exciting discoveries have seldom been directly linked to ecologically relevant behavior, but they are inspiring new hypotheses relating steroids to brain function that promise to make the connection soon.

Steroid Measurement and Dynamics

Many experimental studies of hormones and social behavior measure steroid levels. As in any kind of science, it helps to know something about the assumptions and limitations of the methods used to do this. What are the experimenters up against in trying to figure out what the numbers mean? What should an informed critical reader know?

Steroid levels that are reported are most often amounts of the steroid in plasma separated from whole blood drawn from the systemic circulation, and are most often expressed as weight per unit volume. For example, plasma testosterone in male amphibians ranges from less than 1 to over 200 nanograms per milliliter (ng/mL) and estradiol in female lizards ranges from less than 1 to over 200 picograms per milliliter (pg/mL) (Norris 1996).

Getting a blood sample from a wild animal presents special problems. Imagine trying to study hormonal responses to intermale competition in elephants. Or it might be important on research or conservation grounds to avoid disturbing the animal. For these reasons, there is increasing interest in being able to measure steroids noninvasively. Fecal or (in humans) saliva samples are viable alternatives provided they are stored properly and carefully validated (as they must be) (Goymann et al. 2002; Lynch et al. 2003; Buchanan and Goldsmith 2004). In some digestive systems the hormonal signal has been integrated over a substantial time period, which is either an advantage or a disadvantage depending on the question asked.

The most commonly used method for measuring steroids in samples is some type of radioimmunoassay (RIA), a competitive protein-binding assay in which radioactively labeled steroid added to the sample competes with the animal's own steroid for an antibody (the binding protein). The four criteria for a valid RIA are well established (Midgley et al. 1969). They include (1) specificity (is it measuring what you are trying to measure instead of picking up some other steroid or blood constituent?); (2) accuracy (on average, does it give values that correspond to the actual amount in standard samples?); (3) precision (how variable are the numbers that the assay yields? if the same sample is assayed multiple times, do the numbers agree?); and (4) sensitivity (can it detect the steroid at the levels present in the samples?). It is always reassuring to see these criteria addressed in the methods section of a paper, especially when an assay is being used in a new species where its performance has not already been established. For example, it is wise to look hard at sensitivity whenever samples are reported as below the limit of detectability. This does not always mean that the levels are biologically low. An assay that is sensitive enough for one species' normal range of values may not be for another's. Where steroids have been extracted from the sample prior to assay, an additional consideration is recovery, that is, what percentage of the steroid actually present was successfully extracted? The RIA values have to be corrected for this recovery percentage in arriving at a final number. Assays are not perfect, and the final results are estimates of true but unknown values.

More recently, enzyme immunoassays (EIA), another type of competitive assay, have been developed that have sufficient sensitivity for a range of animal sizes. These have the important virtues of being faster than an RIA and not requiring radioactive substances or licenses. As with any assay, they must be validated first.

Assuming that the assay is valid, what has been measured? Some of the steroid molecules in the blood are attached to binding proteins, including nonspecifically binding albumins, which are much larger molecules than the steroids themselves. For example, glucocorticoids in the blood of many vertebrates are bound to corticosteroid-binding globulins (CBGs), which in spite of their name bind some other steroids as well. Because the function of these binding proteins has been a bit of a mystery, they are often ignored, but this threatens to be a perilous strategy (Breuner and Orchinik 2002).

Furthermore, what has been measured is the amount of steroid in the blood at the one point in time when the sample was taken. However, steroids, like other hormones, are subject to a variety of biological rhythms over timescales of minutes to months. Their levels are dynamic, not static. Some are secreted and released in a pulsatile manner, so that blood levels rise suddenly and then fall gradually. This phenomenon is well known in mammals but is also seen when birds are cannulated so that blood samples can be taken frequently (Ottinger 1983; Bacon 2001). The rate of decay of the pulse and the half-life of the steroid in the blood reflect clearance through binding to receptors at the targets and metabolism by the liver. Steroid half-lives are on the order of minutes to an hour or more, and vary depending on the steroid and the animal's metabolic rate.

Several steroids tend to show pronounced daily rhythms. Glucocorticoids in mammals and in some birds peak around the time of the onset of the daily period of activity (awakening from sleep), while in other birds they peak during the night when the birds are inactive (Carsia and Harvey 2000). Testosterone in male mammals, including humans, shows a similar rhythm, that is, peaks at awakening, contrary to our cultural notions that night is the sexy time. Given this dynamism, how can one sample adequately capture an animal's steroid level? In some sense it can't. The practical solution to this problem used by many researchers is to take all the samples from all the animals at the same time of day, usually the time when the steroid level is highest or in the middle of its daily range, so that results of experimental manipulations are not confounded by daily hormone rhythms.

Finally, what is important for the production of behavior is the amount of steroid that the brain and other targets "see," not necessarily the amount in the general circulation. The steroid receptors in the brain and elsewhere usually have a higher affinity for steroids than any binding proteins in the blood. Furthermore, even if the target and blood amounts are positively correlated, the target's active role in converting the steroid to another form means that we almost never know how much of the locally active steroid the target cells actually saw.

Neuropeptides and Prolactin

Peptides are amino-acid-based molecules and thus very different from steroids both chemically and biologically. Along with other amino-acid-based molecules, but unlike steroids, their structure (exact sequence of amino acids) differs among species. The version of the peptide found in one species won't necessarily work when given to another, especially if the two are unrelated. Even if it does work, the effect (the function thus altered) will not necessarily be similar, because similar molecules can be employed for different purposes in different taxa. The good news is that peptides are closely linked both to interesting social behavior and to genes, and hold great promise for aiding the search for molecular mechanisms of such behavior. Each amino acid in the peptide sequence is coded by a triplet of DNA nucleotides. Peptides are in a profound sense close to the genome.

A large number of peptides function as chemical messengers and regulators in both the nonneural periphery and the nervous system of metazoans. The same exact peptide molecule can be both a neuropeptide (the category of primary interest here) and a peripheral hormone. Like many other chemical regulators of neural activity, neuropeptides are produced by discrete brain cell groups. The brain is a source of peptides as well as a target, but because peptides do not readily cross the blood–brain barrier, some brain region targets "see" only those peptides that were produced in the brain, not those in the peripheral circulation. This is why research on peptides and social behavior seldom measures peptide levels in the general circulation or administers peptides systemically, but instead relies on manipulations and measurements of peptides in the brain itself.

Table 1.1 includes some of the large and ever-increasing number of neuropeptides that have been discovered. Many of these fall into families, that is, groups of peptides that have high but not complete overall amino-acid sequence resemblance. The families that will appear the most frequently in this book are the oxytocin family of nonapeptides (a relatively conserved family of peptides containing nine amino acids, named for a mammalian version), the gonadotropin-releasing hormone (GnRH, formerly called luteinizing-releasing hormone or LHRH) family of decapeptides (containing 10 amino acids), and the corticotropin-releasing factor (CRF, also called corticotropin-releasing hormone or CRH) family. These families are very widely distributed among animals, suggesting that they are quite old, possibly appearing early in metazoan evolution (Peter 1983; Sherwood and Parker 1990; Lovejoy and Balment 1999; Gorbman and Sower 2003). Functions related to reproductive behavior appear to be old as well, predating vertebrates. For example, lys-conopressin, a member of the oxytocin family, regulates sexual behavior in a snail (van Soest and Kits 2002).

Peptides of the GnRH family have an essential role in reproduction in all vertebrates. At least 14 forms have been found altogether, and each vertebrate lineage produces one or more of them. Molecules with GnRH-like activity also occur in invertebrates such as *Aplysia* (a mollusc) and cnidarians, and it will be interesting to see if they turn out to have reproductive functions in these animals as well (Tsai et al. 2003). The discovery of multiple GnRH forms has produced a confusing naming situation. Initially forms were named for the type of animal in which they were discovered, for example, "mammalian GnRH." They still have these names even though they have since been found in other kinds of animals; the names are historical accidents that no longer make sense (Fernald and White 1999). Chapter 7 will take another look at the GnRH family but until then "GnRH" will refer to those that regulate the gonadotrophic hormones of the vertebrate pituitary. In mammals GnRH is released in a pulsatile manner by a "pulse generator" in the hypothalamus. The GnRH message to the anterior pituitary lies in the frequency of the pulses, not their amplitude. The higher the frequency, the more gonadotropin is released into the circulation.

The neuronal cell groups that produce the neuropeptides linked to social behavior are located in characteristic brain regions. For example, in several mammals arginine vasopressin (AVP) is produced by cells in the supraoptic nuclei of the hypothalamus, the medial nucleus of the amygdaloid complex (MeA), the bed nucleus of the stria terminalis (BNST, the output from the MeA to the hypothalamus), and the lateral septal region (de Vries and Miller 1998). When different vertebrate lineages are compared, there is some conservation but also some diversity in the anatomical distribution of a neuropeptide such as AVP (Moore and Lowry 1998; Goodson and Bass 2001). The working hypothesis is that neuropeptides have different functions depending on where they are produced (in what neurons), where the projections from those neurons go, and whether the source or projection is steroid regulated. In a wide array of vertebrates, AVP neurons in the BNST and MeA and their projections to the lateral septum are steroid regulated.

Prolactin is an amino-acid-based hormone, but the number of amino acids is much greater than in neuropeptides such as GnRH or AVP. It is produced by the anterior pituitary and in the brain. Its name comes from its stimulating effect on milk production in mammals. In spite of this sex- and taxon-specific name, it is produced by both sexes, and molecules resembling mammalian prolactin are found in other vertebrates. Because prolactin is a large molecule with an amino-acid sequence that varies significantly across taxa, it is difficult to measure and to manipulate. An RIA antibody that recognizes mammalian prolactin will not necessarily recognize avian or other prolactins. Prolactin that is taken from pituitaries of mammals such as sheep and given to birds or fish will not always mimic the biological effects of their own prolactins.

Where and How Do Steroids Act to Alter Behavior?

Steroids, like other hormones, have targets, that is, groups of cells that contain the machinery to respond. The brain is a key steroid target for understanding social behavior, because it contains a lot of steroid target cells and is responsible for making behavior responsive to the environment, purposeful, and intelligent (fitness enhancing). Other steroid targets also impact social behavior, such as deer antlers or sonic muscles of vocalizing fish. With the explosion of interest in neuroscience, the brain has taken center stage, but the periphery cannot be ignored. Among other reasons, hormone-dependent external morphological characters are clearly important signals to other animals.

The brain targets, like peripheral targets, express steroid-metabolizing enzymes that regulate what the active steroid is and how much of it the neurons see. Targets are dynamic, not passive, a key concept for understanding behavior. This "supply and demand" aspect of local steroid metabolism is essential for a signaling system. The molecular mechanisms of the regulation of the genes for these enzymes are a critical link between steroids and behavior. One consequence of the role of local steroid conversion in brain and other targets is that circulating steroids aren't necessarily the active steroids for behavior. This means that even steroids that are thought to be behaviorally irrelevant because they are only "weakly" androgenic or estrogenic should not be ignored, because they could be converted in the brain to behaviorally active steroids. This is why the abundant but supposedly "weak" androgen DHEA is receiving renewed attention (for example, Soma et al. 2002).

Studies of aromatase provide strong support for this principle that local conversion is critical for steroid action. It is present in all vertebrate brains, and its regional distribution (especially in the diencephalon and limbic system) is reasonably conserved (Callard 1984; Saldanha et al. 1998; Naftolin et al. 2001). Levels of aromatase activity are especially high in brains of birds and teleost fish. Aromatase is expressed in ovaries but not testes of these animals, but nonetheless males have circulating estradiol. All of that estradiol has most likely come from the brain. The initial discovery of aromatase activity in the brain led to the radically new hypothesis that some of testosterone's role in the development and expression of maleness might actually be due to local conversion to estrogens (Naftolin and MacLusky 1984). The counterintuitive hypothesis that estrogens might be the "real" male sex hormones for some components of masculine behavior has received resounding experimental support for several species of birds and mammals (Balthazart and Ball 1998).

Steroid molecules can readily pass through the cell membranes of neurons and glia, but they can just as readily pass back out. To become concentrated in sufficient quantity and for long enough to do anything there needs to be some mechanism to grab and hold on to them. This is what intracellular steroid receptors do. Six types are thought to occur in most vertebrates: progesterone receptor (PR), androgen receptor (AR), glucocorticoid receptor (GR), mineralocorticoid receptor (MR), and two types of estrogen receptors, estrogen receptor alpha (ERα) and estrogen receptor beta (ERβ) (de Kloet 1995; Carsia and Harvey 2000; Breuner and Orchinik 2001a,b; Sloman et al. 2001; Thornton 2001). In spite of their names, they are not always specific to a single steroid or steroid category. Androgen receptors often bind both testosterone and 5α-DHT and their role in behavior is well established. In mammals, and probably birds as well, MRs (type I corticosteroid receptors) have a high affinity for (bind strongly) glucocorticoids as well as mineralocorticoids such as aldosterone, and regulate basal (resting) levels and daily rhythms of glucocorticoids. There are lots of them in the brain, mainly in hippocampal and septal neurons. GRs (type II corticosteroid receptors) are also found widely in the brain, including the hippocampus, and in both neurons and glia. They have a lower affinity for glucocorticoids than type I receptors and appear to function primarily when glucocorticoids are elevated, for example, during a stress response. The role of ERα in responses of targets related to behavior is well established, and the term "estrogen receptors" usually refers to this receptor type. The behavioral or other functions of ERβ and of the recently discovered teleost ERγ are not yet clear, but research in these areas is being actively pursued (Hawkins et al. 2000; Ogawa and Pfaff 2000; Temple et al. 2001; Ábrahám et al. 2003).

The locations of the sex steroid target cells in the brain are somewhat but not entirely conserved in vertebrates. Forebrain targets for circulating steroids include specific nuclei or regions within the preoptic area, the hypothalamus, a portion of the amygdaloid complex and its homologs, the BNST and its homologs, and the septum (fig. 1.2). The gross distribution of sex steroid targets also overlaps the distribution of GnRH family peptides, especially in the septal and preoptic areas (Demski 1984). The steroid target areas in the brain shown in fig. 1.2 are targets because neurons (or glia [Jordan 1999]) there express these receptor proteins. The figure is oversimplified because the different kinds of receptors (and possibly receptor subtypes as well) don't have exactly the same anatomical distributions (Bernard et al. 1999). The receptors themselves are highly but not totally conserved proteins (Baker 1997). There

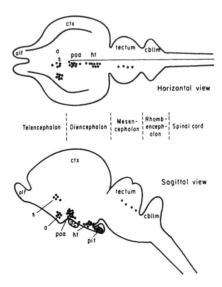

FIGURE 1.2. Conserved brain targets for sex steroids. This schematic of a generalized vertebrate brain shows where groups of labeled neurons (each group indicated by a black dot) have been found following injection with radioactively labeled testosterone or estradiol in most species studied. In some species additional target areas are seen using this or other methods (see, for example, fig. 2.11 showing the telencephalic steroid targets unique to songbirds). a, amygdala or its homolog; cbllm, cerebellum; ctx, cortex; ht, tuberal region of the hypothalamus; oc, optic chiasm; olf, olfactory bulb; pit, pituitary; POA, preoptic area; s, septum. Reprinted from Morrell and Pfaff (1981). © 1981 Plenum Press (Kluwer Academic/Plenum Publishers).

is some anatomical overlap between the expression of enzymes that produce steroids in the brain and the expression of the receptors for those steroids, although not as much as used to be assumed (Ball et al. 2002).

How can a steroid molecule bound to an intracellular receptor accomplish anything as powerful as making an antler grow, a bird sing, or a fish spawn? The answer is wonderful: steroids "tickle the genome." Steroid receptors are what are called ligand-dependent DNA-binding transcription factors. The steroid is the ligand. The steroid–receptor complex binds to hormone response elements located in the promoters of steroid regulated genes (fig. 1.3). In combination with co-activators and co-repressors they alter gene transcription and regulate the amount of mRNA transcript emanating from the target genes. Those transcripts are in turn translated into peptides and proteins. Exactly what genes are affected, and how the target tissue will respond, depends on the steroid receptor co-factors and downstream mechanisms (Charlier et al. 2002). At peripheral targets such as antlers of deer or clasping muscles of amphibians, these proteins might result in cell multiplication and tissue growth. In the brain, the protein products can include enzymes for steroid synthesis and metabolism, steroid receptors (steroids often regulate their own receptors), enzymes for

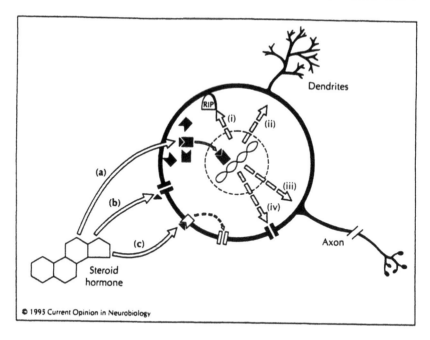

FIGURE 1.3. Mechanisms of steroid action on neurons. (a) Steroid hormones bind to intracellular receptors that alter gene expression, resulting in (i) programmed cell death (apoptosis), (ii) growth or regression of dendrites, (iii) changes in synaptic function, and (iv) synthesis or modulation of ion channels. Two classes of nongenomic effects are also shown: (b) direct binding of steroids to ion channels, and (c) binding of steroids to steroid receptors on the cell surface and modulation of ion channels via second messengers. The cell surface and intracellular receptors are structurally dissimilar. Reprinted from Weeks, J. C. and Levine, R. B. 1995. Steroid hormone effects on neurons subserving behavior. *Curr. Opin. Neurobiol.* 5: 809–815. © 2000 Elsevier Science.

production of neurotransmitters, receptors for neurotransmitters, neuropeptides and their receptors, ion channels, proteins for building and repairing axons, dendrites, and synapses, and substances that increase the number of newly proliferated neurons (Harding 1992; Saligaut et al. 1992; Micevych and Hammer 1995; Young and Crews 1995; Zakon 1998; Fowler et al. 2003). That's a lot of behavior-impacting gene products! Through their intracellular receptors steroids alter neural activity now and in the future, alter their own production and reception and that of other steroids, and regulate some of the other neural signaling systems important for social behavior.

The fact that steroids change gene activity puts them at the center of any effort to bridge the gap between genes and social behavior. It tells us that the connection between genes and hormonal mechanisms is bidirectional, not unidirectional. Furthermore, social interactions cause changes in steroid levels, which in turn cause changes in the gene activity of both participants, a point that will be developed in chapter 2.

Steroid receptors are not the only gene transcription factors and steroids are not the only cellular messengers that activate genes. Rather, they are part of one superfamily containing over 60 related transcription factors. This superfamily includes receptors for thyroid hormones and for vitamins such as retinoic acid, and a vast number of orphan receptors such as SF-1, whose ligands have not yet been discovered.

One of the hallmarks of this gene transcription mechanism of action is that the whole process, especially the DNA transcription and mRNA translation, takes significant time, on the order of hours. Tissue growth or neural remodeling would require gene transcription over an even more extended period. This time course of hours to days is consistent with many well-known behavioral effects of sex steroids. For example, if a castrated male quail is given testosterone, he begins to crow in 3–5 days and to mate in 5–7 days (Beach and Inman 1965; Adkins and Adler 1972).

Is this the only pathway by which steroids can act to alter behavior? There is a great deal of interest in multiple signaling (transduction) pathways for steroids. A particularly exciting prospect that could help solve several puzzles about social behavior is fast nongenomic mechanisms for altering neuronal activity, located in cell surface membranes (Zakon 1998; Moore and Evans 1999; Makara and Haller 2001). Possibilities include additional categories of steroid receptors, interactions between steroids and neurotransmitter receptors, and effects of steroids on the ion channels that control neuronal firing (fig. 1.3). The downstream pathway from such cell surface mechanisms would likely be a second messenger pathway instead of the gene transcription pathway of the intracellular steroid receptors.

There is already some evidence for rapid actions of these kinds on social behavior. Here "rapid" means a steroid effect seen in less than 15–20 minutes (Orchinik 1998). Corticosterone rapidly terminates mating behavior in newts through a membrane corticosteroid receptor (not yet isolated and identified) that constitutes a third category of corticosteroid receptor along with types I and II (Moore and Orchinik 1991). Corticosterone or cortisol treatment rapidly increases locomotor activity in mammals, birds, and turtles (Sandi et al. 1996; Breuner et al. 1998; Cash and Holberton 1999). Although membrane receptors for corticosterone are widespread in bird brains, it is not yet known if they are the mechanism of action for the rapid increase in activity. Rapid effects of steroids on the mating behavior of quail and rats and on investigation of conspecific chemicals by male rats have been obtained (Cross and Roselli 1999; Frye 2001; Balthazart et al. 2004). A membrane receptor for estrogens is suspected, but aside from its location it is not yet clear whether it constitutes a third estrogen receptor type along with ERα and ERβ. Goldfish and pigs have steroidal pheromones that result in rapid change in the behavior of conspecifics when detected by their chemosensory systems. This presumably means that there are membrane receptors for these steroids in these systems (Signoret 1967; Melrose et al. 1971; Sorensen and Stacey 1999).

The discovery of steroidogenic enzymes and neurosteroids in parts of the brain that are not classical targets (that lack intracellular receptors) suggests that these neurosteroids might act mainly through rapid membrane-based mechanisms (Mellon and Griffin 2002). Furthermore, membrane-located mechanisms might have different steroid affinities. If so, then steroids (either neurosteroids or weak circulating steroids such as DHEA) that bind poorly to intracellular receptors could nonetheless have significant effects on neuronal activity. For example, allopregnanolone, a brain-produced metabolite of progesterone, is a potent ligand at the $GABA_A$ receptor (Majewska et al. 1986). Because $GABA_A$ is the principal inhibitory neurotransmitter in the brain, this discovery helps explain an old but previously puzzling phenomenon: progesterone in large dosages has anesthetic and barbiturate properties.

These newly discovered or hypothesized mechanisms of steroid action provide a potential means for rapid facilitation as well as inhibition of behavior, thus increasing its flexibility and adaptability. It will be important for research to establish connections between these mechanisms and naturally occurring social behavior of ecological significance, as Moore and Orchinik's work with newts has done.

Steroid Manipulation

What methods are available for studying steroid effects on behavior, and what are their limitations? The classic experimental design for showing that a behavior is steroid dependent is the "remove and replace" design of endocrinology. The source of the steroid is removed to see if the behavior disappears, and then the steroid is administered to see if the behavior reappears. Removing the source, for example, gonadectomizing an animal, has its problems, however. The surgery may be difficult or (in the case of adrenalectomy) cause illness or death. There may be another source for the steroid that cannot be removed at all, such as the estrogen-producing caudal telencephalon of songbirds. Fortunately, there is another experimental approach, using drugs developed for use in treating people with steroid-sensitive cancers. Some of these drugs work by inhibiting a key step in a steroidogenic pathway. Fadrozole, letrozole, and vorozole are aromatase (estrogen synthesis) inhibitors, and finasteride (now used by millions of men to slow down male pattern baldness) inhibits 5α-reductase, which converts testosterone to 5α-dihydrotestosterone, the baldness culprit. Others work not by reducing the amount of steroid, but as competitive antagonists ("blockers") at the intracellular receptors, thus preventing the steroid from acting at the target. For example, tamoxifen is an estrogen receptor antagonist, flutamide is an androgen receptor antagonist, and RU-486 is both a progesterone and a glucocorticoid receptor antagonist. As this last example implies, these drugs are not always as specific as either clinicians or researchers would like. Also, they seldom completely inhibit or antagonize, and can even have mixed agonist/antagonist effects. For this reason, the fuzzier term "selective receptor modulator" is increasingly preferred to "antagonist" or "blocker," and even "selective" can be suspect. Furthermore, because of the negative feedback involved in regulation of steroid levels, these drugs may cause circulating steroid levels to rise. Also, membrane receptors won't necessarily be blocked by intracellular steroid receptor antagonists; they will be blocked only if the ligand binding site is similar (Schmidt et al. 2000). These are important considerations in interpreting results of experiments that use such drugs.

Replacing the steroid is often accomplished by regular injections to achieve good control over dosage, but administration by inserting an implant is usually preferred if disturbing the animal (or catching it, if it's free-living) is a problem. Peripheral administration works well because steroids readily cross the blood–brain barrier. In working with a new species it is desirable to determine the circulating level produced by the treatment to see if it is in the physiologically normal range. One of the future challenges in testing hypotheses about neurosteroids and ecologically relevant behavior is how to manipulate those steroids in the brains of free-living animals.

Mechanisms of Peptide Action

Because neuropeptides and protein hormones such as prolactin are such different molecules from steroids, we would expect their mechanisms and time course of action to differ as well. As a general rule they are relatively large hydrophilic molecules that do not readily cross a blood–brain barrier or enter into cells. For this reason their receptors are membrane (cell surface) receptors and not intracellular receptors. The receptor molecule amino acid sequences are not as conserved as those of the peptide ligands. Most neuropeptide receptors are members of the G-protein-coupled receptor superfamily (Darlison and Richter 1999). Peptide binding stimulates synthesis of cyclic AMP, a second messenger (the peptide is the first) that triggers the cascade of enzyme activity resulting in proteins that change the cell's physiology (Platt and Reynolds 1988; Norris 1996). A GTP-binding membrane protein (G-protein) mediates between the peptide receptor protein and the enzyme for cyclic AMP synthesis, adenylate cyclase. None of this takes a long time, and peptides can act fairly fast.

The peptides involved in social behavior each have different receptors and some, such as vasopressin, have more than one receptor subtype. There is diversity between receptor types and between taxa in the distributions of these receptors in the brain (Goodson and Bass 2001). That makes it difficult to summarize in any simple way where the principal peptide targets related to social behavior are located, but a few regions stand out. The hippocampus is a major target for CRF. Oxytocin receptors are found in the medial and central nuclei of the amygdaloid complex and in the olfactory system and septum. In mammals $V1_a$ receptors (the type of AVP receptors linked to social behavior) are found in the paraventricular and supraoptic nuclei of the hypothalamus, in the medial nucleus of the amygdaloid complex and bed nucleus of the stria terminalis, and in some species in the septum and the diagonal band of Broca (de Vries and Miller 1998).

The mechanisms that regulate peptides and their receptors include steroids. Peptide and protein hormone actions sometimes require a background of steroid priming in which the steroids, acting through their intracellular receptors, stimulate transcription of the receptor genes, upregulating the peptide receptors (Norris 1996). In addition, peptides and their receptors are a key part of the downstream pathways for steroid action on the brain and behavior. This cross-talk between steroids and peptides and peptide receptors is why peptides must be included in thinking about steroids and social behavior (Witt 1997; Albers and Bamshad 1998).

Multiple Messengers, Multiple Behaviors

Chemical messenger regulatory systems interact with each other at several levels, such as when neurotransmitters and neuromodulators are regulated by steroids in some brain regions and peptide and protein hormone actions require prior steroid priming. These modulatory chemicals provide the flexibility and temporal and spatial coordination necessary for adaptive behavior sequences (Bicker and Menzel 1989). This makes it unlikely that any of the steroid or peptide chemical messengers in table 1.1 will have a specific (one-to-one) relationship to a given behavior. That is, any neurotransmitter, neuromodulator, or hormone functions in more than one type of behavior, and a single type of behavior is based on multiple chemical messengers, not one chemical regulator (Orchard et al. 1993; Weiger 1997). Each chemical regulator is functionally versatile. Structural genes code for some of the messenger molecules and for messenger receptors and so we would not expect any one-to-one correspondences between genes and behaviors either.

As ethologists and neuroscientists have long appreciated, inhibitory mechanisms are just as important as excitatory ones for behavior. Each act, however simple, results from a mix of excitatory and inhibitory inputs, each of which is modulated by multiple chemicals; the balance between them determines the output, allowing more precise regulation than a simple on–off switch would (Katz and Harris-Warrick 1999). Decisions between competing behaviors are outcomes of changing balances between excitatory and inhibitory mechanisms. A behavior can be increased not only by increasing excitatory tone but also by decreasing inhibitory tone. Copulation by male newts (*Taricha granulosa*) illustrates this beautifully (Moore and Orchinik 1991). The peptides AVT and GnRH plus sex steroids (the excitatory mechanisms), glucocorticoids (which are inhibitory), and a host of other neuromodulators work together to produce an animal that mates in response to female stimuli but not if a predator threatens.

The importance of steroid priming for actions of other steroids and of peptides means that synergism occurs in experimental studies of hormones and behavior. "Synergism" in everyday language is often used incorrectly to refer to additive effects of two treatments. True synergism means that the effect when two hormones/peptides are given is greater than the sum of the effects of each given alone. The prototypical example is the combined effect of estrogen and progesterone on female rat and guinea pig lordosis. Neither a low dose of estradiol alone nor an injection of progesterone alone are particularly effective, but the combination given in the right sequence (first estrogen, then progesterone) brings on the behavior of a naturally estrous female. The synergism occurs because the estrogen priming increases gene transcription for the progesterone receptor protein, enabling the progesterone to act (Blaustein and Erskine 2002).

An old principle of comparative endocrinology is that the functions and effects of hormones have changed more in evolution than the hormones themselves. Current understanding of steroid and peptide actions has illuminated the mechanistic basis for diverse actions of the same hormone at different targets. We should look to those mechanisms (enzymes, receptors, and the cascades that they initiate) to see what has supported behavioral change over evolutionary time, a point that will be developed in chapter 5.

Hormones, Plasticity, and Development

Adult brains are like celestial objects or continents—more dynamic and plastic than most scientists used to imagine. No new neurons were born (it was believed), learning and memory happened somewhere somehow but only at the biochemical, not structural, level, and hormones somehow acted on neurons but without visibly altering their structure. It is fascinating to see how much this formerly canonical view of the adult nervous system has changed. Now it is clear that new neurons are formed and used throughout life in many vertebrates and invertebrates (Leonard et al. 1978; Easter et al. 1981; Gould et al. 1999; Harzsch et al. 1999). Now it is known that some forms of learning change the anatomy of neurons. Now it is clear that steroids can change the number, size, form, and dendritic structure of individual neurons and stimulate the formation of new connections between neurons (Luine and Harding 1994; Ball et al. 2002; Woolley and Cohen 2002). It seems quite likely that these steroid-induced structural changes are causally related to important real-world phenomena such as seasonal changes in behavior or a bird's memory for the songs it heard as a juvenile. Social interactions often alter individuals' hormone levels, and the social experiences that an animal has might change the structure of its brain through hormone-related mechanisms, a profoundly significant insight.

Many aspects of social behavior are subject to modification by experience, such as by social learning (for example, mate choice copying) or Pavlovian conditioning (Heyes and Galef 1996; Hollis 1997; Domjan et al. 2000; Woodson 2002). Even hormone levels themselves are subject to Pavlovian conditioning. For example, when a male mouse smells a female, his levels of luteinizing hormone and testosterone rise within minutes. The same hormonal response can be elicited by previously neutral stimuli that have come to signal the imminent appearance of a female (Graham and Desjardins 1980). Pavlovian conditioning, a universal property of nervous systems, enables animals to anticipate what is to come and enhances reproductive success (Hollis et al. 1997; Domjan et al. 1998; Adkins-Regan and MacKillop 2003). Research also indicates that some hormones, including testosterone and estradiol, induce states that the animal can detect and discriminate between, and that have rewarding or aversive properties (Alexander et al. 1994; Frye et al. 2001; Wood 2004). Responses of the immune system are also subject to Pavlovian conditioning (Ader and Cohen 1992), which may be relevant to their contribution to potential costs of hormones (to be addressed later in this chapter) and their role in social behavior evolution (to be addressed in chapter 2).

The developing brain is the epitome of plasticity. Although adults learn a lot, the capacity for drastic change is reduced. This toning down of brain plasticity is a mechanistic basis for the phenomenon of critical periods in development, special times when experiences have long-term consequences of a sort that are not possible outside of an early "window of opportunity." Filial and sexual imprinting are well-known examples of early learning with this special property.

The critical period concept is central to the important organizational hormone theory first proposed by Phoenix, Goy, Gerall, and Young (1959, see also Young et al. 1964). The theory distinguishes between organizational and activational hormone effects and posits three important differences between them. First, organizational effects establish the substrate for the future behavioral sex of the animal, whereas activational effects merely stimulate (activate) the substrate that has already developed. Second, organizational effects are permanent, lasting for the life of the animal even though the hormonal condition that produced them may be long gone, whereas activational effects are reversible, disappearing if the hormonal condition subsides. Third, organizational effects are only possible early in development, during a relatively limited critical period, whereas activational effects normally occur in adulthood.

Nowadays "activate" sounds a bit strong given the permissive nature of hormone actions on social behavior. Also, subsequent research has shown that it is not always possible to make a clean distinction between organization and activation (Arnold and Breedlove 1985), as, for example, when hormone treatment in adulthood turns out to have a long-term effect on behavior. The theory predates the discovery of structural plasticity in the adult brain, but that discovery itself does not undermine the original distinction between organization and activation. What is essential to those concepts is not whether structural change occurs, but whether such structural change is permanent and whether it can happen only during an early critical period.

The organizational hormone theory is still extremely useful and continues to be an important stone in the conceptual foundation of the field of hormones and behavior. The idea that hormones acting early in life can have different effects from those acting later has important implications not just for understanding sexual differentiation of social behavior (chapter 4), but also for thinking about hormones as mechanisms of condition-dependent signaling, as mediators of trade-offs, and as architects of life histories. Steroids with potential organizational effects include not only those produced by the young individual, but also those to which it is exposed through the mother (via internal gestation or egg yolks), siblings (intrauterine position effects), or both. Organizational effects are relevant to insects and possibly other invertebrates as well as vertebrates (Nijhout 1994; Elekonich and Robinson 2000). Because organization is permanent, such effects add to the increasing concern about exposure of wild animals to anthropogenic endocrine disrupters in the environment (Guillette et al. 1995; Palanza et al. 1999; Ottinger and vom Saal 2002).

The mechanisms underlying organizational effects of steroids are not yet well understood. What is clear is that steroid metabolites produced in the brain are key in understanding organizational as well as activational effects (Negri-Cesi et al. 2000). Studies of the regulation of the P450arom gene in young and adult brains may help explain why early hormone effects are so qualitatively different during organization compared with activation (Lephart 1997). It is also clear that binding of steroids to intracellular steroid receptors is involved, because steroid-receptor antagonists that work in adults to inhibit steroid-dependent behavior also work in young animals to prevent normal development of the behavior. But how do these steroid actions permanently sculpt the circuitry for future behavior and why is this sculpting only possible at young ages? These questions will be taken up in chapter 4.

How the Necessary Control of Steroids by the Environment Is Achieved: The HPG and HPA Axes

Animals do not achieve fitness in a vacuum but in a physical and social environment to which their hormones must respond appropriately. In many temperate zone birds, the increasing daylengths of spring stimulate massive growth of the gonads and a corresponding increase in sex steroid production (Wingfield and Farner 1993). A stressful stimulus causes release of adrenal cortico-sterone, which interrupts mating in a male newt (Moore and Orchinik 1991). A dominant female coral reef fish turns into a male when the prior male of the group disappears (Shapiro and Boulon 1982).

How is it possible for the stimuli of the world outside an animal's body to have such a huge impact on what goes on inside it? Those external stimuli are detected by sensory organs that then convey signals to the brain or head ganglion. When the relevant hormones are neurohormones produced by brain neurosecretory cells, as in many invertebrates, the answer is straightforward: neurons that know about the stimulus tell those other neurons to produce or release the hormones. But in vertebrates, the gonadal and adrenocortical steroids of interest come from nonneural tissue located far from the brain. These hormones are regulated by yet other hormones from the anterior pituitary (fig. 1.4). Now that we're in the pituitary we're physically closer to the brain, but there's still a problem receiving information about the outside world, because the anterior pituitary is nonneural tissue. In tetrapod vertebrates, it is not even innervated. So how does sensory information get translated into the endocrinology of the anterior pituitary to then regulate the gonads and adrenals?

The solution lies in a set of discoveries that rank as great achievements in organismal biology (Harris 1955; Scharrer 1959; see also Raisman 1997). The story has been best told in mammals. A special little portal circulatory system (the hypothalamo-hypophysial portal system) connects the hypothalamus and the anterior pituitary. The hypothalamic neurosecretory cells produce a set of peptides, some excitatory and some inhibitory, that are released into this system to reach the cells of the anterior pituitary to regulate their hormone production. This allows communication between brain and anterior pituitary so that the external physical and social world can influence the animal's gonads and adrenals. Hypothalamic peptides also regulate the production of prolactin by other cells of the anterior pituitary. In mammals dopamine (which is not a peptide, and which acts as a neurotransmitter elsewhere) acts as a prolactin inhibitory factor.

GnRH is the hypothalamic releasing peptide that increases the levels of the polypeptide hormones FSH (follicle-stimulating hormone) and LH (luteinizing hormone), the gonadotrophic hormones that stimulate increases in gonadal sex steroids. LH is released in a pulsatile manner as a result of rhythmic activity in what is called the GnRH pulse generator (a set of hypothalamic neurons). Higher frequency pulses result in higher blood levels, as when LH is triggering ovulation. A set of largely negative feedback loops between the gonadal hormones (the sex steroids along with some nonsteroidal hormones such as inhibin and activin), the anterior pituitary, and the hypothalamus then maintains hormone levels within some reasonable range. If testosterone gets too high, less GnRH and LH is produced, which turns down the testosterone level. If it is too low, GnRH and LH go up, raising testosterone. Positive feedback occurs

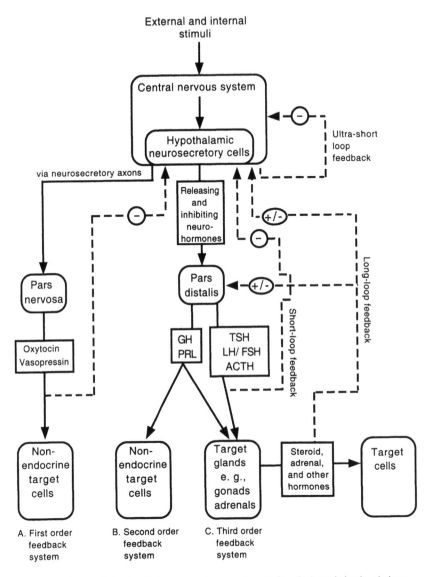

FIGURE 1.4. The hypothalamic–pituitary axes of mammals, including the hypothalamic–pituitary–gonadal (HPG) and hypothalamic–pituitary–adrenal (HPA) axes. The releasing and inhibiting hormones are transported via the hypothalamo-hypophysial portal system. At each level in the axis, hormones or neurohormones exert feedback on one or more other (usually higher) levels. Such feedback loops can be positive (+), negative (–), or both, depending on other factors. Many features of these axes are seen in other vertebrates. GH, growth hormone; PRL, prolactin; TSH, thyroid stimulating hormone; LH/FSH, luteinizing hormone/follicle-stimulating hormone; ACTH, adrenocorticotrophic hormone. The pars distalis is the anterior pituitary. The pars nervosa is the posterior pituitary (neurohypophysis). Redrawn from Norris (1996). © Academic Press (Elsevier).

when rising estrogen levels from developing ovarian follicles cause an increase in the GnRH pulse frequency, thus producing the spike of LH that triggers ovulation. This three-tiered system is called the hypothalamo–hypophysial–gonadal axis or hypothalamic–pituitary–gonadal axis (HPG axis). Levels of GnRH that are either too high or too low will cause suppression of the rest of the axis.

A three-tiered system analogous to the HPG axis, the HPA axis, connects the hypothalamus, anterior pituitary, and adrenal cortex, forming the pathway for the steroid hormone limb of the classic stress response (fig. 1.4). In response to an environmentally challenging (homeostasis perturbing or "stressful") situation, peptides in the paraventricular nuclei (PVN) of the hypothalamus such as CRF are released into the portal system. Upon reaching the anterior pituitary, these stimulate release of ACTH and other hormones into the circulation, which in turn stimulate glucocorticoids from the adrenal cortex. Negative feedback loops bring the hormones back to baseline levels after the challenge is over, but positive feedback can also occur, as when a very aversive event produces sustained glucocorticoid elevation. The sites where glucocorticoids exert feedback include the PVN but also regions outside of the hypothalamus, such as the hippocampus for negative feedback (via type II receptors) and portions of the amygdaloid complex for positive feedback (Lathe 2001). The adult "tone" of the HPA axis (its resting state and response threshold) is influenced by organizational actions of glucocorticoids resulting from experiences in early life (Levine and Mullins 1968; Carsia and Harvey 2000; Walker et al. 2002).

As the site of the releasing hormones, PVN is a control center of the top tier of the HPA axis. When HPA activation occurs in response to simple physical stressors such as cold, blood loss, or pain, the information that there is a problem comes up from the spinal cord or lower brainstem to the PVN in the upper brainstem, the categorization of the stimulus as meriting HPA activation is automatic and "hard-wired," and CRF is the principal releasing hormone for the response. Activation of the HPA in response to stimuli related to social behavior (sights and sounds of other individuals, social events, and psychological states) requires a more nuanced interpretation than the brainstem can provide. Is the stimulus a predator and therefore dangerous, or a conspecific presenting a golden reproductive opportunity? If it's a conspecific, is it a familiar individual or a stranger? Whether conspecifics are dangers or delights depends on who they are and what they have done to the observer in the past. As a receiver of processed sensory information from all modalities, the forebrain amygdaloid complex is well positioned to provide the necessary top-down interpretation, "deciding" how the animal should respond (flee? approach? court?). The mammalian "amygdala" is a complex of nuclei. Homologs of at least some of them are present in all jawed vertebrates (Butler and Hodos 1996). The discovery in mammals that some of these nuclei work together to determine whether a stimulus is dangerous (fear inducing) enough to merit a stress response is an important chapter in contemporary brain science (LeDoux 1995). If the decision is that a stimulus is dangerous, the central nucleus of the amygdaloid complex then sends signals to the PVN and to other brainstem loci responsible for the physiological reactions to a fear inducing stimulus that enable the animal to deal with it.

Multiple hypothalamic peptides regulate the HPA axis, and there seem to be a greater number of them involved in regulating the axis in birds than in mammals (Romero and Sapolsky 1996). When the HPA is stimulated from the top down in mammals, the response at the hypothalalmic tier involves less CRF and more of the oxytocin family of peptides. The exact peptide signature (ratio of CRF to AVP/AVT to OT) depends on the psychological state resulting from the stimulus. Psychological state is widely recognized as a crucial intermediary in humans. Bungee jumping is a thrill for some but a trauma for others. The psychological component is almost certainly critical for nonhumans as well. In addition to individual differences, there are big seasonal and species differences in whether a stimulus is a stressor and activates the HPA axis (Orchinik 1998; Silverin 1998). Seasonal changes raise another important issue as well, whether "stress" is the right word to use for HPA activity. Characterizing adrenal activation as a "stress" response doesn't seem to capture what is happening when baseline glucocorticoid levels change markedly on a seasonal basis (Romero 2002).

As the final common signaling molecules from the brain that reflect integration of multiple inputs and orchestrate the responses of the lower tiers, peptides like GnRH and CRF have pivotal roles in reproductive success and survival (Ball 1993; Bonga 1997; Aguilera 1998). They regulate the HPG and HPA axes that ensure that steroid levels are appropriate for current and predictable future environmental and social conditions (Wingfield and Silverin 2002). These two axes can influence each other as well. Chronic stress, including social perturbation, nutritional stress, inflammation, parasites, and endotoxins, and glucocorticoid administration depress the HPG axis of many mammals and lizards, an effect that is caused by the increase in CRF and glucocorticoids (Sapolsky 1993; Dunlap and Schall 1995; Rivest and Rivier 1995; Kalra et al. 1998; Schneider and Wade 2000). The HPG axis of nonmammals seems to require more extreme HPA activation to be depressed, and moderate glucocorticoid elevation is often characteristic of the breeding period (Wingfield and Silverin 1986; Romero 2002; Moore and Jessop 2003). Sex steroid treatment tends to lower glucocorticoid levels in mammals but it elevates them in some birds (Hillgarth et al. 1997). Both GnRH and the HPA peptides (CRF, AVP/ AVT, OT) are produced by other parts of the brain in addition to the hypothalamus and have other functions in addition to their roles in the top tiers of the HPG and HPA axes, including some direct behavioral functions such as augmentation of anxiety by CRF (Habib et al. 2000) and facilitation of lordosis by GnRH (Dudley and Moss 1991).

The HPG and HPA scenarios spelled out earlier are based largely on research with mammals and to a lesser extent birds. Nonmammalian vertebrates also have hypothalamic regulation of the pituitary and have a similar array of hypothalamic peptides, but what these peptides do and whether they are involved in pituitary regulation is seldom understood (Norris 1996). In those birds studied so far, the peptide VIP seems to be the primary releasing factor for prolactin regulation (Sharp et al. 1989; El Halawani et al. 1990; Maney et al. 1999; Vleck and Patrick 1999). Fishes other than teleosts, lampreys, and hagfish have a vascular link between the hypothalamus and pituitary (Sower 1998). Teleosts, however, have direct innervation of the anterior pituitary by the preoptic area (the region anterior to the hypothalamus) (Peter et al. 1990). This is the mechanism for external regulation of the gonads in response to changes in the physical and social environment and of the glucocorticoid-producing interrenal tissue in response to stimuli that threaten survival (Bonga 1997; Norris 1996). Teleosts have GnRH neurons, but they have a wider distribution outside the hypothalamus compared to other vertebrates. In spite of these special derived anatomical features of the teleost HPG and HPA axes, GnRH and CRF are still the key molecules for the top tiers (Lovejoy and Balment 1999).

It is an oversimplification, of course, to imply that the problem of how steroid production is regulated adaptively is solved now that some links between the hypothalamus and the pituitary are known. Although a remarkable number of brain regions send some kind of neural projection to some part of the hypothalamus, nonetheless there are many gaps in understanding how sensory information about the social world reaches the hypothalamic cells producing the releasing peptides. Also, the hormone levels at the three tiers of the HPG and HPA axes aren't always well correlated. Behavioral effects of GnRH and CRF independent of their stimulating effect on peripheral steroids complicate the picture in potentially interesting ways.

Insect endocrinology shows some interesting parallels with the neuroendocrine axes of vertebrates (E. Scharrer 1959; Acher 1986). Insect hormones are regulated by external as well as internal cues via the brain by means of two- or three-tiered cascades with feedback loops that begin with brain neurohormonal peptides and end at peripheral target organs (Nijhout 1994). In one "axis," brain neurosecretory cells produce PTTH (prothoracicotrophic hormone, a neuropeptide), which is stored in and released from the corpora cardiaca; PTTH in turn stimulates the prothoracic gland, which makes the prohormone ecdysone, which is converted to ecdysteroids when it reaches its targets (e.g., epidermis). In another axis, brain neurosecretory cells produce the peptides allatotropin and allatostatin, which regulate the production by the corpora allata of juvenile hormones, their sole product, which in turn act on multiple targets including the nervous system. Communication between tiers occurs mainly by direct innervation. There are no special portal systems to get the neurohormones out of the brain (there is no need given how insect circulation works), and are no gonadal steroids, but in other ways the organizational scheme is similar, reflecting a common need for hormonal systems to be adaptively regulated by the outside world. Some of the same peptides are used as well (Ottaviani and Franceschi 1996).

Diversity in Mechanisms

Many hormonal mechanisms are rather conserved in vertebrates but there is interesting diversity nonetheless. The special way that the hypothalamus communicates with the anterior pituitary in teleost fish is just one of several notable examples of derived features appearing only in one lineage, in this case a very large one. Diversity can arise through evolved adaptations to particular environmental circumstances (natural selection), through the behavior of conspecifics (sexual selection), or through the sheer passage of long periods of time with no selection against the occasional mutation (neutral evolution). The following are additional notable examples of diversity resulting from these processes:

1. Juvenile hormones (which are not steroids) appear to be a family unique to insects and crustaceans.

2. Insects cannot make steroids de novo because they don't make cholesterol. The ecdysteroids are made from sterols obtained from the diet (Rees 1985).

3. Vertebrate groups differ in whether the predominant glucocorticoid is cortisol (teleost fishes, primates) or corticosterone (rats and mice, birds, lizards).

4. 11-Ketotestosterone and other 11-oxygenated nonaromatizable androgens are characteristic of teleost fish, where they are the predominant circulating androgen and replace DHT as the other important androgen (Borg 1994). Testosterone is present as well but 11-ketotestosterone is often more potent than testosterone for male sexual characters and behavior (Brantley et al. 1993; Borg 1994). It is not clear how these steroids act or why teleosts have them and not other vertebrates. The recent discovery of two androgen receptor types in both teleosts examined may help answer the first question (Sperry and Thomas 1999a,b).

5. The presence in the brain of the steroid metabolizing enzyme 5β-reductase, which converts testosterone into 5β-dihydrotestosterone, is unique to birds. 5β-DHT has little or no biological activity and levels of it are highest in nonbreeding individuals and seasons. Thus, 5β-reductase seems to function as an inactivation shunt (Massa et al. 1979; Balthazart 1989). This ex-

ample illustrates the important principle that regulation involves inhibition or inactivation along with facilitation and upregulation.

6. The telencephalon of the speciose oscine passerine lineage of birds ("songbirds") has several striking features. In addition to the song system itself (a set of interconnected nuclei dedicated to song and song perception), there are intracellular androgen receptors in several of the telencephalic song system nuclei, there are estrogen receptors in one of the song system nuclei (HVC), and the caudal nidopallium has a large field of aromatase positive cells (Metzdorf et al. 1999; Schlinger and Brenowitz 2002). Intracellular androgen and estrogen receptors and aromatase are not seen in homologous parts of the telencephalon in other birds or in mammals and are unusual in other vertebrates as well.

How "Costly" Are These Hormonal Mechanisms?

The dominant theoretical framework in behavioral ecology emphasizes costs and benefits to individual animals of engaging in a behavior. Benefits must outweigh costs in the currency of fitness for a novel behavioral solution to a problem to be maintained and increase in frequency over evolutionary time. Also central is the concept of trade-offs that limit what is possible (Krebs and Davies 1997). These trade-offs occur at all levels including internal, behavioral, and life history (trade-offs between different fitness components), and are based not only on energetic costs but also on any other external or physiological resources that are in limited supply and on time limitations. If the brain uses more oxygen than usual, less is available for some other internal function. Two different behaviors cannot usually be done at the same time. Investment in the present could reduce investment in the future.

Behavior has direct energetic costs and some behavior that is good for social and reproductive goals increases predation risk. It is increasingly recognized that the costs of behavior also might include the costs of the mechanisms, including hormonal mechanisms, that support the behavior. Determining costs of mechanisms is difficult, and the concept of "costs" can be slippery. However, because it is an important part of some research on steroids and social behavior that will come up in several chapters, it is worth taking a look here at what the costs of steroids might be.

Some of the physiological costs of HPA activation and of chronic glucocorticoid elevation are well established, including atheroschlerosis, ulcers, immunosuppression resulting in increased disease risk, and, in some taxa, HPG inhibition (Cooper and Faisal 1990; von Holst 1998; Salzet et al. 2000; Jasnow et al. 2001). Such immunosuppression also occurs to natural stressors, including social conditions (Nelson et al. 2002). HPA activation is part of an emergency response system whose costs can be borne because the alternative is likely to be death.

In a captive study simulating natural predation, estrous deermice (*Peromyscus maniculatus*) were more likely to suffer predation by a weasel than nonestrous deermice, and this was because of their odor, which is sex steroid dependent (Cushing 1985). Because this odor results in part from excreted steroid metabolites and serves to attract males, this is a particularly interesting illustration of the costs and benefits of steroids.

In categorizing potential costs of testosterone in male songbirds, Wingfield et al. (2001) consider both direct costs (those of the testosterone itself) and indirect costs (those resulting from the behavior that testosterone stimulates). Their list includes higher energetic costs, reduced fat stores, oncogenic effects of estrogenic metabolites of testosterone, increased mortality, interference with pair bonds, increased injury, interference with parental care, and immunosuppression. They refer to an idea first proposed by Naftolin (see Naftolin and MacLusky 1984) that steroidogenesis at the targets might be a mechanism to avoid costs of that steroid in the circulation. Because different behaviors involve somewhat different brain regions, limiting a potent steroid to the relevant target not only avoids systemic costs such as oncogenesis or immunosuppression, but also reduces costs resulting from inadvertent and unwanted stimulation of other behaviors. Another way to reduce the costs of a steroid might be to increase behavioral sensitivity to it (increase receptor numbers or sensitivity) (Hews and Moore 1997).

Steroid-stimulated behaviors, including general locomotor activity, can be energetically demanding and so steroids would be expected to increase metabolic rate (oxygen consumption) indirectly, because of these behaviors (Vehrencamp et al. 1989; Emerson and Hess 2001). Estradiol increases activity level and metabolic rate in female rats, whereas progesterone has the opposite effect (Wade and Schneider 1992). Brain tissue is energetically expensive with respect to oxygen consumption and the forebrain song system nuclei of songbirds are more metabolically active when they are testosterone stimulated (Wennstrom et al. 2001).

Do sex steroids also increase resting or basal metabolic rate (RMR or BMR), the "cost of living" (Hulbert and Else 2000)? In principle, steroids could have a direct effect on BMR if they change core body temperature or (because the brain is a big oxygen user) change neural activity during behaviorally inactive states such as sleep (Hänssler and Prinzinger 1979). In practice, measuring BMR so that it is unconfounded by locomotor activity (including minor locomotor activity like "fidgeting") and by the stress of confinement is challenging, especially for highly mobile animals with large home ranges. So is figuring out how to correct for the changes in body mass that usually occur when steroid levels are experimentally manipulated.

Any direct effects of sex steroids on BMR might be most likely to impact fitness in animals with small body size, rapid metabolism, and low fat stores, such as many small birds. There have been several attempts to see if testosterone affects BMR in birds (for example, Lynn et al. 2000); very few, however, have yielded positive results (Buchanan et al. 2001). The evidence from lizards is more positive. Female *Sceloporus virgatus* had lower BMRs than males when adjusted for body mass (Merker and Nagy 1984). Both testosterone and estradiol increased oxygen consumption in *Chalcides ocellatus* (al Sadoon et al. 1990).

In mammals the positive evidence comes mainly from humans, perhaps because it is easier to measure BMR in our species. Hamilton (1948) observed that castrated men have lower metabolic rates and live longer, and concluded that "a price is paid for a beard" (p. 315). Both testosterone and estradiol increase metabolic rate (Lyons 1969). Men have a higher BMR than women of the same height and weight, but this sex difference doesn't necessarily generalize to other species. Comparative analyses of BMR (for example, Bennett and Harvey 1987) do not mention the sex of the animals, as if there are no sex differences. Even glucocorticoids have not been shown to affect BMR in birds and mammals in spite of the fact that stress itself increases oxygen consumption. All told, the major physiological costs of sex steroids, if there are any, must lie elsewhere.

Folstad and Karter (1992) proposed that testosterone is costly in part because it suppresses the immune system, leaving animals more vulnerable to pathogens, and that this cost helps keep testosterone dependent male signals honest. Their proposal was based on rather little evidence, most of it from humans and a few laboratory mammals, although it is consistent with observed sex differences in immune function and parasitism rates (Nelson et al. 2002). It had the beneficial effect of inspiring a number of experimental attempts to test the hypothesis in birds and other nonmammals and a greater interest generally in the immune systems of nonmammalian vertebrates. Support for the hypothesis is mixed and seems to depend on the species, whether the animals are in captivity or free-living, and whether acquired or innate immunity is measured (Hillgarth et al. 1997; Hasselquist et al. 1999; Roberts et al. 2004). It is innate immunity that is seasonally regulated, has high nutritional costs, and is phylogenetically old (Lee and Klasing 2004). If immune function is redistributed (adaptive reallocation of energy) rather than suppressed overall, there would not necessarily be unavoidable costs (Raberg et al. 1998; Braude et al. 1999). In small rodents, immune system functioning is enhanced on short days (Nelson and Demas 1996), but this is due to melatonin, not steroids. The overarching conclusion that testosterone suppresses the immune system in animals appears to be premature. Furthermore, in those cases where positive evidence has been obtained, it is likely that other hormones, for example, glucocorticoids, and not the testosterone itself, are the culprits. Administering testosterone increases corticosterone in songbirds and corticosterone is a well-established immunosuppressant (Hillgarth et al. 1997; Evans et al. 2000; Casto et al. 2001). Regardless of what the relevant hormone is, immune status clearly has fitness consequences, not just for survival (disease and parasite resistance), but also for the likelihood of being chosen as a mate (Nelson and Klein 2000).

Little attention has been paid to potential costs of estrogens, even though they are oncogenic in mammals, males of some species have surprisingly high circulating levels, and some male courtship behavior is based on estrogens as well as androgens (see chapter 2). Experimentally administered estrogen has aversive properties (Ganesan 1994). Birds are sensitive to toxic effects of estrogens administered systemically, so much so that there is a fine line between a behaviorally effective dose of estradiol and a lethal dose (Warren and Hinde 1959). Small wonder then that estradiol usually circulates in amounts a hundred times lower than testosterone, with aromatization of androgens at the target serving to augment the local level. In mammals estradiol can either enhance or depress immune function, and again these effects could be mediated by changes in other hormones, including peptides (Grossman 1985; Whitacre et al. 1999; Geary 2001; Nelson et al. 2002). Both DHEA and prolactin are immune enhancing in those mammals studied (Nelson et al. 2002).

What about the steroids themselves and their mechanisms of action? Are these likely to be costly? Steroids are small molecules containing abundant elements, but both the enzymes required for their synthesis and conversion and their receptor proteins would have similar costs to other bodily proteins. Glucocorticoids are highly oxidized molecules of the sort that tend to be toxic and mutagenic. The GnRH and oxytocin families of peptides are relatively small as peptides go but prolactin is a much bigger molecule. In many species of animals and especially in females, gonadal steroids quickly drop when animals are food limited (e. g., Aubret et al. 2002), but this could be because food scarcity is a signal that it is a bad time to think of investing in reproduction rather than because the hormonal mechanisms themselves require more calories than are currently available. In birds and some other vertebrates the gonads (not just the gamete-producing tissue, but also the steroid-producing tissue) shrink to nearly nothing during the nonbreeding season, as if something about this tissue, its products, or their consequences is too costly to afford in the off-season. For birds, the cost of carrying gonadal weight while flying might be significant, but are there other costs as well?

2

Mating, Fighting, Parenting, and Signaling

Behavior such as mating and parenting has fairly obvious fitness consequences but those consequences are not always so obvious for other social behaviors. Nor is it obvious when mating or parenting are done in bizarre ways, for example, by forced copulation that causes injury or by eating the babies. Animal behaviorists have made great progress understanding the functions of many social behaviors, thanks to research that tests hypotheses about whether and how they promote reproductive success. This chapter examines how the steroid and peptide mechanisms of chapter 1 contribute to adaptive social behaviors and whether they impose costs or set limits to what is possible. It begins with some basic findings, concepts, and principles, then considers rhythms in hormones and behavior, and, finally, looks at bidirectional relationships between hormones and behavior from a signaling perspective. Along the way, some overly simplistic notions will get debunked, such as equating testosterone with aggressive behavior and equating prolactin with parental behavior.

Courtship and Mating

As the behavior that serves to bring female and male gametes together, mating is critical for reproductive success and under continual natural and sexual selection at all levels from the behavioral to the molecular (fig. 2.1) (Civetta and Singh 1998). It is also the most extensively studied and best understood behavior with respect to underlying hormonal mechanisms. Courtship behaviors, those actions performed at close range that determine whether the interaction will proceed to actual mating, often show a similar relationship to circulating hormones as mating does. Some of the principles derived from the study of mating and courtship can be generalized to other behaviors that indirectly serve reproductive functions.

Invertebrates have a staggering array of ways of getting eggs and sperm together, including emission of enormous numbers of gametes into the water by sessile organisms, release of sperm by parasitic males located inside the female's body, transfer of large spermatophores, and internal fertilization by copulations lasting many hours. Although these behaviors occur when the gonads are in the right state, they are seldom dependent on any hormones from the gonads. Invertebrate gonads, unlike vertebrate gonads, usually do not com-

FIGURE 2.1. Mating behavior of (A) Japanese quail, (B) gopher tortoises, and (C) big brown bats. In many species of vertebrates that have been studied, mating behavior is dependent to some degree on gonadal steroids. Big brown bats are nocturnal and mate in the dark, here in an "upside down" position. Photo A is by the author, B is courtesy of Valerie Johnson and Craig Guyer, and C is courtesy of Mary Mendonça and Kristen Navara.

bine the dual functions of gamete and hormone production, and so removal of the gonads won't necessarily alter mating behavior even though it renders the animal infertile. For example, *Octopus* males whose testes are removed continue to copulate, passing nonexistent spermatophores to females (Wells and Wells 1972).

Not only do vertebrate gonads produce both gametes and hormones, but the same gonadal sex steroids that regulate gamete production and release also regulate the behavior of both parties that will bring the gametes together. In a wide array of species, removal of the gonads results in a decline in courtship and mating behavior, and treatment of gonadectomized individuals (those whose gonads have been surgically removed) with testicular or ovarian sex steroids restores the behavior to its previous level. A positive result from this remove-and-replace experimental design has long been the gold standard of evidence for hormone dependence. The results of such experiments can be quite dramatic. For example, male Japanese quail are enthusiastic copulators that grab the female's neck feathers almost immediately upon presentation, mount, and achieve cloacal contact and insemination in a matter of seconds (fig. 2.1A). If the testes are removed the frequency of this behavior falls to zero over the next week, never to recur again. Once testosterone replacement is begun, attempted and successful matings magically begin to reappear during the next week or two (Beach and Inman 1965). Strutting, a male courtship and threat display, and crowing, a loud male vocalization, follow the same pattern. On the female side, it is impressive to see how a single injection of less than a milligram of estradiol into a 100-kg ovariectomized female pig produces all the normal estrous behavior for the normal duration, not only the receptive stance (standing with rigid legs and ears cocked) that allows the male to copulate, but also an intense fascination with the smell, sound, and feel of a male (Signoret 1970).

These relationships between circulating hormone level and behavior make it very tempting to assume that individual differences within a species in the behavior of unmanipulated animals are correlated with their hormone levels—that a male lizard that displays to females intensely and frequently has more testosterone than one that doesn't, or that a female monkey that mates with every male in the group has more estradiol than one that mates only once. It often comes as a surprise to newcomers to the field of hormones and behavior to discover that, more often than not, this is not true. The correlation can be near zero or even negative (Batty 1978). How can this absence of a correlation in a group of individuals be reconciled with the obvious association between hormone level and behavior in remove-and-replace experiments and "experiments of nature" in which marked changes in sexual behavior occur at puberty as sex steroid levels rise?

The explanation lies in the form of the dose–response relationship between hormone level and behavior and the concept of a hormonal threshold for a

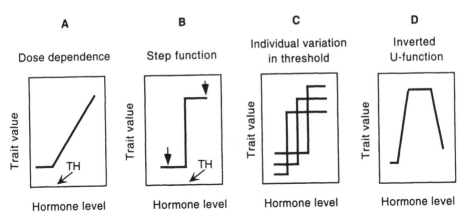

| A | B | C | D |
| Dose dependence | Step function | Individual variation in threshold | Inverted U-function |

FIGURE 2.2. Schematic diagrams of relationships between steroid level and behavioral or morphological traits. The first three are the models presented in Hews and Moore (1997) and are redrawn from that source (© 1997 Chapman & Hall [Kluwer Academic/Plenum Publishers]). The threshold dosages for a response are indicated for models A and B. Model A is also called a dose–response or graded function. See figs. 2.3 and 2.4 for experimental data consistent with these models. In model B the unlabeled arrows indicate where gonadectomized animals (left arrow) and gonadectomized animals receiving hormone replacement (right arrow) would fall. Model C is a step function with individual differences in thresholds. When averaged across individuals, it gives the appearance of a dose-dependent function. Model D is an inverted U function where intermediate hormone levels produce the greatest response.

step function (Meisel and Sachs 1994; Hews and Moore 1997) (fig. 2.2). For nonbehavioral characters, including many hormone-dependent aspects of physiology and morphology, there are graded dose–response relationships. For example, as the amount of testosterone administered to castrated animals is increased above zero, the redness of a rooster's comb or the size of a male frog's forelimb clasping muscles increases. Any threshold dose for the response to start occurring lies below the normal circulating hormone range of intact animals. While this occasionally happens with behavior, more often a step function is seen, for example, in rabbits and sheep (D'Occhio and Brooks 1982) (figs. 2.3 and 2.4). Once the dose is at or above the threshold (4 mg of testosterone propionate per day for sheep), the behavior occurs at a similar level regardless of the dose. When gonadectomized animals receiving no hormone replacement are compared with gonadectomized animals given hormone replacement designed to mimic normal breeding levels, or when pre- and postpubertal animals are compared, or when animals outside of the breeding season are compared with those currently breeding, the comparison is between "doses" on opposite sides of the step-up. But when unmanipulated postpubertal animals are compared during their breeding period, all their sex steroid levels fall safely above the threshold in the part of the step function that is flat. "Safely above" isn't surprising, because genes producing hormone

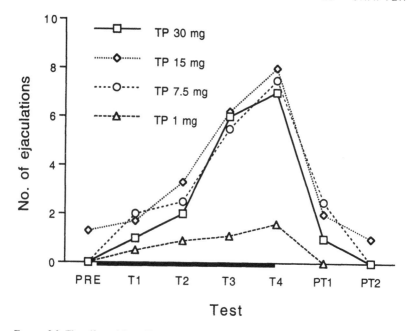

FIGURE 2.3. The effect of four different doses of testosterone propionate (TP) on completed matings (ejaculations) in castrated male rabbits. Injections were given for 120 days (black bar on X axis), with mating tests at four times during that interval (T1–T4) as well as before (PRE) and after (PT1 and PT2) the injection series. There was no dose–response relationship above the 1-mg/day dose. Instead, there was a substantial effect of how many injections had already been given (the duration of treatment). Such results are better described by a step function (model B in fig. 2.2) than by a dose-dependence function (model A). Redrawn from Ågmo and Kihlström (1974). © 1974 Elsevier Science.

levels below the threshold for mating would of course be subject to very strong negative selection.

Occasionally the dose–response relationship is an inverted U function (fig. 2.2). For example, the middle range of doses of corticosterone stimulated the greatest amount of locomotor activity in sparrows (Breuner et al. 1998). Romeo et al. (2002a) reported inverted U functions relating both estradiol and testosterone to sexual behavior in male hamsters. What might cause behavior to fall off at the upper part of the dose range? An uninteresting possibility that fortunately doesn't apply to either of those examples is toxicity. Much more intriguing would be a change in another hormone, for example, suppression of GnRH by the higher doses of sex steroids in the case of the hamsters.

Where do the individual differences in behavior come from then if not from circulating hormone levels? Research has identified several interesting sources for mating behavior such as the animal's genotype (see chapter 5) and its past sexual experience. Organizational effects of early hormones are another potential source and might mediate an effect of genotype. But regardless of whether

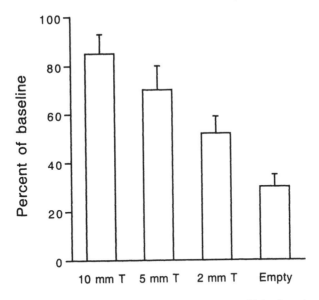

FIGURE 2.4. The dose–response relationship between testosterone (T) implant size and penile erection reflexes in castrated male rats (model A in fig. 2.2). (Here and elsewhere, error bars represent standard errors of means.) The males were spinally transected and had previously received larger testosterone implants that were then replaced with those on the X axis. The decline in penile reflexes was closely related to the size of the second implant. In contrast, the smallest implants were sufficient to maintain copulatory behavior in castrated spinally intact males, with no further increase in performance with larger implants (a step function). The penile reflex results also confirm that testosterone can act at the level of the spinal cord as well as the brain to facilitate motor components of mating. Redrawn from Hart et al. (1983). © 1983 Academic Press (Elsevier).

animals are "duds" or "studs" and why, they remain duds and studs even when gonadectomized and given the exact same amount of hormone (Grunt and Young 1952; Cunningham et al. 1977). The duds were not duds because of insufficient circulating hormone. This and the threshold principle have been known for many years, and yet the "dose–response fallacy" continues to be widely assumed (Hews and Moore 1997; Fusani and Hutchison 2003).

Another potential source of individual differences is brain steroid metabolism. Because of the way steroid metabolizing enzymes are regulated in some brain regions, brain and blood levels of steroids are somewhat independent. In principle, a lack of correlation between behavior and blood precursor steroid level could occur simultaneously with a positive correlation between behavior and brain target steroid level. Testosterone is the predominant circulating testicular steroid in nonteleost vertebrate males but some of this testosterone is turned into other steroids in the brain that are the behaviorally active forms. One of the more amazing results of research on steroid regulation of courtship and mating behavior by male birds and mammals is the importance of brain-

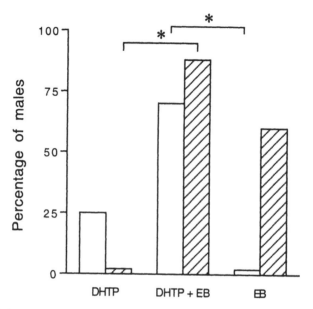

FIGURE 2.5. Contrasting effects of injections of two testosterone metabolites, dihydrotestosterone (administered as dihydrotestosterone propionate, DHTP) and estradiol (administered as estradiol benzoate, EB), on the percentage of male Japanese quail with regressed testes showing courtship displaying (strutting, clear bars) and mating (hatched bars). DHTP activated strutting but not mating, whereas EB activated mating but not strutting. The combination of these two metabolites of testosterone was even more effective and restored both behaviors. In breeding males, circulating levels of these metabolites are low, and both brain enzymatic pathways (5α-reductase and aromatase) are required for circulating testosterone to produce the complete repertoire necessary for successful male reproduction. In this and other figures, asterisks indicate statistically significant differences at $p < .05$. Redrawn from Adkins and Pniewski (1978).

produced estrogens in some species (Adkins 1981; Balthazart and Ball 1998). For example, in quail estradiol treatment by itself can stimulate mating behavior in castrated males or males with regressed testes (fig. 2.5). Testosterone does not stimulate sexual behavior or sexual interest effectively if it is combined with an aromatase inhibitor or an estrogen receptor antagonist (Watson and Adkins-Regan 1989b; Balthazart et al. 1995). For other behaviors, such as crowing, conversion of testosterone to estrogen is not as important as conversion to dihydrotestosterone (Adkins and Pniewski 1978; Deviche and Schumacher 1982). The complete male-typical behavioral repertoire of the birds that have been studied results from a combination of androgenic and estrogenic metabolites of testosterone (Hutchison 1990; Harding 1991). Mammalian brains have lower levels of aromatase. While aromatization is not quite as important for male behavior as in birds, nonetheless, the general conclusion that conversion of testosterone in the brain to other steroids contributes to male behavior still holds (Meisel and Sachs 1994).

Why are step-function relationships to circulating hormones more common for behavior than for other characters? For example, in human males, testosterone doses producing circulating levels spanning the normal range produce graded increases in muscle strength but a flat (step-) sexual behavior response function (Bhasin et al. 2001a,b). Why do some behaviors show a step-function relationship while others show a graded relationship? For example, the copulatory behavior of male rats shows a step-function relationship to testosterone at the same time that penile reflexes and aggressive behavior show a more graded function (fig. 2.4) (Hart et al. 1983; Albert et al. 1990). These contrasts also serve as a reminder that the actual dose–response relationship must be determined experimentally rather than assumed. One possibility is the nature of the behavior and how it is defined and quantified. If the behavior is scored as yes, it occurred, or no, it didn't, the data won't allow a graded response. If the behavior is observed infrequently, as in many free-living animals, there isn't much possibility of extensive variation in frequency either. By measuring separately components of behavior that can vary in a graded manner, such as the pitch of a crowing male's voice, at least the measurement allows a graded relationship to be seen if it exists. In a similar vein, behavior that is not a single act but a whole set of different but functionally connected acts is more likely to vary in a graded manner if it is summarized as a composite score based on how many of the components occur. Such composite scores may, however, obscure important differences between components.

From a social behavior perspective, the most interesting reason that hormone levels above the threshold are not correlated with behavior is that social context has a big influence on the likelihood that hormone-dependent behavior will occur (Lincoln et al. 1972; Crews and Moore 1986; Keverne 1992; Wallen 2001). Just because a male has hormone levels (androgens or estrogens, in the circulation or in the brain) sufficient to mate doesn't mean that he will be allowed to mate by other males, will be accepted by females, or will think it is wise to try. This seems obvious for humans, but it's equally applicable to other species. Because behavior is socially modulated, hormones are permissive (necessary but not sufficient). This principle probably extends to peptide-behavior relationships, too (Trainor et al. 2003). The concept of "permissive" is sometimes contrasted with "activational" (necessary and sufficient), although in the context of sexual differentiation "activational" more often means "not organizational" without necessarily implying sufficiency.

In addition to the usual reasons common to all of science for being skeptical about the value of correlational studies for testing causal hypotheses, one should be doubly skeptical when a correlation between blood hormone level and behavior is presented as evidence that the hormone is producing the behavior. The reverse direction of causation, where engaging in the behavior causes a hormonal change, is just as likely to have produced the correlation.

Individual and Species Variation in
Hormone Dependence of Mating Behavior

Removing the gonads of animals does not always eliminate mating and other sexual behavior. One of the classic findings of experiments with domestic cats was that castration of males before they had acquired any sexual experience with females had a greater effect on mating than castration after they had acquired sexual experience (Rosenblatt and Aronson 1958). This effect is familiar to cat owners who are dismayed to discover that neutering their adult male does little to prevent him from escaping the house in search of females. A role for experience in the retention of behavior after castration is not limited to cats, but is seen in some other vertebrates such as the lizard *Cnemidophorus inornatus* (Sakata et al. 2002). Why does sexual experience make the behavior of individuals less hormone dependent? What is the aspect of the experience that is critical? What has changed about the mechanisms of hormone action? Does this phenomenon contribute to the increases in reproductive efficiency and success with age that have been documented in many wild species? These questions remain largely unanswered, but in the years since the research with cats it has been learned that experiences of many kinds can change brain structure permanently, suggesting that a neurobiological explanation might eventually be discovered. New hypotheses are likely to emerge from studies of steroid receptor knockout mice and of ligand-independent steroid receptor activation (Phelps et al. 1998; Auger 2001, 2004).

Castrated male big brown bats (*Eptesicus fuscus*) are capable of mating in dyadic encounters but are not seen to mate in group settings, as if they can't outcompete intact males for access to females (fig. 2.1C) (Mendonça et al. 1996), another way that hormones are necessary but not sufficient because of social context. Castration eliminates mating in lizards of several species unless the males are left in their home cage or on their territory (Moore and Lindzey 1992). Here a familiar environment seems to help maintain the behavior independently of circulating gonadal steroids, as if the environmental cues are serving as conditioned stimuli for brain level mechanisms (neurosteroid production? ligand-independent steroid receptor activation?).

In addition to individual and situational differences, vertebrate species seem to differ in the degree to which mating behavior (especially male mating behavior) is dependent on gonadal hormones. While castration nearly always results in some decrement by some measure, there is enormous variation in the magnitude of the decrement. For example, castrated cichlid fish (*Sarotherodon* sp.), spadefoot toads (*Scaphiopus couchii*), white-crowned sparrows (*Zonotrichia leucophrys*), and common marmosets (*Callithrix jacchus*) are reported to continue right on courting females and attempting copulation for weeks or months, whereas castrated sticklebacks (*Gasterosteus aculeatus*), lizards (e. g., *Uta stansburiana*), Japanese quail, and talapoin monkeys (*Miopithecus talapoin*) stop mating shortly after castration (Aronson 1959; Liley and Stacey 1983; Moore and Kranz 1983; Moore and Lindzey 1992; Meisel and Sachs 1994; Harvey and Propper 1997; Dixson 1998).

How should this diversity be explained? Over the decades a variety of hypotheses have been proposed, some of which have stimulated important research leading to new discoveries (Meisel and Sachs 1994). Because these hypotheses address the question from different viewpoints (different Tinbergian questions) they are not competing alternatives. Reviewing them is a good way to see how researchers have thought comparatively about hormone behavior relationships over the years and will preview themes in later chapters.

One set of hypotheses focuses on the endocrine mechanisms themselves. Perhaps not all of the gonadal tissue was successfully removed by the castration surgery or the gonads regenerated. This hypothesis is readily rejected by microscopic examination at the end of the experiment or by confirming that circulating sex steroids are low to undetectable. While some of the cases of teleost fishes that continue to court and mate after surgery could be due to incomplete castrations, this probably doesn't explain all of them (Liley and Stacey 1983). Perhaps the behavior is supported by gonadotropins from the pituitary such as LH, which increase following castration due to the loss of negative feedback by gonadal steroids. Although a gonadotropin hypothesis has been proposed repeatedly to explain several different puzzles in behavioral endocrinology, there is remarkably little evidence for any effects of these or any other pituitary hormones on social behavior with the exception of prolactin's role in parental behavior (Harding 1983). Perhaps adrenal production of sex steroids is supporting the behavior in the absence of the gonads. Independently of its merits for explaining mating by castrated males, studies of a few female mammals, such as horses and musk shrews (*Suncus murinus*), have found a significant contribution of adrenal steroids to female receptivity or proceptivity (female solicitation), although in musk shrews and probably mares as well the ovaries also need to be present (Asa et al. 1980; Fortman et al. 1992).

Frank Beach, a comparative psychologist and one of the "founding fathers" of the modern field of hormones and behavior, proposed a brain phylogeny hypothesis to explain diversity in the hormone dependence of male mating behavior (Beach 1947). According to this hypothesis, "higher" mammals such as primates are more encephalized (have more forebrain and cortex) than "lower" mammals such as rats, and the more encephalized the species, the less hormone dependent the behavior. This hypothesis was grounded in a *scala naturae* concept of phylogeny and brain evolution prevalent at that time. Also, the evidence at that time suggested that the sexual behavior of humans (thought to be the most encephalized species) might not be very hormone dependent, because some castrated men, unlike rats, seemed to be able to function surprisingly well sexually. We now know that the true phylogeny of mammals should be conceptualized as a tree-like structure, that large brains with a lot of forebrain and cortex have evolved more than once in the mammals, not just in the primate lineage, that the sexual behavior of men is dependent on gonadal testosterone (Bagatell et al. 1994), and, as Aronson pointed out back in 1959, that so-called "lower" vertebrates have just as much diversity in the hormone dependence of behavior as mammals do. Yet claims that "higher" animals are more hormonally emancipated still crop up from time to time.

Although Beach's hypothesis turned out to be incorrect, the idea of using phylogenetic hypotheses to make progress in understanding diversity in hormone–behavior relationships was an important insight (Hart 1974). An up-to-date use of this approach would draw on currently accepted phylogenetic trees, would include some kind of allometric approach to relative sizes of different brain components, and would apply a comparative method that controls for correlations due to shared ancestry. The remaining challenge is whether the phenomenon itself (sexual behavior following castration) can be pinned down well enough to serve as the character in such a serious comparative analysis. For example, what should the social context be for its assessment? Wouldn't all the species have to assessed in the same way by the same research team? Just thinking about such an attempt is fatiguing!

More recent hypotheses have taken an ecological and functional approach as proposed by Hart (1974), asking when the behavior occurs during the year relative to actual breeding and whether it seems to serve strictly reproductive functions rather than multiple functions (Asa et al. 1980; Crews and Moore 1986; Wingfield 1994b). Regardless of their form, behaviors that occur throughout the year and serve functions in addition to reproduction are hypothesized to be less sex steroid dependent than behaviors confined to the mating season that serve to get gametes together. Primate observers have long recognized that mounting of another animal seems to serve multiple functions, occurs outside of the breeding season, and is performed by juveniles as well as adults and by both sexes. These are not the hallmarks of a highly hormone-dependent behavior. In a similar vein, Barth (1968) proposed that in female insects receptivity is endocrine regulated in longer-lived species where the behavior will go on and off in relation to egg cycles but not in short-lived species that mate only once and then die. These kinds of hypotheses are an essential part of any effort to better integrate behavioral endocrinology and behavioral ecology.

Female Mating Behavior and Sex Differences in Hormone Dependence

Reproduction is often more energy and time limited for females than males, with greater present and future costs. This asymmetry has ramifications for the evolution of many other sex differences, including the greater condition dependence of female hormone levels (Aubret et al. 2002). It is now widely appreciated that females and males exercise different strategies for when to mate, with how many partners, and with what kind of partners (Andersson 1994; Shuster and Wade 2003). Parental behavior per se is only a source of asymmetry when it limits reproductive rate (Clutton-Brock and Parker 1992). For example, many teleosts have male parental care yet egg guarding and fanning don't limit the male's ability to attract and mate with females (Berglund and Rosenqvist 2003).

Females are not necessarily fertile all the time during the breeding period. Many mammals ovulate on a periodic basis and most cannot store viable sperm for significant periods. A female wolf ovulates only once a year (Asa 1997). A female green anole lizard (*Anolis carolinensis*) ovulates just one ovum every two weeks, and during mating she is immobile under the male for 20 minutes, a likely predation risk for both parties (Crews 2002). Some female insects mate only once in a lifetime.

Thus, for some females a critical problem to be solved is coordinating mating behavior with the availability of fertilizable eggs. The solution for many is tight regulation of mating behavior by the same hormonal milieu that accompanies ovum maturation and impending ovulation. With tight hormonal control of mating, sexual behavior drops fast and far after removal of the ovaries, with little variation between individuals or species. Viewing the relationship between hormones and mating behavior from a functional perspective that acknowledges sexual asymmetries leads to deeper understanding of the strong tendency (with some exceptions) for male mating behavior to be less strictly dependent on the gonads than female mating behavior (Aronson 1959). It also suggests the hypothesis that mating behavior might be more hormone dependent in males than females in species where males invest more in reproduction.

The classic case of ovulatory hormones as mechanisms of female mating behavior is estrus in many female mammals. Female golden hamsters (*Mesocricetus auratus*) are a good example (Lisk and Reuter 1980; Lisk et al. 1983). Like many terrestrial mammals, they are nocturnal and relatively solitary. Hamsters are spontaneous ovulators, like humans and other primates. That is, they do not require a mating stimulus to ovulate, in contrast to induced ovulators, which ovulate in response to the stimulus of mating. When ovulation is not imminent, the female is aggressive toward males and does not let them into her underground burrow. As ovulation approaches, however, the female begins depositing a vaginal secretion that is very attractive to males around her burrow entrance. This is an important component of her proceptive behavior. Eventually a male enters and is allowed to sniff the female's anogenital region and then copulate for about an hour. During the male's sniffing and mating the female is in an immobilized lordosis posture, her receptive behavior. Afterward the female turns aggressive again and drives the male away from her burrow.

This pattern will then repeat every four days (the length of her ovulatory cycle) unless she gets pregnant, which in the wild she probably always does.

What produces this marked change in the female's behavior? Vaginal marking is activated by the rising levels of estrogen (principally estradiol) coming from the maturing ovarian follicles to be ovulated. Once this estrogen priming has occurred, rising ovarian progesterone induces sexual receptivity. Shortly after ovulation these steroid levels fall and the female's behavior reverts back to aggression. Ovariectomized females never show vaginal marking or receptivity and are aggressive all the time. If injected with estradiol followed by progesterone, all the normal proestrous (vaginal marking) and estrous (mating) behavior occurs a predictable time later.

It is very common to find that ovarian hormones are critical for female vertebrate mating behavior, either estrogens alone, estrogens followed by progesterone, or progesterone followed by estrogens (sheep). Yet there are interesting exceptions and variations on this theme. Female musk shrews (*Suncus murinus*) have no spontaneous ovarian cycle, and interactions with males rapidly induce receptivity even when there are no developing follicles (Rissman 1990). Mating then stimulates follicular development and ovulation. Ovariectomized female big brown bats still show proceptive behavior and mate even more than intact females (Mendonça et al. 1996). Ovariectomized and adrenalectomized female common marmosets (*Callithrix jacchus*) continue to show cyclic proceptivity (Kendrick and Dixson 1984). Female Jordan's salamanders (*Plethodon jordani*) ovulate weeks or months after mating and insemination, and so presumably mating has a different hormonal basis than ovulation (Rollmann et al. 1999). In some species the hormones stimulating female sexual behavior are not steroids, as when prostaglandins are responsible for turning off a female frog's mate rejection calls (Diakow et al. 1978). These frogs have external fertilization, as do many teleost fishes. Liley and Stacey (1983) formulated a general principle along these lines, proposing two hormone–behavior patterns for female vertebrates: (1) external fertilization, where mating occurs at oviposition, not ovulation, and oviposition hormones like prostaglandins are the stimulators for the female's behavior; and (2) internal fertilization, where mating occurs near ovulation and estrogens from maturing follicles serve to stimulate female mating behavior.

Sex can be costly for female invertebrates as well as vertebrates and mating is often very closely tied to ovulation. Here the tight coordination is achieved not by ovarian hormones but by having the gonadotrophic hormones, which are juvenile hormones, stimulate sexual behavior. Rising levels of juvenile hormones from the corpora allata have been shown to stimulate receptivity in females of relatively long-lived species in several insect orders (Nijhout 1994).

Among nonprimate mammals, females of most species don't show any sexual interest in males or engage in copulatory behavior except when they are fertile (or soon will be, as in the musk shrew case), which is why estrus (behavior associated with ovulation) is so striking and obvious. Female porcupines (*Hystrix africaeaustralis*) seem to be an exception, for they are reported to be surprisingly sexually active outside the periovulatory period (Morris and van Aarde 1985). Female Old World primates were thought to be exceptions as well, and indeed some are. Human females have an ovulatory cycle that is indistinguishable hormonally from the cycles of their closer primate relatives, but without estrous behavior or an estrous cycle. Nor are there any obvious external signs of ovulation. Primates with these characteristics are said to have concealed ovulation. Researchers have been quite creative in coming up with hypotheses to explain why some primates have concealed ovulation while other primates and most other mammals show behavioral and other signs of ovulation (Heistermann et al. [2001] is a recent example).

In the meantime, there have been significant developments in knowledge of the behavior of female primates (Dixson 1998; French and Schaffner 2000; Wallen 2000). Experiments with rhesus macaques (*Macaca mulatta*) show that when females are given a means to control whether to engage in a sexual interaction, so that it is their interest and not the male's that is measured, their sexual motivation is more closely tied to ovulation than was previously believed. Furthermore, the mating behavior of female monkeys and apes differs from that of hamsters and other mammals. There is no immobilization, lordosis, or other special reflex that limits the ability to copulate to the fertile period and requires hormonal activation. Hormones regulate desire, not ability (Wallen 1990). Historically, humans have not been consciously aware of when ovulation is occurring, but there are subtle changes in some women's behavior that are detectable in data summarized over whole groups of subjects (Morris and Udry 1982; Stanislaw and Rice 1988; Hedricks 1994). Ovulation is not completely concealed; nonetheless, humans and some other female primates do not show the phenomenon of estrus. Perhaps the question is not why they don't, but rather why some animals have evolved such powerful morphological or behavioral signals of ovulatory status (Nunn et al. 2001). And, of course, what is key is whether conspecifics can tell when ovulation is occurring, not whether human researchers can (Dixson 1998).

In a number of domestic and wild mammals, females in estrus show an increased tendency to mount other females (Dagg 1984). When female–female mounting occurs during estrus it tends to be stimulated by periovulatory changes in gonadal sex steroids (Goy and Roy 1991). Is this behavior merely a functionless epiphenomenon of the estrous hormonal state, or a behavior with some adaptive function? Such a question is not easy to answer. There is no dearth of hypotheses for possible functions, but satisfying tests of those hypotheses are missing. For example, the social organization of some ungulates that display female–female mounting is sexually segregated (Ruckstuhl and Neuhaus 2002). Perhaps female–female mounting serves as a long-distance visible signal that the mounter is ready to ovulate, making male approach worthwhile (Billings and Katz 1999). The hypothesis that males are attracted from a distance to a mounting female is testable. It is harder to test the hypothesis that this is why female–female mounting evolved—that male attraction was the selective pressure for its origin and maintenance. Such a test would require experiments on male attraction to mounting females in multiple species varying in degree of sexual segregation, followed by a comparative analysis based on a known phylogeny of those species to see if female mounting is associated with distant males. That's a tall order.

Female *Cnemidophorus uniparens* lizards mount other females. The best predictor of this behavior is a postovulatory endocrine state on the part of the mounter in which estradiol is low and progesterone is high (Crews 2002). Experimental manipulations confirm that progesterone stimulates the male-typical copulatory pattern in these females. What is fascinating about this case is that the species is an all-female parthenogen. Crews' hypothesis for the function of this behavior is that females of the ancestral sexual species needed stimulation from mating in order to develop and ovulate multiple eggs (a reasonably common phenomenon in animals), and that the parthenogen has given up the need for sperm but not the need for that stimulation, which now can be provided only by other females (Crews et al. 1986). Even if correct, there are still mysteries here. First, this might explain the maintenance of female mounting, but what was its evolutionary origin? A functionless epiphenomenon of progesterone later co-opted to substitute for males? Second, being mounted may be good for the mountee, but what's in it for the mounter that has maintained it? Is some kind of reciprocity at work?

Just as male behavior can be based on estrogens as well as androgens, we can ask if female behavior is ever based on androgens from ovaries, adrenals, or both, acting either directly or through brain aromatization. Circulating androgens, like estrogens, peak at ovulation in many female vertebrates. Peripheral androgens that are aromatized in the brain support receptivity in the musk shrew (Rissman 1990). Although some female lizards, snakes, and turtles have substantial circulating testosterone, in most cases the behavioral significance of endogenous androgens, if any, has not been investigated (Whittier and Tokarz 1992; Staub and DeBeer 1997). Testosterone stimulates female receptivity in the green anole, an effect that is blocked by simultaneous treatment with the estrogen synthesis inhibitor fadrozole (Adkins and Schlesinger 1979; Winkler and Wade 1998), but it is not known whether blocking the action of endogenous testosterone would reduce receptivity in ovulatory females. In leopard geckos (*Eublepharis macularius*) females receiving long-duration implants of testosterone (mimicking the male-typical testosterone pattern) were unreceptive and aggressive to males, whereas females receiving short-duration implants (mimicking the ovulatory cycle levels of females) were the most receptive to males (Rhen et al. 1999).

Aggressive Behavior

Compared to mating behavior, aggressive behavior has a more complicated relationship to sex steroids, peptides, and, indeed, almost any other chemical messenger acting in the brain (see Parmigiani et al. 1998 for an example). One important reason is that aggressive behavior is a very heterogeneous category in form, function, and motivation (Moyer 1976). Aggression can be offensive or defensive. Acts of aggression can serve to fend off would-be predators or aggressive conspecifics, to subdue a large prey item, to compete for limited resources essential for survival, such as food, water, and shelter, to cause an irritating conspecific or heterospecific to stay further away, as well as to defend a territory and to compete for nest sites or mating opportunities. We would not expect sex steroid regulation of kinds of aggression that serve survival rather than reproductive functions, that occur year-round whether breeding or not, and that are used by both sexes and all ages. Instead, we would expect connections between reproductive hormones and aggression that serves direct or indirect reproductive functions, especially if the behavior is most frequent or intense during the breeding season and more so in one sex. Thus male-typical fighting and displaying that occur mainly during the mating period, such as the ritualized combat of male red deer (*Cervus elaphus*) during the rutting season, decrease following castration and are restored by testosterone treatment (Lincoln et al. 1972). A similar pattern of results shows up repeatedly in the literature on sex steroid hormones and aggression across a broad array of vertebrates and across a diverse array of aggressive acts ranging from vocalization at a distance to close range injurious acts like biting (fig. 2.6A) (Svare 1983; Barfield 1984; Archer 1988). The function of the behavior better predicts its hormonal basis than its form does. The kinds of male aggressive behavior with the closest connections to sex steroids seem to be those that probably have been under sexual selection.

A corollary of this principle is that aggressive acts such as biting that serve multiple functions will not be eliminated by castration. Contests between males over females might subside, but biting that serves other functions will continue. When both mating and aggressive behavior are measured in castrated males, more often than not mating behavior decreases more than aggressive behavior (for example, in the mouse *Peromyscus californicus* [Trainor and Marler 2001]).

The concept of hormones as permissive depending on context is especially important in aggressive behavior. An animal's position in a dominance hierarchy plays a large role in the expression of aggression (chapter 3), and residents of an area are more likely to be aggressors than intruders. No matter how testosterone laden a male is, he would be stupid to take on an opponent who will surely win or will retaliate to lethal effect. In addition to the risk of injury

A

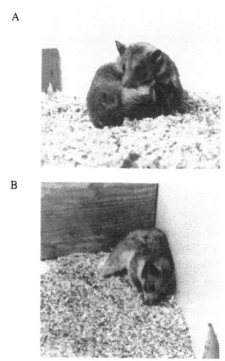

B

FIGURE 2.6. Fighting (A) and flank marking (B) by golden hamsters. Flank marking is a scent-marking (signaling) behavior with a predominantly agonistic function. Both behaviors are hormonally regulated. Courtesy of Robert Johnston.

or death, aggressive encounters and especially defeat produce the kind of HPA activation that takes a heavy physiological toll (Haller 1995; Haller et al. 1998). Energetic costs also put a brake on the aggression promoting properties of testosterone (Marler et al. 1995).

Male aggression tends to get equated with testosterone but this equation is just as oversimplified for aggression as it was for mating behavior. The failure of castration to invariably reduce aggression is but one reason. In some species aggression is greater outside the breeding season and it is decreased, not increased, by testosterone (for example, the hamster *Phodopus sungorus* [Jasnow et al. 2000]). Estrogens produced by brain aromatization of testosterone contribute to aggression in Japanese quail and some domestic mouse strains (Hasan et al. 1988; Schlinger and Callard 1990). Acute elevations in glucocorticoids stimulate aggressive behavior in mice when testosterone is also present (Leshner 1983; von Holst 1998).

The brain peptides arginine vasopressin and arginine vasotocin (which from here on will be called simply vasopressin and vasotocin) have been repeatedly linked to male–male aggressive signaling and fighting in a wide variety of vertebrates, especially those peptides produced in the septal region and hypo-thalamus (Sheehan and Numan 2000; Goodson and Bass 2001). For example, vasopressin action in the anterior hypothalamus is a critical mechanism under-lying flank marking by hamsters, in which odors are deposited that serve ag-gressive functions (fig. 2.6B) (Ferris et al. 1994). In a field experiment, vaso-tocin treatment enabled male gray treefrogs to acquire calling sites from resident males (Semsar et al. 1998).

Female aggression and its hormonal basis have been relatively neglected (Floody 1983). This is unfortunate because many females are very aggressive, sometimes more so than males, aggression among females is an important dyadic level process driving the spacing patterns and social systems of many animals, and in mammals the fitness consequences of rank in a dominance hierarchy are better established for females than for males.

It should be possible to predict in advance the relationship between female aggressive behavior and hormones by looking at when the behavior occurs relative to the breeding season and ovulation and what functions it serves. Female golden hamsters are a good example (Floody 1983). Like other solitary mammals, they are aggressive toward both females and males all the time except that they are not aggressive toward males when they are in estrus. Being aggressive is the "default" that does not require stimulation by sex steroids. The function of sex steroids is to turn off the aggressive behavior so that mating can occur. Manipulation experiments confirm this hypothesis. Ovariectomy has little effect on female hamster aggression. Instead, the combination of estrogen plus progesterone inhibits aggression toward males, rendering the female peaceable enough to get her ova fertilized. On the other hand, in bank voles (*Clethrionomys glareolus*) and mountain spiny lizards (*Sceloporus jar-rovi*), ovariectomy reduces female aggression and sex steroids stimulate it (Ka-pusta 1998; Woodley and Moore 1999). This pattern suggests the hypothesis that female aggression is sexually selected in these animals. Female lab rats, descendents of a group living social species, continue to direct aggression at other females after ovariectomy (DeBold and Miczek 1984), as if aggression is more related to competition for food or shelter.

There are limits, however, to how well one's intuitions predict the hormonal basis of female aggression. Mammal mothers with babies are notoriously nasty. The aggressive defense of the young coincides with lactation, inspiring hormonal hypotheses. However, extensive research has shown that this aggres-sion is actually less hormonally based than other forms of aggression (Svare 1983; Lonstein and Gammie 2002). At best, hormones prime the system during pregnancy to be able to respond to stimuli from the young by behaving nastily toward anyone that gets near them. The critical stimuli priming the behavior once the babies are born come from suckling, not from hormones. The depen-dence on stimuli from the young is characteristic of other components of parental behavior that are not aggressive, another reminder that function, not form, is a better predictor of the relationship of aggression to hormones.

When both sexes engage in similar kinds of aggression, we can ask if the hormonal bases are also similar. There are several hypotheses, including the possibilities that (1) estradiol (from ovaries in females, from brain aromatization of testosterone in males) stimulates the behavior in both sexes; (2) testosterone or progesterone stimulates it in both; (3) testosterone stimulates it in males via androgen receptors, whereas estradiol stimulates it in females via estrogen receptors; and (4) sex steroids are not the basis in either sex. Testing such hypotheses requires manipulations in both sexes to block actions of endogenous hormones, which have seldom been done. Looking at correlations between hormone levels and aggression, or administering the same hormone treatment to both sexes, are not critical tests. We can also ask if peptide mechanisms are similar. An antagonist to the V1a subtype of vasopressin receptor reduces aggression in both sexes of prairie voles (*Microtus ochrogaster*) (Stribley and Carter 1999).

When increased female aggression might enhance reproductive success (might be selected for), females can't necessarily accomplish this over the short term by evolving higher circulating testosterone levels if higher levels will interfere with ovulation, because that would be too costly. Some other mechanism, such as greater sensitivity of brain receptors to the existing circulating steroids, is a more likely route. Over longer periods of evolution, however, increased testosterone is clearly possible (for whatever selective reason), because some female vertebrates have quite high testosterone levels yet ovulate perfectly well.

Parental Behavior

Parental behavior differs from sexual behavior in ways that matter for an understanding of its connections to hormones. It is not universal in vertebrates. It occurs in mammals, birds, crocodiles, a number of teleost fish, and a few lizards, snakes, and amphibians, but is absent in other vertebrates. This scattered distribution suggests multiple independent evolutionary events, so that neither the behavior itself nor the underlying mechanisms can be assumed to be homologous (Clutton-Brock 1991; Reynolds et al. 2002). Parental sex roles are also diverse. Females are the sole parental care agent in many mammals, in some birds and fish, and in snakes and lizards that have parental care. Males are the sole caretaker of the young in a few birds and some fish, and both sexes care for the young in a few mammals, many birds, and some fish. Parenting is an entire repertoire of behaviors, not a single behavior. It occurs over an extended period of time, often against a complex and changing hormonal background. Parental behavior changes as the needs of the offspring change. There is a substantial role of experience, both within and between litters or clutches, which likely contributes to increases in reproductive success with age. An error in parental behavior can easily result in loss of all the offspring, whereas an error in a sexual interaction is usually less catastrophic. There is a need for plenty of redundant backup mechanisms.

The hormonal basis of offspring care is best understood in a few mammals such as rats, rabbits, and sheep, but some of the lessons learned are probably applicable to a wider array of taxa (Rosenblatt and Snowdon 1996). The endocrinology of pregnancy, birth, and lactation provides a list of potential candidate hormones for female parents, including gonadal sex steroids, especially estrogens, which are high during pregnancy, prolactin, which stimulates milk production, and the peptides oxytocin and vasopressin, which increase in the brain as well as in the circulation. Experimental evidence supports a role for each of these three hormone/peptide categories in at least some mammalian species (Numan 1994; Bridges 1996; Pedersen 1998). Estrogens not only are important in their own right, but also prime the brain to be able to respond to prolactin and oxytocin. Progesterone tends to be inhibitory, but the exact relationship between gonadal sex steroids and maternal behavior probably depends on whether the female normally mates during a postpartum estrus. The sites in the brain where these hormones act to facilitate maternal behavior overlap considerably with sites for sexual behavior, and include the medial preoptic area and the bed nucleus of the stria terminalis (BNST) (Numan 1994).

Female sheep that have given birth not only care for the lamb in the usual mammalian ways (licking, letting the lamb nurse) but also form a selective attachment to their own lamb during the first few hours after birth. This is highly adaptive because the females live in groups that move constantly, so that lambs could easily get mixed up. What opens up this "window of opportunity" to accept a lamb as her own is a combination of steroid priming during gestation followed by an increase in oxytocin in the paraventricular nuclei of the hypothalamus (fig. 2.7) (Lévy et al. 1996). Prolactin is not essential for any of this. In spite of its importance for mammalian lactation, prolactin is by no means "THE maternal hormone."

Steroids and peptides are more important for the onset of maternal behavior (when the offspring have just arrived and need immediate attention) than for its continued maintenance. The evidence for this important distinction is most extensive for rats but the principle is very likely to apply to many taxa. Once female rats have been exposed to young for a few days, the behavior will continue even if all relevant endocrine glands are removed, and females that have never given birth (or even juveniles) will begin to care for pups after a few days of exposure (Rosenblatt et al. 1979). Furthermore, after the first litter, hormones are not even required for rapid maternal behavior onset following birth (Bridges 1996; Fleming et al. 1996; Lévy et al. 1996). Instead, there is a

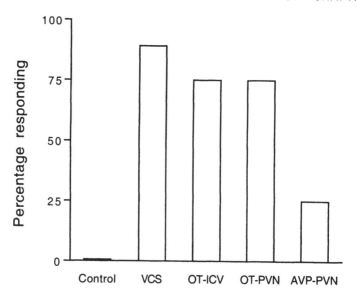

FIGURE 2.7. The percentage of female sheep responding to a foster lamb with maternal behavior following different treatments. Maternal behavior was assessed by both acceptance behaviors (for example, licking and letting the lamb suckle) and rejection behaviors (for example, butting the lamb). The females were primed with progesterone and estradiol to mimic hormonal changes of pregnancy and then were given vaginocervical stimulation (VCS, mimicking birth and causing oxytocin release), oxytocin in the cerebroventricular fluid (OT-ICV), oxytocin in the paraventricular nucleus (OT-PVN), or vasopressin in the PVN (AVP-PVN). Controls received neither VCS nor peptide. The results support the hypothesis that oxytocin released during birth induces maternal interest in the lamb. Data of Da Costa et al. (1996). © 1996 Blackwell Science Ltd.

lasting effect (a permanent sensitization) caused by the first birth such that stimuli from the young are sufficient to induce the behavior thereafter. This picture that has emerged, in which a special peripheral hormonal condition is not always needed for maternal behavior, is a far cry from the popular idea that maternal love is endowed by nature with strong hormonal determination. Instead, what is impressive (as ethologists know well) is the nurture evoking power of very young animals. Female rabbits visit the young for only five minutes a day to nurse them, and yet the stimulation the female receives during this time is sufficient to keep her in the maternal state (González-Mariscal and Rosenblatt 1996). There are many reports of females that have not given birth recently "adopting" young and caring for them. Maternal behavior offset at weaning is due more to a change in the stimuli given off by the young than to any hormonal change.

The power of offspring as stimuli shifts the mechanistic focus onto the sensory and perceptual aspects of parenting (Stern 1990). The selective attachment of the female sheep to her lamb is a chemosensory phenomenon in which she imprints to and remembers the lamb's odor through actions of norepinephrine in the main olfactory bulb (Kendrick et al. 1992). The reason that female rats that have not given birth require some exposure to pups to begin behaving maternally is that pup odors are initially aversive but become attractive with time. To some extent behaving maternally is the "default" for a rat that will occur unless some other competing tendency inhibits it (such as an aversion to their odor), and hormones serve as the disinhibitors, curbing fear or avoidance to permit the behavior to occur before the pups have died of neglect (Fleming 1986).

When male mammals behave parentally, we can ask whether the hormonal support mechanisms for behavioral onset are similar to those of females (Schradin and Anzenberger 1999; Wynne-Edwards and Reburn 2000; Ziegler 2000). In common marmosets, a biparental species, males show increases in prolactin and steroids before the birth while the female partner is pregnant. After birth the drug bromocryptine lowers both prolactin levels and infant retrieval in both sexes (Roberts et al. 2001). Thus far this is the only experimental evidence for the increasingly popular hypothesis that prolactin is a paternal behavior hormone. Male Djungarian hamsters (*Phodopus campbelli*) don't even have to be in the same room as the female prior to birth to spring into action as little midwives as soon as the pups appear, provided they mated with her at some point (Jones and Wynne-Edwards 2001). Paternal behavior in the biparental California mouse (*Peromyscus californicus*) decreases following castration and is restored by testosterone replacement. Testosterone has this effect through brain aromatization, making estrogen a key parental steroid in these males (and probably in the females as well, based on work with other rodents) (Trainor and Marler 2002). In male gerbils (*Meriones unguiculatus*), on the other hand, castration increases paternal behavior (Clark and Galef 1999). Castration has no effect on the paternal behavior of male prairie voles, and even sexually inexperienced males are spontaneously parental without prior exposure or experience (Lonstein and de Vries 1999). In cases like this, nonhormonal mechanisms such as peptides that can gear up rapidly are more likely candidates to be mechanisms. These species differences in the hormonal basis of paternal care may be related to mating systems, especially whether they create trade-offs for the males between mating and parental effort, a hypothesis that is discussed in chapter 6.

The hormonal basis of postincubation care of offspring by birds has been studied experimentally in very few species, principally ring doves (*Streptopelia risoria*), chickens, and turkeys. Chicks provide stimuli that induce parental behavior in birds, just as baby mammals induce parental behavior. In chickens and New World quail such induction causes sex steroids to decrease (Richard-Yris et al. 1983; Sharp et al. 1988; Vleck and Dobrott 1993). Consistent with this decrease, ovariectomy doesn't prevent chick care by chickens (Lea et al. 1996). Ring doves belong to a lineage of birds, the order Columbi-

FIGURE 2.8. A ring dove regurgitating crop milk to one of its two squabs. Among pigeons and doves, both parents feed the squabs in this manner. Actions of prolactin on the brain are instrumental in stimulating the behavior. Courtesy of John Buntin.

formes, that is quite special as parents (Silver 1984). Both sexes have crops lined with tissue that responds to prolactin by producing crop "milk," which is regurgitated to feed the chicks (fig. 2.8). In addition to this indirect pathway between prolactin and chick care, in which prolactin stimulates a peripheral organ that then generates sensory feedback to stimulate the act of feeding chicks, prolactin also acts directly on the brain to facilitate the behavior in both sexes (Buntin 1996). The only other experimental evidence for a role for prolactin in the care of posthatching offspring in birds is a study in which free-living female willow ptarmigan (*Lagopus lagopus*) who were given prolactin remained closer to their chicks when flushed compared to controls (Pedersen 1989). In light of the picture that has emerged in mammals and the special nature of doves, it is premature to assume that prolactin is the hormone responsible for postincubation care of offspring in birds generally. Finding that prolactin levels are elevated during the period of chick care is suggestive, but because stimuli from the young can cause prolactin elevation in some species, such correlations are not evidence that prolactin causes chick care (Sockman et al. 2004). Another reason for caution is that in birds with precocial young, such as galliforms and waterfowl, prolactin falls when the chicks hatch and is low during the period of chick care (El Halawani et al. 1984; Buntin 1996). If prolactin is doing anything for these parents, it's more likely to be priming the onset of the behavior than maintaining it.

Males of several species of temperate zone passerine birds become poor parents when their testosterone levels are experimentally elevated, a phenomenon that is important for the discussion of mating systems in chapter 3 and life history trade-offs in chapter 6 (Silverin 1980; Hegner and Wingfield 1987; Dittami et al. 1991; Ketterson et al. 1992). The treated males are in effect too busy with other priorities like competing with other males and attracting females to have any time left for their chicks. The steroid receptor antagonist cyproterone acetate increased male parental care in spotless starlings (*Sturnus unicolor*) (Moreno et al. 1999). As in galliforms, care of posthatching offspring is associated with low sex steroid action. Nothing about prolactin can be concluded from these experiments. In male dark-eyed juncos (*Junco hyemalis*), who reduce feeding of chicks when treated with testosterone, the treatment doesn't reduce prolactin levels or interfere with prolactin binding in the brain (Schoech et al. 1998). Some other mode of action seems to be at work. Is testosterone causing a change in the level of some hormone other than prolactin that facilitates feeding? What are the nonhormonal mechanisms underlying the setting of behavioral priorities that testosterone is impacting?

When teleost fish care for the young, it is more often the male who is the parent. A common pattern is for the male to build a nest, court and mate with females who approach the nest, and then care for the eggs the females lay in the nest while also continuing to court and mate with new females. Parental care often ends with hatching, but even when it doesn't little is known about the underlying physiology. Male sticklebacks (*Gasterosteus aculeatus*) glue their nests together with a secretion from the kidneys and male Siamese fighting fish (*Betta splendens*) blow bubble nests with mucus. These secretions, along with the nest building and egg care behavior, are androgen dependent. Castration reduces nest building and egg care, and 11-ketotestosterone treatment is more effective than testosterone for restoring it (Liley and Stacey 1983; Borg 1994). The mechanism of action of the 11-ketotestosterone on the behavior is unclear. One possibility is that it is converted in the brain to 5a-dihydrotestosterone (Borg 1994). A prolactin hypothesis has been entertained for parental behavior in some fish. Prolactin treatment induces egg fanning behavior (Blüm and Fiedler 1965; de Ruiter et al. 1986). On the whole, however, it is a difficult hypothesis to test for technical reasons, and so the jury is still out (Liley and Stacey 1983; Schradin and Anzenberger 1999).

Steroids are not known to be causally related to care of posthatching young in any invertebrates. Instead, other hormone categories serve such functions. The onset of parental behavior in burying beetles (*Nicrophorus orbicollis*) is associated with an increase in juvenile hormones. Stimuli from the young both influence hormone levels and prolong parental behavior, a remarkable parallel with the induction phenomenon in birds and mammals (Trumbo 1996).

How Hormones Alter Behavior: Circuits, Networks, and Processes

Male cardinals break into song as their sex steroid levels rise in late winter. As estrogen levels rise, female tigers that have lived solitarily for weeks suddenly start roaring and allow a male that responds to approach and mate. How do hormones produce these dramatic fitness-enhancing changes in behavior? There are two complementary ways this question has been approached. One is to ask about the neural and biochemical mechanisms of hormone action. The models guiding this approach are often anatomical "wiring diagrams" that trace pathways from one cell group and region to another (e. g., Pfaff et al. 1994; Newman 2002). The models are combined with the type of information presented in chapter 1 about steroid targets and receptors and the molecular consequences of steroid action. The other approach is less biologically reductionistic and asks about the underlying processes (for example, motivational or cognitive) that might have been affected, focusing on the conceptual level between the nervous system and the observable behavioral outcome (see Haller et al. 1998 for a recent example). This is the traditional domain of psychology, but it is increasingly relevant to behavioral ecology as well.

An old way to conceptualize behavior is to view it as motor output that occurs in response to stimuli. This reflex approach, while outdated in some respects, nonetheless suggests timeless and important questions about whether and how hormones act on peripheral sensory components, on motor components, and on the central nervous (brain and spinal cord) components that process and interpret sensory inputs, organize motor outputs, and link the two appropriately. There was a time when lively debate occurred over whether central nervous system or peripheral hormone targets were more important for behavior (Adkins-Regan 1996). Hormones clearly stimulate peripheral structures such as the red rumps of female baboons, the crops of doves, and the scent glands of salamanders in parallel with their stimulating effect on the animals' behavior. "Peripheralists" thought that these structures might be the primary hormonal targets and that the behavior was stimulated by sensory feedback from these peripheral structures and not by hormones acting centrally. That debate has long since been resolved as both central and peripheral actions are clearly important, operating in concert and requiring careful experiments to disentangle. The research summarized earlier on central and peripheral actions of prolactin on ring dove parental behavior is a beautiful modern example (Buntin 1996).

There is abundant evidence that motor organs and systems can be a target for steroid action and many such examples are now classics of neuroethology. Androgens (testosterone, 5α-dihydrotestosterone, or 11-ketotestosterone) increase the size of the muscles used by males to produce sounds and other signals that attract females and warn other males, to clasp females, to produce the penile movements for fertilization, and (in some mammals) to be strong generally (Bass 1989; Kelley 1997; Bhasin et al. 2001a,b). Further upstream, the brainstem and spinal cord motor neurons controlling these muscles can also be androgen sensitive (Arnold et al. 1976; Arnold 1984).

In these examples the motor system is sexually dimorphic and is used in a special way for reproductive purposes. This is exactly when we expect to find steroid regulation. In contrast, sensory organs of vertebrates are not usually sexually dimorphic in any obvious way, are used for many nonsocial purposes, and understandably have been relatively neglected as potential steroid targets. Males and females often differ in their responses to the same stimuli, but this is usually assumed to reflect sex differences in perception or motivation, rather than peripheral sensory ability. There are important exceptions, however, where the sex difference begins at the periphery (Gandelman 1983; Moffatt 2003). Estradiol increases the size of the receptive field of the somatosensory nerve innervating the perineal skin of the female rat even when the nerve is disconnected from the central nervous system, an effect that in an intact female is likely to increase her propensity to adopt the lordosis posture when tactile contact occurs there (Komisaruk et al. 1972; Kow and Pfaff 1973/1974). The retinal, optic tectal, and retinotectal neurons of goldfish (*Carassius auratus*) contain aromatase, and the developing primary visual cortex of rats contains androgen receptors, raising as yet untested hypotheses about a role in regulation of visual functions related to reproduction and suggesting a potentially fruitful hybrid zone between behavioral endocrinology and sensory ecology (Gelinas and Callard 1993; Nunez et al. 2003). The inner ear of female midshipmen fish (*Porichthys notatus*) contains estrogen receptors, and steroid manipulations alter auditory sensitivity (Sisneros et al. 2004). The sheer existence of steroidal pheromones in animals such as pigs and goldfish means that olfactory receptor tissue is a kind of steroid target (Signoret 1967; Melrose et al. 1971; Sorensen and Stacey 1999). Steroids influence responses of olfactory receptor neurons to social odors in goldfish and mice (Cardwell et al. 1995; Halem et al. 1999).

Much of the brain is involved in higher-level sensory processing or motor organization and both steroids and peptides are known to act at this level to facilitate social behavior. Vasotocin facilitates the courtship behavior of male rough-skinned newts through effects on sensorimotor integration (Thompson and Moore 2000; Rose and Moore 2002). Steroid receptors are present in the main olfactory bulb of ferrets (*Mustela putorius furo*) and testosterone increases the activity of olfactory bulb neurons in response to female odors in ferrets (Kelliher et al. 1998). In golden hamsters there are steroid receptors in the projections of the accessory olfactory bulb (the bulb receiving input from the vomeronasal organ) to a subcircuit of the extended amygdala (a portion of the medial amygdaloid nucleus and a portion of the medial part of the bed nucleus of the stria terminalis). Through manipulation experiments Wood and Newman (1995) have shown where the hormone-sensitive and chemosensory circuits communicate with each other, providing an anatomical explanation for why steroids and female odors are both required for males to copulate.

There is abundant evidence that steroid actions on the anterior hypothalamus, preoptic area, extended amygdala (medial amygdaloid nucleus and bed nucleus of the stria terminalis) and their homologs, and septal region underly the expression of steroid regulated social behavior. This is best established for mating behavior. In most vertebrates that have been studied, gonadectomized animals that fail to mate will begin to mate when a small amount of testosterone or estradiol is implanted into the anterior hypothalamus or preoptic area. One of the earliest such experiments was carried out in a bird, the domestic fowl (Barfield 1969). The "hot spots" are different for male-typical and female-typical behavior. The medial preoptic area (see fig. 1.2) works best for male mounting, whereas the ventromedial hypothalamus works best for female receptivity. The behavior is not necessarily as complete or as vigorous as in the intact animal (steroid action at other sites and at the periphery is also required), but it is remarkable that steroid action at a single site is effective at all.

Network and computational models are increasingly important in neuroscience. In mammals there is much convergent evidence for a social behavior brain network that includes the sex steroid-sensitive regions listed in the previous paragraph (Newman 2002). These regions function as nodes that all project to each other and contain vasopressin as well as steroid receptors. Activity of the same network underlies social behaviors such as affiliation, sexual, aggressive, and parental behavior, each of which reflects a different pattern of relative activity across the nodes (Newman 2002). Much less is known about the nodes of the social behavior network in nonmammalian vertebrates aside from the contribution of the hypothalamus and preoptic area to mating behavior.

Now let's ask what processes are enhanced by steroid administration and by naturally increasing steroid levels. One important conceptual distinction is between motivation, arousal, and desire versus ability and motor performance (Wallen 2001). Experiments using implants, lesions, and other techniques for manipulating brain regions tend to support the proposal that the anterior hypothalamus–preoptic area (AH-POA) and extended amygdala (MeA and BNST) have somewhat different roles in steroid-stimulated mating behavior (Everitt 1990), roles that are not unique to mammals but might be a more general vertebrate characteristic (Thompson et al. 1998). The AH-POA seems to affect the performance of the behavior, for example, the ability to execute sexual reflexes. The MeA and BNST are actually two subcircuits, one for general arousal and the other (the subcircuit with the steroid receptors) that determines that sexual behavior will occur (Newman 2002). Lesions of MeA often affect sexual arousal and interest in the sexual partner as secondary consequences of a general interference with arousal. Studies of the behavior of ERα knockout mice suggest a role for estrogen action in generalized arousal (Garey et al. 2003). Arousal and motivational effects are not limited to steroids; the effect of vasotocin on vocal behavior in gray treefrogs (*Hyla versicolor*) seems to be motivational (Tito et al. 1999).

In social behavior there are at least two parties. This allows an indirect process by which hormones can alter an animal's behavior, namely by changing the stimuli that the animal gives off, either passively or actively (behaviorally). These in turn change how other individuals react, which in turn changes the target individual's behavior. A female that is ovulating is more likely to mate not just because she is more interested; males are more interested in her, because her stimulus configuration is more attractive. A male with reproductive levels of hormones gets into more fights not only because he is more aggressive, but also because the stimuli he gives off signal his adult breeding male status and inspire attacks by others. In humans there is yet another highly interesting pathway for hormone effects on behavior, via self-perception. For example, at puberty hormones change a person's body in ways that are easily perceived by the individual, and this bodily perception itself can lead to changes in gender-related behavior (Money 1987). It is fascinating to wonder if such a process also could happen in nonhumans, for example, through self-perceived hormonally regulated odors.

With respect to cognitive processes, steroids have well-documented effects on performance on a variety of laboratory learning and memory tasks. Much of this literature is rather distant from the social behavior realm or from ecologically relevant situations. An important exception is the research on song learning by oscine passerines (fig. 2.9). Because song learning in some of these species occurs only during particular stages of development, it has been hypothesized that changing steroid levels during development are a mechanism for opening and closing the sensitive period. This has been a surprisingly difficult hypothesis to test, partly because castration (the obvious experimental manipulation) does not reduce the levels of circulating estrogens in finches or sparrows, the usual subjects, and estrogens are important for singing (Marler et al. 1988; Adkins-Regan et al. 1990). There is some positive support nonetheless. Male zebra finches that were both castrated and given steroid receptor antagonists (to block both androgens and estrogens) at 20 days of age grew up to sing abnormal songs (Bottjer and Hewer 1992). They still sang, but the songs contained too few learned syllables, as if the sensitive period window for learning them had shut too early. The same treatment of adults did not affect song quality. White-crowned sparrows given testosterone after song acquisition but before song production sang abnormal songs, suggesting that a

FIGURE 2.9. A juvenile zebra finch (just left of center) with its parents (right of center) and two other adult males (left). Juvenile males learn song syllables from their father and other adult males in the group with which they interact. Perturbing the normal steroid levels of juvenile males results in poorer song learning and song quality. Photo by the author.

premature increase in steroid level is as deleterious as insufficient steroid (Whaling et al. 1995).

Testosterone increases persistence of attention in male domestic chicks (Andrew 1991). This process may enable an adult rooster to better resist distracting stimuli during an interaction with a female or a male rival (Andrew and Jones 1992; Jones and Andrew 1992). The same process might underlie the stimulating effect of testosterone on the observable vigilance behavior of male grey partridges (*Perdix perdix*), a behavior that is attractive to females (Fusani et al. 1997). The hypothesis that sex hormones might selectively increase attention to sexual stimuli in humans is also receiving attention (Alexander and Sherwin 1991, 1993).

Female rats that become mothers show improvements in learning and memory that are probably adaptive given the increased foraging demands they face (Kinsley et al. 1999). They also show steroid-produced changes in neurons in the preoptic area, such as an increase in cell body volume (Keyser-Marcus et al. 2001), suggesting that neuroplasticity in this region might underlie this cognitive improvement. Steroid-regulated peptides have also been linked to socially and ecologically relevant aspects of memory. For example, septal vasopressin is important for the recognition of familiar individuals by rodents (Sheehan and Numan 2000).

The discovery of steroid receptors and other steroid mechanisms in parts of the isocortex has expanded the array of potential learning, memory, and other cognitive processes that might be affected by steroids. Together with some of the classical targets such as the medial amygdaloid nucleus, these discoveries provide a bridge to some of the concepts important to behavioral ecologists such as decision making. Different behaviors cannot be performed simultaneously, and so to function adaptively, animals' nervous systems are continually making decisions about what behavior should dominate at any particular moment. The correct decision depends on many factors that must be integrated, including the state of the physical environment, the condition of the animal, its past history, and what others are doing. Steroids and peptides can be mechanisms underlying these important decisions.

Some of the most compelling evidence comes from studies in which acute glucocorticoid elevation from HPA activation quickly throws behavioral and well as physiological switches that redirect behavioral priorities away from reproduction and toward survival (Silverin 1998; Wingfield et al. 1998; Orchinik et al. 2002). Male white-crowned sparrows living in the northern Rocky Mountains are likely to experience a blizzard during the spring breeding period. This is associated with a rise in corticosterone (fig. 2.10A) and a behavioral shift from territorial advertisement and defense in exposed locations toward finding shelter and foraging to avoid starvation. The confirmation that the steroid elevation actually causes the behavioral shift comes from an experiment with song sparrows (*Melospiza melodia*) in which a corticosterone implant reduces the male's response to a simulated territorial intrusion (a speaker playing song with a stuffed male model on top) (fig. 2.10B) (Wingfield 1988).

In male rough-skinned newts (*Taricha granulosa*), courtship and mating is a prolonged process. Testosterone and vasotocin work together to stimulate this behavior but corticosterone is inhibitory (Coddington and Moore 2003). Corticosterone is elevated by the threat of predation and acts quickly to interrupt mating, putting the male in a less vulnerable position. The mechanism for the effect is a membrane corticosteroid receptor that alters the responses of brainstem and spinal cord neurons to cloacal stimulation (Rose and Moore 1999; Lewis and Rose 2003). Interrupting sex when danger threatens is a widespread phenomenon and a similar mechanism may be at work elsewhere. Estrous female mice love the odor of males, but corticosterone treatment inhibits this mate preference in minutes (Kavaliers and Ossenkopp 2001). The trick for the animal is to avoid shutting down the HPG axis unnecessarily over a minor problem that will go away soon, especially if the reproductive period is a very brief opportunity (Wingfield and Sapolsky 2003). In fact, neither LH nor testosterone levels change in the white-crowned sparrows during the blizzard. An older animal with less reproductive value left would also be predicted to avoid shutting it down, a hypothesis that has not yet been tested with natural stressors. Another way to switch behavior quickly in addition to glucocorti-

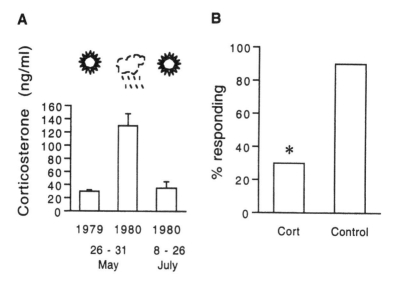

FIGURE 2.10. (A) Free-living breeding male white-crowned sparrows captured during bad weather (a spring snowstorm) had higher corticosterone levels than those captured during better weather. They also spent less time singing and defending the territory and more time "lying low" and foraging. LH and testosterone levels did not differ. (B) Free-living breeding male song sparrows given an implant of corticosterone (Cort) were less likely to respond to a simulated territorial intrusion than control males. Taken together, these results suggest that increased corticosterone resulting from HPA activation is a decision-making mechanism that rapidly shifts the males' behavioral priorities away from territorial defense (mating effort) and toward survival. Redrawn from Wingfield, J. C. 1988. Changes in reproductive function of free-living birds in direct response to environmental perturbations. In M. H. Stetson, ed., *Processing of Environmental Information in Vertebrates*, 121–148. Berlin: Springer-Verlag. © 1988 Springer-Verlag.

coids that is even less likely to reduce gonadal hormones would be to have CRF do the work centrally. This hypothesis is supported by an experiment in which CRF infused into the brain suppressed copulation solicitation in female white-crowned sparrows (Maney and Wingfield 1998).

Not all animals would be expected to inhibit mating during environmental threats. Different breeding systems produce different trade-offs between reproducing and risk taking and thus different decision points about when to switch from mating. This is thought to be why male *Taricha* mating is so easily disrupted by glucocorticoids, whereas male *Bufo marinus* mating is not. *Taricha* has an extended breeding season, and a single mating episode is a small part of the season's success, whereas *Bufo marinus* has a briefer explosive breeding period in which a single mating is a larger contribution to fitness (Orchinik et al. 1988; see also Moore et al. 2000). As we will see in chapter 6, extreme HPA activation doesn't disrupt mating in salmon or marsupial mice either, both of which have "do and die" breeding systems.

Daily and Seasonal Rhythms of Social Behavior

Daily rhythms of behavior and physiology are ubiquitous in animals. Both overall activity level and frequency of occurrence of social behaviors vary markedly between day and night. Inside the animal, levels of hormones such as glucocorticoids or (in some mammals) testosterone also show pronounced daily rhythms. These facts have been known for several decades or more, but it is still unclear whether the hormonal rhythms are a mechanism producing the behavioral rhythms. A chronic difficulty with this hypothesis has been the long latency required for steroids of peripheral origin to act via intracellular receptors to alter gene expression and thereby behavior. More recently discovered rapid actions occurring through membrane receptors could rescue the hypothesis, but the right experiments to test it are rare. Corticosterone and CRF rapidly stimulate locomotor activity in some vertebrates (Moore et al. 1984; Breuner et al. 1998) but this is not by itself direct evidence that the normal daily rhythm of corticosterone actually has behavioral consequences. Also, in some species glucocorticoids peak around the time of the daily activity peak, whereas in others (such as some birds) they peak when the animal is inactive (Carsia and Harvey 2000). Similar species differences have been discovered for diurnal rhythms in testosterone (for example, Foerster et al. 2002). The dawn chorus of birds is a lovely behavioral rhythm, but it remains unknown whether it has anything to do with steroid rhythms (Staicer et al. 1996).

Annual rhythms of social behavior are another matter, for here the causal link between correlated changes in behavior and hormones as the seasons progress is supported by much experimental evidence. The bulk of this comes from species living in north temperate regions whose annual cycles of reproductive behavior are regulated by changes in daylength (photoperiod). The subjects of the experiments include both animals that breed in spring/summer, such as birds for which increasing daylengths stimulate reproduction, and those breeding in the fall, such as sheep and elk for which decreasing daylengths stimulate reproduction. The two sibling species of *Coturnix* quail (*Coturnix coturnix* in Europe and western Asia and *Coturnix japonica* in eastern Asia) provide a simple yet striking example. Males are heard crowing and nests with eggs are found only during the spring and early summer. Males housed outside and exposed to a naturally varying photoperiod show dramatic changes in testis size, with peaks corresponding to the breeding season in free-living birds (Boswell et al. 1993). The summer vs. winter gonadal and behavioral states can be completely mimicked by housing the birds indoors on summer and winter photoperiods. If the lights are on for 14 hours a day or more males will crow and mate, females will lay eggs and mate, and both sexes will have well-developed gonads (Sachs 1969). If the lights are on for 10 hours a day or less, however, the gonads will be completely regressed. The behavior of birds with regressed gonads is the same as that of gonadectomized birds; neither display any reproductive behavior. If birds housed on short days are treated with sex steroids, they assume full reproductive behavior, just as gonadectomized birds treated with sex steroids do (Adkins and Nock 1976). Sex steroids alone turn a winter bird into a summer bird with respect to crowing, courtship, mounting, solicitation, and receptivity. Such studies suggest that for *Coturnix* the annual cycle of reproductive behavior is due largely to the annual cycle of gonadal sex steroid production. This cycle in turn is due to regulation of the HPG axis by daylength (Tanaka et al. 1965; Follett 1973). The only messy part of this otherwise tidy story is that *Coturnix coturnix* males are still producing substantial testosterone (how? where?) even on short days (Boswell et al. 1995).

Nonetheless, in this and many other north temperate birds, the whole HPG shuts down for the fall and winter and the gonads shrink drastically, not just in gamete-producing tissue but also in hormone-producing tissue. The HPG of seasonally breeding mammals is also toned down in the off-season but not as drastically as in birds (Dawson et al. 2001). It is not known whether this taxonomic difference is due to flight (the need to be as light as possible) or to other avian-specific costs of large gonads and their products.

In photoperiodic vertebrates gonadal sex steroids can be a mechanism for turning behavior on and off seasonally. Presumably the behavior needs to be turned off when reproduction is not possible (when animals are not fertile because of food supply and climatic factors) because the behavior itself is costly (uses energy, attracts predators) and might be incompatible with other priorities like molting or migrating. Yet some animals do court and mate at times of the year when no fertilization is possible. For example, socially monogamous ducks and geese may court and select mates in the autumn even though no egg laying occurs until spring. In some species but not others the behavior coincides with an autumnal elevation in circulating testosterone (Bluhm 1988).

Other puzzles are cases where the correlation across the year between behavior and peripheral sex steroids isn't so good. Wild rooks (*Corvus frugilegus*) breed in spring and also have a period of autumn sexuality (aerial display, carrying pine cones) but without any autumnal increase in circulating testosterone (Lincoln et al. 1980). The autumnal elevation in mockingbird (*Mimus polyglottos*) aggression occurs at a time when both testosterone and LH are low (Logan and Wingfield 1990). Male musk turtles (*Sternotherus odoratus*) mate in the spring when the gonads are regressed and again in the fall when they are enlarged (Mendonça 1987). Castrated dusky-footed wood rats (*Neotoma fuscipes*) continue to show a dramatic increase in intermale aggression at the time of year when breeding would normally occur (Caldwell et al. 1984). Male garter snakes (*Thamnophis sirtalis parietalis*) court and mate upon emergence from hibernation in the spring even if castrated the year before (Crews et al. 1984).

What behaviorally relevant mechanisms might be changing in response to environmental cues instead of circulating hormones that could account for such seasonal patterns? One early hint that other mechanisms awaited discovery was that the behavioral consequences of sex steroid administration sometimes differ depending on whether the animal is in a winter short-day state vs. a summer long-day state. Mountain spiny lizards, musk turtles, and domestic canaries, hamsters, and sheep all respond to sex steroids more at some times of year or on some photoperiods than others (Mendonça 1987; Moore and Marler 1988). For example, male mountain spiny lizards mate in the fall when circulating testosterone is highest, and show increasing levels of territorial aggression and testosterone as summer progresses to fall. So far this is unexceptional. However, testosterone implants given to castrated males stimulate breeding season levels of aggression if given in the fall but not if given earlier in the year (Moore 1988). If sex steroids cause seasonal changes in aggression, in effect turning a nonbreeding animal into a breeding animal (behaviorally speaking), why should it make any difference what time of year or photoperiod it is when the hormone is given, especially if the animals are gonadectomized? What is being affected by the environment that would alter the animal's sensitivity to steroids in this way?

One possibility is that the chemical mechanisms for steroid action themselves might be varying seasonally (Wennstrom et al. 2001). In species ranging across several animal groups (including bats, birds, frogs, goldfish, and crabs) seasonal changes have been documented in brain steroid receptors, in neuropeptides such as GnRH and VIP, in sensitivity to neuromodulators of reproductive behavior, and in the activity of brain steroid synthesizing and metabolizing enzymes (Callard et al. 1983; Pasmanik and Callard 1988; Wood et al. 1995; Gahr and Metzdorf 1997; Foidart et al. 1998; Takase et al. 1999; Soma et al. 2003). In amphibians and some fish, androgenic steroids only stimulate male mating behavior if the levels of vasotocin are also those typical of the breeding period (Moore and Zoeller 1979; Liley and Stacey 1983; Penna et al. 1992). Some of these seasonal changes are regulated by peripheral steroids, so that those steroids are still driving seasonal changes in behavior. But if these brain mechanisms vary independently of peripheral steroids in some anatomical regions because of a more direct (non-gonadally mediated) effect of photoperiod, then behavior can change seasonally even when steroid levels don't. Melatonin has been hypothesized to vary seasonally independently of gonadal state in songbird brains, and to interact with steroids to produce annual rhythms in both singing and the volumes of the song system nuclei (Bentley and Ball 2000).

The dose–response models discussed earlier are also relevant to understanding annual rhythms in hormone-regulated behavior. For behaviors that have a step-function (threshold) relation to steroids, seasonal changes in brain mechanisms (receptors, peptides) would mean that the threshold is changing. Also, if the threshold for a behavior is low, then a seasonal drop in peripheral steroid level would not necessarily cause a change in behavior.

A complementary set of hypotheses to explain these seeming dissociations between peripheral steroids and social behavior in annual cycles is more ecological and functional. It asks about the purposes hormones serve in the lives of wild animals and under what circumstances hormonal regulation of a behavior would be selected (Crews and Moore 1986; Wingfield et al. 1990). Hormonal regulation of the production of mature eggs and sperm is a basic fact of life, and so we would always expect gonadal sex steroids to be relatively higher when actual reproduction (fertilization) is occurring. It has already been emphasized that hormonal regulation of behavior is needed if the behavior must be turned on and off seasonally. When will this be so? Some stages of reproduction depend more on food supply or some other limiting resource than others. If there is substantial seasonal variation in food supply (due, for example, to annual cycles of temperature or rainfall), then both hormones and reproductive behavior will need to vary seasonally as well, driven by environmental cues that predict in advance when food will increase (Wingfield et al. 1992). This is the associated pattern in which hormones and reproductive behavior peak at the same time of year and there is significant gonadal control of the behavior (Moore and Lindzey 1992). From these basics we can ask when dissociations would be expected.

Several interesting hypotheses have been proposed. They are difficult to test; the few tests so far have focused largely on males and on testosterone, and most are explanations after the fact rather than predictions in advance. First, if ecology is the ultimate cause of seasonal patterns in behavior and hormones, then ecology should be a better predictor of whether a species has a dissociated or associated pattern than phylogeny (Moore and Lindzey 1992). Few dissociated patterns have been confirmed both behaviorally and endocrinologically to date (mainly garter snakes and a few bats [Crews 2002]). Information is not yet available from their close and distant relatives to give this hypothesis a comparative test in a phylogenetic context.

Second, if the critical limiting resource varies in abundance unpredictably, the animals may need to be behaviorally "on" and ready to "go" all the time. With such an opportunistic breeding strategy, behavior would not vary seasonally in a regular and predictable manner (it might occur in any month of the year if conditions are right) and would not be expected to be as dependent on peripheral sex steroids. There might be a chronic state of readiness (either hormonal and permissive or nonhormonal) with a fast-acting hormonal or peptide mechanism ramping it up (or releasing it from tonic inhibition) in response to rainfall, seed abundance, or a social cue (Crews and Moore 1986; Wingfield et al. 1992; Harding and Rowe 2003). Zebra finches are well-studied opportunistic breeding strategists (Zann 1996). They are native to the arid Australian interior, a region of highly unpredictable rainfall, and are permanently pairbonded. Gonads are developed all the time, up to the point of readiness to produce yolked follicles by females. Nest building and egg laying begin quickly after rainfalls. Hormones modulate the frequency and intensity of behavior such as singing but castration does not eliminate any of their reproductive behavior (Arnold 1975; Harding et al. 1983). Once again, a comparative test of the hypothesis would require experimental manipulations on an array of species that differ with respect to environmental predictability, analyzing the results (one data point per experiment!) in a phylogenetic framework. This would require a heroic effort by an army of researchers.

Third, if food supply and other resources for reproduction are relatively constant year-round, individuals would not have the same need for hormonal regulation of their behavior. While some tropical regions have pronounced wet and dry seasons, others enable animals to breed in any month of the year so that behavior is never "off." With no need for a mechanism to turn behavior on and off, the behavior should be less closely linked to gonadal steroids (Levin and Wingfield 1992). A scenario in which adult breeding behavior is inhibited during the juvenile period and the inhibition is removed at puberty would suffice. Rather than maintaining high levels of sex steroids year-round, individuals could maintain their hormones at lower levels, those known to be sufficient for gamete production and mating. Regardless of whether higher levels are costly, if they are not beneficial there will be no selection for them. Consistent with this constant environment hypothesis acting over a long enough time period to produce species differences, tropical aseasonal frogs have lower testosterone levels than temperate species in the same families (Emerson and Hess 1996). In those tropical birds that have been studied thus far, circulating levels of testosterone seem to be relatively low compared with temperate zone species (Levin and Wingfield 1992). It is not clear whether estrogen levels are lower as well. In the case of the birds, however, the temperate and tropical species that have been compared are not usually relatives, and without some way of taking into account phylogeny and biogeographical (continental) history the comparison has an apples-and-oranges flavor to it. A stronger case would be made if the same species with an extensive north–south range had lower peak sex steroid levels and less hormone-dependent behavior at more tropical latitudes. This assumes that the relationship can evolve relatively rapidly, which is likely to be the case (see chapter 5). In two species of *Zonotrichia* sparrows, no such pattern was seen in males' testosterone levels (Moore et al. 2002). There is still much to be learned here about when and why tropical birds have lower steroid levels.

A fourth hypothesis is that male behavior will only be associated with a pronounced elevation in peripheral testosterone and be highly testosterone dependent if and when (in the annual cycle) it functions in mate competition (Ketterson and Nolan 1994; Wingfield et al. 2001). Here high testosterone levels are viewed as a product of sexual selection. Behavior that occurs year-round in a seasonal breeder presumably has a function other than mate competition outside the breeding season. Consistent with prediction, in male song sparrows (*Melospiza melodia*, which defend territories in autumn as well as the spring breeding period) castration in autumn (when testosterone is low and no mating is going on) had no effect on response to simulated territorial intrusion (Wingfield 1994b). In European stonechats (*Saxicola torquatus*), administration of the androgen receptor antagonist flutamide plus the estrogen synthesis inhibitor ATD reduced male aggressive responses to simulated territorial intrusions during the breeding season but not during the nonbreeding season (Canoine and Gwinner 2002). Supporting evidence also comes from a comparative study of six species of tropical passerines living in east Africa (Dittami and Gwinner 1990). Whether there was an annual cycle of testosterone in males was better predicted by territorial behavior than by when breeding occurred. There was no cycle when territories were defended year-round but there was if territoriality was limited to the breeding period.

It might be necessary to break this hypothesis into two parts, one for how high the testosterone needs to be and the other for how hormone dependent the behavior needs to be. Behavior can have a low threshold but still be quite dependent on that low steroid level. Even if mating serves only to fertilize gametes and has little sexually selected component (for example, when it occurs in a genetically monogamous mating system) it still might need to be coordinated with mature gametes and regulated by sex steroids if it is costly because of predation risk. Following this logic, the mounting behavior of male white-crowned sparrows, which is not testis dependent, might have some other function besides fertilization (Moore and Kranz 1983).

Social behavior that occurs at times when peripheral sex steroid levels are low and is unaffected by gonadectomy could still be steroid based. Peripheral glucocorticoids vary seasonally in vertebrates, with unknown consequences for social behavior (Romero 2002). Fatty acid esters of steroids that hang around a long time might function at targets even after steroid production ceases and circulating levels fall (Hochberg 1998). The discovery of substantial brain steroid synthesis, either de novo or as an amplification of low levels of a peripheral precursor such as DHEA, provides another hypothesis for cases of social behavior related to reproduction occurring when peripheral steroids are low. In male starlings (*Sturnus vulgaris*), aromatase levels in the telencephalon are highest, not lowest, in the winter (Riters et al. 2001). The best evidence that brain steroids are supporting behavior even if the gonads aren't comes from studies of free-living song sparrows (*Melospiza melodia*). The autumnal aggression (territorial defense) of males, which is not affected by castration, is reduced by flutamide plus ATD and by the estrogen synthesis inhibitor fadrozole, supporting the hypothesis that nongonadal estrogens are stimulating the behavior (Soma et al. 1999, 2000). (Recall that estrogen is not a "female" sex hormone and can be important for the stimulation of male behavior.)

Male spotted antbirds (*Hylophylax naevioides*), a neotropical species, have low levels of testosterone but nonetheless steroids are supporting singing and aggression that are associated with breeding activity, because the behavior is reduced by treatment with flutamide plus ATD (Hau 2001). In the antbird case, it is not known if the behaviorally relevant steroids are coming from the gonads or are neurosteroids.

The neurosteroid hypothesis has several attractive features. It implies that the brain mechanisms downstream of steroids for social behavior might be conserved, whereas what varies in different birds and different seasons is the source of the steroids. This is probably a more realistic way to think about evolutionary change in types of seasonality than assuming that all the steroid-related mechanisms are built from scratch or discarded altogether each time such an evolutionary shift occurs. A role for neurosteroids would mean that evolutionary change in whether females also display the behavior could occur without a change in peripheral circulating steroids, and that estrogens could support the behavior in both sexes. Support by neurosteroids might avoid some of the costs of peripheral steroids (Wingfield et al. 2001). The idea that birds are less able to tolerate these costs might shed light on why birds have higher brain aromatase levels than other tetrapod vertebrates.

Hormones and Signaling

Animals engage in a rich variety of behaviors that serve to communicate with others. Signals come in many physical and chemical forms and are detected by nearly every kind of external sensory organ. In many terrestrial vertebrates, signals include chemicals released into the environment that can be detected later on when their producer is no longer around. Some signaling behavior occurs mainly during the breeding period and differs between the sexes, strong hints of reproductive functions and regulation by sex steroids. Conversely, looking for sex differences and a relation to hormones sometimes reveals communicative functions of behaviors that aren't at first glance obviously social, such as sneezing by squirrel monkeys, which apparently is a scent-marking behavior (Hennessy et al. 1980), an example of how studies of mechanisms can lead to new hypotheses about functions.

Hormones are one mechanism for coupling signaling behavior with morphology (Hews and Quinn 2003), as when testosterone stimulates vocal behavior in birds and frogs through effects on the brain and also on the syrinx or larynx, stimulates display of the dewlap in lizards at the same time that it increases the size and color of the dewlap, and inspires head-crashing contests in deer while also supporting antler growth and hardening. It is not safe, however, to assume that all morphological structures used in signaling have the same hormonal basis as the behavior that uses them, an important lesson learned from birds and their plumage physiology (Owens and Short 1995; Hillgarth et al. 1997). The male peacock's magnificent display behavior may be testosterone dependent, but the elaborate feathers he shows off are not.

Many displays and other signaling behaviors that are sexually dimorphic are products of sexual selection, serving mate attraction or intrasexual competition functions or both. Table 6.A in Andersson's (1994) survey of the sexual selection literature lists examples from over 130 vertebrates of traits that quantitative studies show are under sexual selection. Much of the list consists of male signaling behaviors that are already known to be dependent on gonadal sex steroids.

The croaks, chirps, and hums of male frogs and fish result from actions of sex steroids and vasotocin working together on vocalization systems (Diakow 1978; Boyd 1997; Goodson and Bass 2001). While this might sound reminiscent of mating behavior, sex and signaling mechanisms are not entirely overlapping either anatomically or hormonally. Nor are the key peptides the same, as the Great Plains toad (*Bufo cognatus*) illustrates (Propper and Dixon 1997). Breeding aggregations of males produce ear-blasting sounds and clasp females that approach a caller. Treatment with GnRH stimulates clasping but not calling; treatment with vasotocin stimulates calling but not clasping, a clean double dissociation between the effects of the two peptides.

In some species of birds, singing is a male-specific behavior that occurs mainly during the breeding season. Where such cases have been studied, singing is regulated by testosterone acting on androgen receptors, estrogen receptors (through aromatized metabolites of testosterone), or both (Harding 1991; Balthazart and Ball 1998). The brain circuits underlying steroid-induced vocalization in birds other than oscine passerines are not well understood. Steroid targets such as the intercollicular nuclei of the midbrain are likely components in all birds but the contribution of telencephalic structures in nonoscines is unclear (Balthazart and Ball 1993; Cheng 1993; Beani et al. 1995; Yazaki et al. 1999). In oscine passerines a well-defined set of telencephalic nuclei dedicated to singing and song perception sends output to the brainstem nuclei controlling the syringeal muscles (fig. 2.11). The discovery of this "song system" by Nottebohm et al. (1976) has provided an exceptional opportunity to see how steroids act in the nervous system to affect the learning, production, and perception of signaling behavior. The majority of the song system nuclei contain androgen receptors, and HVC contains estrogen receptors (Schlinger and Brenowitz 2002). Hormonal manipulations and seasonal changes in hormone levels result in changes in the overall sizes of several of the nuclei and also the size and structure of individual neurons (DeVoogd 1991; Tramontin and Brenowitz 2000). Strictly speaking, showing that steroids affect the anatomy and physiology of the song system and showing that steroids affect singing are not sufficient to confirm that the former is the cause of the latter, but the connection seems so likely because of the dedicated function of these nuclei.

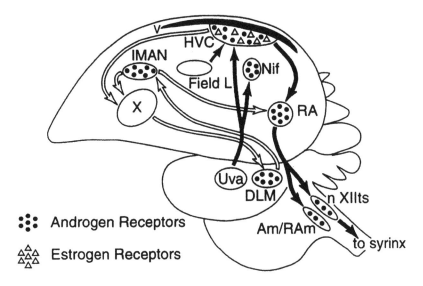

FIGURE 2.11. Schematic of a sagittal view of a songbird (oscine passerine) brain showing major nuclei and projections of the song system and the steroid receptor status of the nuclei. The descending posterior motor pathway from HVC to RA to nXIIts controls the production of song. The anterior forebrain pathway linking HVC, X, IMAN, and DLM functions in song learning. Androgen receptors and/or (in nucleus X) expression of androgen receptor mRNA have been found in all of the nuclei shown, but estrogen receptors have been found only in HVC. Am, nucleus ambiguus; DLM, medial portion of the dorsolateral nucleus of the thalamus; HVC, nucleus HVC of the nidopallium; IMAN, lateral portion of the magnocellular nucleus of the anterior nidopallium; Nif = Nif, interfacial nucleus of the nidopallium; RA, robust nucleus of the arcopallium; RAm, nucleus retroambigualis; Uva, nucleus uvaeformis; V, ventricle; X, nucleus X; nXIIts, tracheosyringeal part of the hypoglossal nucleus. Reprinted from Schlinger, B. A. and Brenowitz, E. A. 2002. Neural and hormonal control of birdsong. In D. W. Pfaff, A. P. Arnold, A. M. Etgen, S. E. Fahrbach and R. T. Rubin, eds. 2002. *Hormones, Brain and Behavior*, vol. 2, 799–840. Amsterdam: Academic Press (Elsevier). © 2002 Academic Press (Elsevier).

Steroid implants placed in single brain regions have not been very effective for stimulating singing in any birds (Phillips and Barfield 1977; Watson and Adkins-Regan 1989a), but this could be because steroid action is needed at multiple sites simultaneously, a difficult experimental manipulation in such small heads.

Some chemosignals are specific molecules that reliably elicit relatively specific, unlearned, rapid behavioral or physiological responses in receivers. These are often referred to as "pheromones," in contrast to complex mixtures of chemicals to which animals learn to respond in the context of other informative cues. Some pheromones are steroids, for example, the chemicals excreted by female goldfish that increase milt (sperm and seminal fluid) volume in males (fig. 2.12) (Sorensen and Stacey 1999). Few pheromones have been identified in mammals; instead chemical signals tend to be complex mixtures

A

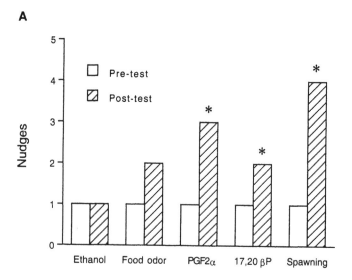

B

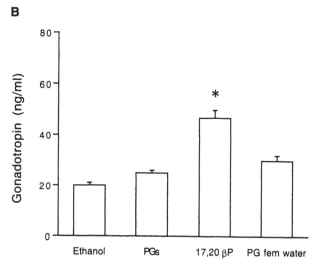

FIGURE 2.12. Responses of male goldfish to chemical cues released by females. A mixture of F prostaglandins (PGFs) is given off by postovulatory females prior to spawning, and 17,20βP is released by preovulatory females. (A) Groups of males were exposed to ethanol (control), food odor (another kind of control), PGF2α, 17,20βP, or a spawning female. PGF2α was more effective than 17,20βP for increasing nudging, a social behavior typical of a spawning situation. (B) 17,20βP increased gonadotropin and milt (sperm and seminal fluid) volume in isolated males. 17,20βP was more effective than prostaglandins (PGs), water that prostaglandin-treated females had lived in (PG fem water), or ethanol (control). Redrawn and modified from Sorensen, P. W., Stacey, N. E. and Chamberlain, K. J. 1989. Differing behavioral and endocrinological effects of two female sex pheromones on male goldfish. *Horm. Behav.* 23: 317–332. © 1989 Academic Press (Elsevier).

that experienced receivers are best able to use. The boar pheromone, the volatile androgen androstenone, is a notable exception and another good example
of a pheromone that is a steroid (Signoret 1967; Melrose et al. 1971). Regardless of whether chemosignals are pheromones, the production, dissemination,
and reaction to chemosignals are often hormone-dependent processes, especially when they serve reproductive functions. For example, male red-bellied
newts (*Cynops pyrrhogaster*) produce a peptide pheromone, sodefrin, that is
highly attractive to female newts; its production is regulated by hormones,
especially androgens and prolactin (Kikuyama and Toyoda 1999). Most male
mammals scent mark, gonadal sex steroids are key stimulators of the behavior,
and animals in a hormonally reproductive state spend more time sniffing scent
marks (fig. 2.6B) (Gandelman 1983; Yahr and Commins 1983; Takahashi
1990; Gosling and Roberts 2001; Moffatt 2003).

Females as well as males engage in hormone-regulated signaling behavior
both as senders and receivers. The copulation-solicitation behavior (courtship, proceptivity) of several mammals and birds is estrogen dependent. Female birds given estrogen and female frogs given vasotocin respond more
strongly to playbacks of male song (Catchpole et al. 1984; Boyd 1994). Scent
marking by female hamsters, in which the perineal region is dragged on the
ground to deposit a vaginal secretion that is very attractive to males, only
occurs on the day of proestrus and is stimulated by rising estrogen (Lisk and
Nachtigall 1988). Pheromones were first discovered in female insects. Female
sex pheromone production is regulated by juvenile hormones in cockroaches
and by pheromone biosynthesis activating neuropeptides in moths and butterflies (Barth 1968; Nijhout 1994). Female humans prefer "masculine" odors
and faces more at the time of ovulation than at other times (Penton-Voak and
Perrett 2001).

Although there are many cases where males and females engage in the same
signaling behavior (for example, duetting birds) there are few where the hormonal basis for the behavior is known in both sexes for comparison. Where
there is no behavioral dimorphism and the behavior occurs year round, it is
not even safe to assume that there is any hormonal basis. Based on studies of
long-calling by laughing gulls (*Larus atricilla*), Terkel et al. (1976) suggested
that in species where females also sing (or long-call), either testosterone is
responsible in both sexes or estradiol stimulates it in females and testosterone
stimulates it in males. This was also an early attempt to think about mechanistic
bases for an evolutionary change in song dimorphism. Testosterone stimulates
similar display behaviors in both sexes of juvenile black-headed gulls (Ros et
al. 2002). Both male and female ring doves perform the wingflipping display,
a nest solicitation behavior. This display is estrogen sensitive in both sexes,
but the ability of different androgens to stimulate it differs in males and females
(Adkins-Regan 1981a; Rissman and Adkins-Regan 1984). Infusion of vaso-

tocin into the brain modulates singing in female sparrows as well as in males (Maney et al. 1997a; Goodson 1998a).

There has been much interest recently in androgens and male signals, especially in the hypothesis that testosterone helps maintain the honesty of those signals. The foundation for this hypothesis begins with the fact that individuals are seldom genetically identical. As with other social behavior, a selfish gene perspective is essential for thinking about signaling and receiving. Whenever two genetically different parties are interacting, conflicts of interest are expected. It is widely thought that in mate choice interactions (intersexual selection) males are trying to sell themselves to be chosen as mates and females are tough sells holding out for the best males. In interactions between male rivals (intrasexual selection), senders are trying to intimidate other males and receivers are trying to call their bluff. What are the females choosing and why? Which male signals will be honest (reliable) indicators of mate and competitive quality? Some attempts to answer these questions have turned to testosterone. The theoretical link is that the costs of signals help maintain their honesty (Zahavi 1977; Hamilton and Zuk 1982; Johnstone 1997; Bradbury and Vehrencamp 1998). These costs include physiological as well as energetic costs. Steroids, especially testosterone, are hypothesized to be among the physiological costs most important for sexually selected signals. For signal strength to be an evolutionarily stable strategy for indicating quality, the fitness cost to the signaler must be greater for lesser quality signalers (those in poorer condition) (Grafen 1990). Costs such as testosterone for production of extravagant male-specific signals are thought to be particularly likely to satisfy this theoretical requirement. This version of indicator mechanisms theory is now known as the testosterone handicap model, and the particular version that emphasizes the immunosuppression costs of testosterone hypothesized by Folstad and Karter (1992) (discussed in chapter 1) is known as the immunocompetence handicap hypothesis (ICHH).

The testosterone handicap model is intended to account for the evolution of behaviorally produced signals such as singing as well as nonbehavioral ones such as head ornaments or bright colors. But the model is not so easy to confirm (Kotiaho 2001; Cotton et al. 2004; Roberts et al. 2004). Let's break it up into its component parts by asking a series of questions that focus on steroids and signaling behavior.

First of all, are signaling behaviors that are hypothesized to function honestly in mate attraction or intrasexual competition actually steroid dependent? On theoretical grounds (as already discussed in this chapter), they are particularly likely to be steroid dependent, because they are often sexually dimorphic and coincide with the breeding period (for example, musth in male elephants, a complex set of chemical and visual signals [Jainudeen et al. 1972; Rasmussen et al. 1984]). That said, steroid regulation should be established empirically rather than assumed for any new species and behavior. Nature has a way of surprising us at unexpected moments.

Second, are these hormonal mechanisms costly? Chapter 1 discussed potential biochemical and physiological costs of steroids, including costs of their production and action, effects on metabolic rate, and effects on immune system function. It is not yet clear how universal or important (fitness impacting) any of these costs are. Folstad and Karter's (1992) emphasis on immunosuppression nicely integrated the immunocompetence handicap model of Hamilton and Zuk (1982) with indicator models of good genes but risks ignoring other costs of steroids and other steroid actions. In regulating behavior, steroids alter the activity and anatomy of neurons. Neurons are big oxygen and glucose users. What might be the fitness related costs of this? How costly is it to maintain extravagant brain nuclei and circuitry dedicated to a sexually selected behavior, as in the case of the oscine passerine song system (Jacobs 1996)? Do birds in poor quality suffer a significant fitness cost by having to maintain and use a song system? Does having one mean that some other behavior will be performed more poorly because of limited brain space and energy? These questions are unanswered.

Third, are the costs (whatever they might be) greater for poor-quality animals? In humans, steroids seem to have greater effects on metabolism when physiologic reserves are reduced (Salhanick et al. 1969). Some of the discrepancies in the avian literature in which experimental administration of testosterone suppresses the immune system yet testosterone is positively correlated with immune function in wild birds could be interpreted to mean that some males can't afford the cost of having more testosterone (Peters 2000). Beyond this the answer to this question is murky. "Cost" is a slippery concept and hard to translate into something that can be measured.

Fourth, does poor condition reduce sex steroid levels? Here the ground is more solid, at least with respect to domestic or laboratory animals, mainly mammals. Chronic stress, including nutritional stress, suppresses the HPG (Pottinger 1999; Schneider and Wade 2000), as does illness (inflammation, endotoxins, cytokines) (Kalra et al. 1998). The hormonal production of the testes is more sensitive to condition than spermatogenesis is (Jainudeen and Hafez 2000; Zhang et al. 2003). Male western fence lizards (*Sceloporus occidentalis*) with malarial parasites have a larger corticosteroid response to capture, have smaller testes and lower testosterone levels, and engage in less territorial and courtship behavior (Dunlap and Schall 1995). Corticosterone treatment of uninfected males lowers testosterone and behavior.

Fifth, is the expression of the behavior condition dependent because steroid levels are condition dependent (rather than because of some other physiological link between condition and behavior)? This could be tested experimentally by giving males in poor condition who are showing low signal expression some steroid treatment (ideally via a brain implant, to avoid peripheral actions that might make condition worse) and seeing if signaling behavior increases in spite of continued poor condition. If not, then some other mechanism is the source of the cost that cannot be borne.

Sixth, in selecting mates do females regard testosterone-based signals as more important (better indicators) than others (Ligon 1999)? Multiple signals provide powerful within-species tests of predictions from the testosterone indicator theory. Tests with galliform birds converge on a "yes" answer to this question. Female jungle fowl (*Gallus gallus*) choose males on the basis of comb size and appearance, not body size or plumage (Zuk et al. 1995; Parker and Ligon 2003). Combs are excellent reflectors of circulating testosterone (and a bioassay for testosterone), whereas body size and plumage are not. Roosters from a genetic line selected for a low antibody response had higher testosterone and larger combs than those from a high antibody response line (Verhulst et al. 1999). Female Gambel's quail (*Callipepla gambelii*) prefer males with high courtship rates, a testosterone-dependent trait, but are indifferent to manipulations of male plumage ornaments, which are not testosterone dependent (Hagelin and Ligon 2001; see, however, Calkins and Burley [2003] for a different outcome with *C. californica*, the California quail). Female gray partridges (*Perdix perdix*) choose males on the basis of vigilance behavior and call structure, which are testosterone dependent, but males are not devalued if their testosterone independent brown breast patch is eliminated (Beani and Dessì-Fulgheri 1995). Male black grouse (*Tetrao tetrix*) with larger combs achieved more matings on a lek, and among males that mated comb size was positively correlated with testosterone level (Rintamaki et al. 2000).

Finally, do females prefer males with higher testosterone levels? Few experiments have given females choices between males with experimentally manipulated testosterone levels designed to mimic different points in the intact range, rather than choices between castrated males with and without hormone replacement (as in Klint 1985). In one of the few exceptions (Zuk et al. [1995] is another), testosterone-treated intact male dark-eyed juncos (*Junco hyemalis*) with high normal levels displayed more and were preferred by females in mate choice tests (Enstrom et al. 1997).

Thus far the discussion has taken the testosterone indicator model of signal evolution and then looked to the empirical literature to evaluate its components. There is some evidence to support it, but some components have not been examined closely and its generality within and beyond the avian lineage is unclear. Now it's time to turn things around, taking some key concepts and facts from behavioral endocrinology and asking what their implications are for the testosterone handicap model and other hormone related indicator theories.

First, step functions between signaling behavior and steroid level are problematic for the model (Hews and Moore 1997; Fusani and Hutchison 2003). It assumes a close graded dose–response relation across the range of testosterone levels produced by males, both between males and within males, yet such relationships are uncommon in hormones and behavior. Testosterone-dependent behavior might be good for telling whether a male is in a breeding versus nonbreeding state, or is in poor versus adequate condition, but how well will it mirror differences in the high normal range to discriminate among the better males? Are the sexually selected behaviors that matter those that do show a dose–response relation to testosterone? Rooster combs (not a behavior) do track testosterone level in dose–response hormone replacement experiments but there is no correlation at the high end of the range in individually housed birds (Parker et al. 2002). Features of song and calls such as acoustic structure or duration that reflect the state of the syringeal muscles might track it reasonably well because syringeal muscles are steroid sensitive and muscles are likely to have a graded relationship to androgens (Luine et al. 1980).

A second consideration is the time course of steroid action and steroid dynamics. Following castration, it may take days for behavior to change. In the case of a sudden illness or food shortage, testosterone-dependent behavior won't necessarily reflect a male's current condition but instead will reflect his condition a few days ago. For some purposes that might be adequate or even preferable. On the other hand, daily and pulsatile steroid rhythms mean that it makes no sense to take one sample for testosterone measurement and expect it to be related to condition or immunocompetence, which are integrated over longer time periods. Aspects of behavior involving more rapid steroid effects or peptide actions have greater potential to reveal current state and track moment to moment changes during a behavioral interaction.

Third, the organization/activation concept has important implications for sexually selected signaling (Crews 2000; Emerson 2000; Hews and Quinn 2003). The testosterone handicap model focuses on testosterone as an expression (activation) mechanism and on the costs of maintaining adult behavior and other signals, rather than the costs of building their neural and morphological apparatus during development. The developmental mechanisms of male-specific behaviors include steroids (chapter 4). These early steroid levels (embryonic, infant, juvenile) will also be affected by condition and some of their effects will be permanent. What is organized early sets the limit for what can be expressed in adulthood. An adult male's hormonally organized behavior and morphological traits are a window into the start he got in life. They might reflect genetic quality just as faithfully or even more so than traits sensitive to current condition. Where there are multiple signals females could use some to judge current condition and others to judge past condition, a highly effective combination. Again, those with no link to a purportedly costly mechanism like testosterone either early in life or in adulthood should be lower priorities in mate choice. Songbirds that learn song during development through hormone-related mechanisms are an interesting case. Learning of song syllables is sensitive to early nutritional condition and females prefer males with more learned song features (Nowicki et al. 2002a,b; Buchanan et al. 2003). In zebra finches the quality of the male's one and only song (the number of learned syllables it contains) is inferior if steroid actions were not right during juvenile life (Bottjer and Hewer 1992). Adult song rate is testosterone sensitive, an activational effect (Arnold 1975). The red beak color of males reflects both adult testosterone status and immune function status (McGraw and Ardia 2003; Blount et al. 2003), whereas their plumage ornaments are impervious to early or adult sex steroid or nutritional status (Birkhead et al. 1999; Arnold 2002). What do females seem to care about in this species? They prefer males with good-quality song (Lauay et al. 2004), high song rates (Houtman 1992), and bright red beaks (Burley and Coopersmith 1987).

Fourth, from the perspective of behavioral endocrinology the nearly exclusive emphasis on testosterone is odd. As already explained, the idea that "maleness = testosterone" is a great oversimplification. In any species in which testosterone administration changes levels of other hormones such as corticosterone, interpretation of the results of testosterone manipulation experiments becomes problematic. There is much better evidence for a connection between adrenal glucocorticoids and immune function than for a link between sex steroids and immune function (chapter 1). Condition (illness, nutritional status) has a big effect on glucocorticoids. Estrogens also need to be considered. Where male signal expression or development depend on brain-produced steroids such as estrogens (for example, songbird singing), behavior will be a poorer reflection of a male's ability to tolerate the peripheral costs of sex steroids. While brain amplification of a peripheral steroid might reduce the male's costs of signal production (Wingfield et al. 2001), that would make the signal a less useful indicator of his quality to females. What turns on a female ring dove is not the male's androgen-dependent bow-cooing or hop-charging, but his estrogen-dependent nest-cooing, a product of brain aromatase (see next section). Peripheral estrogens, on the other hand, could be quite costly. Do females care about a male's peripheral estrogen status? De Bruijn et al. (1988) found that estrous female rats preferred castrated males given estradiol to those given testosterone. Does female choice have anything to do with the high circulating estrogen levels in male pigs and horses? What signals if any are kept honest by them? In teleost fish testosterone supports sperm production but male-specific signaling behavior is more often dependent on 11-ketotestosterone. The costs of 11-ketotestosterone are unknown, but having different steroids for sexually selected vs. other reproductive characters might allow them to evolve somewhat independently. In species where males provide direct benefits in the form of parental care, females might be looking for signals reflecting hormones predicting parental ability.

Finally, from the perspectives of both behavioral endocrinology and behavioral ecology, it is decidedly odd that all the emphasis is on male signals in mate choice. Yes, it is common for sexual selection to operate more strongly on males than females, but there are exceptions and in socially monogamous species both sexes are choosing partners. Female steroid levels are more sensitive to condition than male levels. Females of some species have elaborate signals that are very attractive to males and have a known hormonal basis, for example, the hugely swollen bright red rumps of female baboons or the sex attractant odors of female hamsters and insects. Are these indicating anything important about the individual female's quality because of their hormonal basis, or are they simply indicating the female's ovulatory status and physical location in the world? Also, regardless of which sex is the receiver, we need to ask whether there are any hormonal costs associated with the development or expression of signal reception and interpretation. Female songbirds do have a neural song system, which they use to discriminate songs even if they don't sing (Schlinger and Brenowitz 2002).

Chemical cues that are steroids are fascinating to think about in the context of indicator mechanism theory. There are inevitably steroids in the excretory products of animals at levels that are somewhat correlated with circulating steroid levels. By evolving the ability to detect those steroids with their chemosensory systems, receivers gain a direct window into the circulating steroid levels of other individuals, levels reflecting their condition as well as reproductive status. For example, a female elephant's responses to male urinary odors increase as the male's testosterone level increases (Schulte and Rasmussen 1999). Where males go out of their way to deposit such odors by scent marking or by manufacturing special steroids (for example, boar pheromone), these cues are signals. Signals with a direct material link to the condition of the signaler are relatively cheat-proof (Kodric-Brown 1998; Gosling and Roberts 2001). Thus, steroidal odors should be high priorities in females' assessments of individual male quality. Female pigs are not just attracted to the male's androgenic pheromone, but are more likely to adopt the receptive posture in response to it (Signoret 1970; Melrose et al. 1971).

Any testosterone indicator mechanism model of sexually selected signal evolution suggests that when behaviors that are initially subject only to natural selection become increasingly sexually selected, they should become more hormone dependent and better correlated with hormone level as well. Other selective forces may complicate that trajectory. For example, the same signal may function in both intrasexual competition and mate attraction, and these may have different or even incompatible mechanistic solutions. Testosterone may be an indicator of competitive ability in a man but that doesn't make him interesting to women (Swaddle and Reierson 2002). Brain steroid metabolism could be part of the solution. For example, conversion of testosterone to DHT and binding to androgen receptors might be good for male competition, and aromatization to estrogen and binding to estrogen receptors might be better for appealing to females. That shifts the trade-off for the male to the steroid level, because a testosterone molecule can be sent down only one of the two pathways.

Hormonal Responses to Signals and Cues

The relationship between steroids and social behavior is bidirectional. Not only do hormones affect behavior, but engaging in social behavior can alter the hormone levels of the participants. Chapter 1 described the internal mechanisms allowing external social stimuli to influence hormone levels. Here we need to consider the phenomenon as a form of communication and ask who is benefitting and why that led to the evolution of these hormonal responses to other individuals. The need to understand the consequences for the recipient becomes even more urgent once it is realized that steroid changes will have serious consequences, such as altered gene expression. Even when circulating steroid levels don't change, social interactions can cause changes in gene expression leading to changes in steroid receptor numbers in the brain (Fernandez-Guasti et al. 2003).

Let's start with a relatively straightforward case that appears to be largely mutualistic: chemical signaling between female and male goldfish during mating (fig. 2.12) (Sorensen and Stacey 1999). For aquatic animals with external fertilization, gamete release has to be coordinated very precisely between the sexes. The costs of a mistake are great for males as well as females. In both sexes of goldfish, $17,20\beta P$ ($17\alpha,20\beta$-dihydroxy-4-pregnen-3-one) and other related steroids are involved in production of mature gametes. When the female ovulates, the prostaglandin $PGF2\alpha$ (not a steroid) stimulates her spawning behavior. The female releases both $17,20\beta P$ and $PGF2\alpha$ into the water. When detected by the male, $17,20\beta P$ increases milt volume and fertilizing capacity and $PGF2\alpha$ stimulates his spawning behavior. Thus, there is a preovulatory and a postovulatory signal from female to male that ensures that their gametes and their behavior are ready at the same time. This seems to benefit both the female, who can ensure that her eggs will be fertilized, and the male, who can get ready in time to do that. Many fish produce $17,20\beta P$, but only a few respond to it the way male goldfish do. Thus, it is highly likely that the goldfish's chemical signaling system has been derived from chemical products that were not originally communicative (Sorensen and Stacey 1999).

Next let's consider cases in which males stimulate female ovaries. Male vocalizations, odors, and other stimuli enhance ovarian development in at least some members of every vertebrate group studied (Wingfield et al. 1994). In some species females ovulate only if they are exposed to stimuli from a male, a phenomenon called induced or reflexive ovulation. It has long been known that female pigeons and doves will not normally lay any eggs unless they are housed with a male and that a gonadectomized male doesn't work (Erickson 1985). Female cats, rabbits, and a diverse assortment of other mammals do not spontaneously ovulate the way humans and other primates do, but instead ovulate only in response to the tactile stimulation of copulation with a male (Jöchle

1975). In musk shrews the ovaries remain in the juvenile condition with unde-veloped follicles until a male appears on the scene (see also the discussion of puberty in chapter 6) (Schiml et al. 2000). In this species behavioral interaction with a male or even just exposure to male odors rapidly increases GnRH pro-duction, setting in motion follicular development and eventual ovulation. How can we explain why some species have spontaneous ovulation while others have induced ovulation? What's in it for the female to be one or the other? Why did induced ovulators evolve to be so seemingly manipulated by males?

There are still no clear answers to these questions for mammals but an im-portant lesson has emerged: a phylogenetically based comparative analysis is essential. Only this can reveal which of the two states (or positions on a contin-uum) is ancestral and which is derived (which "came first"). Only then can one see when and where (phylogenetically) changes in states occurred to search for ecological or social correlates. There hasn't been a consensus in the literature about the order in which the states evolved. Schiml et al. (2000) and Conaway (1971) concluded that induced ovulation is ancestral, whereas Milligan (1982) concluded that it is derived. None of these discussions used an explicit modern mammalian phylogeny. Instead, questionable assumptions were made, for ex-ample, that insectivores and rodents are primitive groups. Using Novacek's (1992) cladogram of mammalian orders, the comparative information already published, and MacClade (Maddison and Maddison 2001) to reconstruct the past, induced ovulation emerges as the derived state (Adkins-Regan, unpub-lished). This is the case regardless of whether a conservative or relaxed crite-rion is used to categorize ovulatory states. The same conclusion is reached if the phylogeny of Murphy et al. (2001) is used. The ovulation types of some of the critical basal taxa are unknown, however, and so this conclusion could change as that information becomes available.

Most hypotheses about selective pressures and benefits to the female assume that spontaneous ovulation is the normative default state (because it is what humans have?) and induced ovulation is what needs explaining. Instead, the pros and cons of each ovulation strategy need to be considered. Also, spontane-ous ovulation doesn't necessarily mean cyclic ovulation. Many female mam-mals get pregnant at the first ovulation of the season. At any rate, a common hypothesis is that induced ovulation is an adaptive strategy when males are few and far between and the arrival of one is unpredictable (for example, solitary mammals with low population densities) (see references in previous paragraph). This hypothesis has not yet been adequately tested. In light of contemporary signaling theory, with its selfish gene perspective, one might wonder if induced ovulation is the female's way of testing the male. Perhaps only high-quality males can induce ovulation.

On the surface, the significance of induced ovulation seems clearer for birds. Species such as doves that are induced ovulators are socially monogamous and biparental. The reproductive cycle is a coordinated sequence of stages (courtship, nest building, incubation, chick care) in which behavioral interactions stimulate hormonal changes in both sexes (Lehrman 1965). The eggs are incubated constantly, with the female and male alternating (Ball and Silver 1983). The female should not invest in yolked follicles or ovulate until the nest and the relationship with the male are satisfactory, because without them the eggs will fail. Most ducks are also socially monogamous, and females lay eggs only if they are pairbonded with a mate (Bluhm 1988). In contrast, female galliform birds lay eggs even if housed where there are no males. Social monogamy and paternal care are uncommon in galliforms. With sperm storage of a week or more and mate attraction vocalizations by both sexes, there is probably little chance in the wild that a male would not be around to fertilize eggs when needed. The female should just go ahead and yolk up follicles and plan to ovulate when it is the right time for her. In a similar vein, it seems pretty sensible that animals with external fertilization would be induced ovulators, and that turtles, which store sperm for very long periods (so that fertilization is dissociated from mating), would be spontaneous ovulators.

But is the bird account a just-so story or a hypothesis that can be confirmed in a species rich comparative analysis? And has the female's perspective really been given due consideration? Female ducks do not lay eggs with a randomly assigned male, but only with a self-chosen male (Bluhm 1988). The original interpretation of the ring dove's induced ovulation was that the male's bow-coo courtship display stimulated the female's HPG axis to cause ovulation. Shouldn't the female have more say (physiologically speaking) in whether her HPG axis gets stimulated? Further research revealed that indeed she does (Cheng et al. 1998). The critical male behavior is not his bow-cooing—his attempt to persuade her, which is androgen dependent—but his nest soliciting display, which is performed by both sexes together and is estrogen dependent (Hutchison 1990). The proximate stimulus for LH release and ovulation is not the sound of the male's nest-coos, but the sound of the female's own nest-coos. There is auditory input to the preoptic area, neurons in the preoptic area respond to the female's coos, and those responses are fed to GnRH neurons. It is her (the receiver's) nervous system that makes the decision. Notice how studies of the behavioral and neural mechanisms underlying the phenomenon have helped solve a puzzle about function.

The important conclusion, that the receiver's nervous system makes the decision, may apply to a wide range of female hormonal responses to male social stimuli. Consider, for example, the Bruce effect, in which chemical cues in the urine of male mice cause females to terminate pregnancies produced by mating with a prior male. A male's urine alone won't produce the effect unless he's housed near females (de Catanzaro et al. 1999). On the face of it, this sequence of events seems quite counter to the female's interests. In meadow voles (*Microtus pennsylvanicus*), however, termination is not inevitable and the female's behavior toward the male determines whether disruption occurs (Storey 1994).

This is consistent with Schwagmeyer's (1979) proposal that the phenomenon is a form of mate choice (a choice between the original and the new male), with half of the choice exerted after mating.

Mating-induced termination of receptivity occurs in some insects as well as many vertebrates through mechanisms that include peptides and prostaglandins (Barth 1968; Whittier and Tokarz 1992; Truman and Riddiford 2002). Termination of receptivity is often interpreted as a form of mate guarding. This is fine for the male but whether and how it promotes the female's fitness is less often considered. The phenomenon of induced corpora lutea also reduces a female's receptivity. In some female rodents the stimulation of mating is required for fully functional corpora lutea (Adler 1978). Following ovulation a corpus luteum forms in the ovulated follicle but it is not very functional and doesn't produce the high progesterone levels typical of a functional corpus luteum. Mating causes an increase in prolactin that lasts for several days, which in turn "rescues" the corpus luteum for 12–14 days (Terkel 1988). The high progesterone of the functional corpus luteum allows implantation of fertilized eggs and also inhibits further ovulatory cycling and estrous behavior for 12–14 days. Because this is not a universal mammalian phenomenon, it raises questions about function. One hypothesis noted that the stimulus requirements were species-typical (different rodents have different penile intromission patterns during copulation) and proposed that the induction requirement served as a species recognition vaginal code reflecting co-evolution of male behavior and female response (Diamond 1970; Allen and Adler 1985). This idea has not held up well, because the female stimulus requirements are not that selective (Meisel and Sachs 1994). Another hypothesis notes that species with induced corpora lutea are small with very short cycles and short gestation periods. There is an inevitable trade-off between rapid frequent cycling and a good luteal phase. Fully functional corpora lutea are essential for pregnancy but there isn't much time to get them going. The female needs all the help she can get to have any chance of success (Adler 1978; Terkel 1988; Albert et al. 1992). This scenario implies natural selection both on the males, for the kind of copulation that will produce enough stimulation, and on the females, for the right prolactin response to that stimulation, a cooperative view of the situation. This does not explain the evolution of species-specific mating patterns on the part of the males, however. Perhaps a less cooperative view would be, for example, an alternative hypothesis based on sexual selection.

Females can produce changes in the hormone levels of other females, as when female presence causes puberty to begin at a later age in mice or a dominant female causes a subordinate to fail to ovulate (see chapters 3 and 6). Another famous example is estrous synchrony in rats and its analog in humans, menstrual synchrony (McClintock 2002). Some progress has been made identifying the chemical cues that are the proximate stimuli affecting the recipients. It is often assumed that these are acting via the accessory olfactory system coming from the vomeronasal organ but this seems very unlikely for humans. The evidence that humans have a functional vomeronasal organ is poor (Meredith 2001). The genes for the vomeronasal receptor proteins are a completely different family from the main olfactory receptor proteins, and in humans all but one of the vomeronasal receptor family genes are nonfunctional pseudogenes (Kouros-Mehr et al. 2001; Liman and Innan 2003). The main olfactory system (the receptors in the nasal epithelium) are perfectly capable of serving to receive the cues for cycle adjustment. The mystery is the function of such adjustment. Why should the recipients shift their cycles? How could that have been adaptive, especially in a premodern environment in which females were seldom cycling? Estrous cycling and menstrual synchrony per se are probably neutral reflections of a broader and more adaptive category of hormonal responses to the social and physical environment, such as condition-dependent adjustments in pubertal age (McClintock 2002).

Additional communication puzzles emerge when male hormone levels rise in response to female cues. In the case of induced ovulation, we at least know what the consequence of the hormonal change (increased LH) is for reproduction: ovulation occurs. In a number of male vertebrates female stimuli cause a rise in circulating LH and testosterone within minutes (Taleisnik et al. 1966; Bronson and Desjardins 1982; Wingfield et al. 1994). This happens mainly when the males are not already living with females, for example, upon an initial encounter. It is easy to assume that this is somehow useful for the male. Surely more testosterone must be a good thing! But why? What exactly are the physiological or behavioral consequences of this hormonal change? Do they increase the male's fitness and if so how? Most of the well-established effects of testosterone take quite a while to occur. The female would be long gone before they kicked in. Most "off the top of the head" hypotheses are predicated on rapid consequences.

Let's consider a few possibilities. First, an acute hormonal elevation might increase the male's attractiveness to the female during the premating courtship period, especially if males in poor condition can't mount a very good response. Second, it might improve his mating vigor or stamina or shorten his latency to mate or ejaculate, a plus for a sneaker mating strategy. Third, an acute elevation in testosterone might get the male's sperm mobilized for use, for example, by stimulating contractility of the vasa deferentia or accessory glands. A fourth possibility involves more long-term consequences. An acute hormonal elevation might help the male remember where and when he encountered a female so he will return there in the future.

Plausible though these might be, they have not been adequately tested. Furthermore, there is an alternative hypothesis that elevations in circulating steroids are an epiphenomenal by-product of changes in hypothalamic peptides such as GnRH or CRF that are serving some other more central function but incidentally alter the downstream parts of the neuroendocrine axes (van de

Poll 1991). The reason the functional hypotheses have been poorly tested is understandable, however. It is difficult to experimentally block male hormonal surges without changing baseline levels. Often the best one can do is to compare males that are castrated and receiving testosterone replacement (who can't undergo a testosterone surge) with intact males, who can, but even then it is hard to equate baseline levels and mimic natural variation with time of day. The second hypothesis has received experimental support. Male mice experience a pulse of testosterone when exposed to a female or her urine. Sniffing her urine shortens the male's latency to copulate, and an injection of testosterone mimics this effect (James and Nyby 2002; see Malmnas [1977] and DeJonge et al. [1992] for similar findings in rats). Exposure to a female or her odor also reduces anxiety, as does an injection of testosterone (Aikey et al. 2002). Thus, it seems likely that a testosterone pulse benefits the male by overcoming his fear of the female so he can start mating quickly. The most direct test of the hypothesis, blocking testosterone pulses and seeing if the latency-shortening effect of exposure to a female disappears, is still lacking, however. In addition, it is not known if a short mating latency increases the male's fitness. It could just as plausibly decrease it if the female prefers a male who courts first or the male can't mobilize as many sperm in time if he starts off fast.

The same questions arise when testosterone or other hormone levels rise during aggressive competition. Again, the phenomenon occurs in many vertebrates and in free-living as well as captive animals (Harding 1981; Leshner 1983; Wingfield et al. 1994; Oliveira et al. 2002). Most reports concern adult males, but there are some exceptions. Female dunnocks (*Prunella modularis*) compete for males, and their fecal androgens increase when competition is increased experimentally by removing males (Langmore et al. 2002). Both sexes of black-headed gull (*Larus ridibundus*) chicks show testosterone elevations in response to challenges (Ros et al. 2002). The taxonomic breadth of the phenomenon makes it easy to overlook the fact that its functions are seldom known. Again, there could be both short-term consequences, in case there are rapid effects of testosterone, and long-term consequences based on well-established steroid actions. There is also another side of the phenomenon to consider. While animals that emerge as winners may show a rise in testosterone, both winners and losers may show HPA activation and elevated circulating glucocorticoid levels during the encounter, and in the losers this may last longer, causing testosterone levels to drop (Sapolsky 1993; von Holst 1998). The usual interpretation of HPA activation is that it mobilizes energy to enable the animal to deal with the crisis at hand and either overcome it or escape from the situation. This is obviously adaptive but if it goes on too long it will become very costly. For those bird species in which testosterone treatment causes a rise in corticosterone, any short-term benefits of an aggression-induced rise in endogenous testosterone could be mediated by an energy-mobilizing rise in corticosterone.

What might be the short-term benefits of a testosterone increase during or immediately after an aggressive encounter? If it occurs during the encounter, it might increase the amplitude of honest but intimidating behavioral signals (if the male is in good condition), thus ending the encounter sooner, increase muscle strength (Tsai and Sapolsky 1996), or enable the male to keep on being aggressive for a longer period of time. This last hypothesis was tested in mountain spiny lizards, who show a big post-encounter rise in aggression after a staged encounter (Moore 1988). This also happened in males that were castrated and receiving testosterone replacement, ruling out a surge of gonadal testosterone as the mechanism. The hypothesis was rejected and changes in brain steroids were offered as an alternative mechanism.

A recent model for humans proposes a positive feeedback cycle in which engaging in competitive interactions causes testosterone levels of winners and losers to diverge, which then increases behavioral divergence and so on (Mazur and Booth 1998), but this assumes (without evidence) some relatively short-term activational effects of the changes in testosterone produced by the interaction. In animals that can detect other individuals' hormonal metabolites in urine or other products, winners and losers might smell different after an encounter (Oliveira et al. 1996).

In species such as mice and rats where glucocorticoids stimulate aggression centrally, we can ask if HPA axis activation might be rapidly beneficial during the encounter because of its aggression-promoting properties. Makara and Haller (2001) reported a rapid effect of corticosterone on aggression in rats and proposed that the rapid effects are immediately adaptive in addition to preparing the system for the genomic level effects to come.

Delayed consequences to males of hormonal changes during competitive encounters have been proposed (Leshner 1983). In mandrills (*Mandrillus sphinx*) gains and loss in status are reflected in changes in sexual skin coloration (Setchell and Dixson 2001), as if coloration is an honest and dynamic badge of status. Winners will need to fight again, and the rise in testosterone may prepare them for that (Wingfield et al. 1990). Winners are more likely to be winners in the future and losers are more likely to remain losers—the winner and loser effect. Perhaps these effects are due to the hormonal changes during the winning and losing encounters. When male mice are defeated, the HPA axis is elevated and the HPG axis is turned down. They become less aggressive and more submissive. Preventing the increase in corticosterone (by adrenalectomizing them and providing corticosterone replacement) delays the increase in submission but preventing the change in testosterone (by castrating them and providing testosterone replacement) is without effect (Nock and Leshner 1976). Male California mice maintained on testosterone replacement that are given an injection of testosterone shortly after an aggressive encounter (pro-

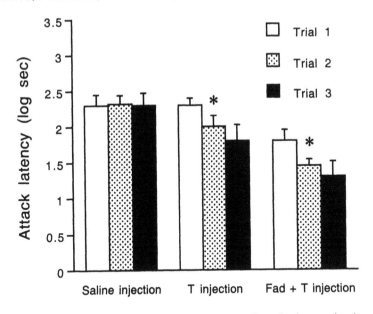

Figure 2.13. Male California mice injected with testosterone (T) to simulate transient increases following aggressive interactions were faster to respond aggressively the next day. This effect was not seen in controls that were injected with saline. The males were castrated and had testosterone implants sufficient to produce baseline levels. They were tested three times and were injected following trials 1 and 2. The testosterone injections elevated circulating levels for about 30 minutes. These results support the hypothesis that an aggression-induced transient increase in testosterone might have a behavioral function at a later time. Males given fadrozole (an estrogen synthesis inhibitor) plus the testosterone injection also responded faster the next day, indicating that it is androgenic, not estrogenic, consequences of testosterone that are responsible. Redrawn from Trainor, B. C., Bird, I. M. and Marler, C. A. 2004. Opposing hormonal mechanisms of aggression revealed through short-lived testosterone manipulations and multiple winning experiences. *Horm. Behav.* 45: 115–121. © 2004 Elsevier.

ducing a 30-minute spike in the steroid) are faster to attack the next day (fig. 2.13) (Trainor et al. 2004).

This last experiment suggests another source of hypotheses for longer-term benefits of hormonal changes in animals of both sexes in response to both sexual and aggressive encounters, namely, research on hormones, learning, and memory. HPA axis activation in response to aversive stimuli has a very well established positive effect on consolidation of a long-term memory for the situation that produced the activation (McGaugh 1989). This is obviously adaptive because the animal remembers for a long time to avoid the place, the individual, or whatever was dangerous. Glucocorticoids are secreted not just in response to negative (aversive) stimuli, including aggressive encounters, but also in response to sexual stimuli (Piazza and Le Moal 1997). It is worth asking

FIGURE 2.14. Winner (fish facing down) and loser (fish facing up) male blue gouramis. Once the males have interacted, such behavior rapidly becomes conditioned to stimuli predicting that an interaction is imminent. Courtesy of Karen Hollis.

if either these glucocorticoids or the elevations in sex steroids serve any analogous memory-enhancing functions. Both hormonal changes and social behavior (aggression and courtship) readily undergo classical (Pavlovian) conditioning and will occur in response to a previously neutral stimulus if it predicts the appearance of a conspecific (Graham and Desjardins 1980; Pfaus et al. 2001). This enables the animal to respond faster and more efficiently (Hollis et al. 1997; Domjan et al. 1998). Winner and loser effects in gouramis (*Trichogaster trichopterus*) and golden hamsters are due in part to Pavlovian conditioning (fig. 2.14) (Hollis et al. 1995; Huhman et al. 2003). Is the conditioned anticipatory hormonal change in any way causing the behavioral anticipation? Finally, hormones and peptides have rewarding or punishing properties and their consequences can be detected and discriminated. Animals will seek out or avoid a place if being there is followed by an experimentally manipulated change in hormone. GnRH is rewarding for male rats even when they are castrated and receiving testosterone (de Beun et al. 1991). The latency for the rewarding effect is short, less than an hour. Testosterone is rewarding in rats (Alexander et al. 1994; Frye et al. 2001; Wood 2004). The mechanism mediating this effect is unclear. Possibilities include interoceptive cues and neurosteroid metabolites. Regardless of the mechanism, hormonal responses to encounters might change the probability that the animal will go to that place or approach that animal again.

All of these hormonal responses to others raise important questions about both function and mechanism. It seems very likely that progress on either front would benefit the other. The discussion has emphasized puzzles about function, but the pathways upstream of the HPG and HPA axes responsible for social effects on steroid levels are usually unknown. The work with ring doves (Cheng et al. 1998) is one of the few cases where the outlines of a pathway are in place for a response of the HPG axis to a species typical signal. Acknowledging the receiver's assessment and decision-making role in its hormonal responses to others suggests involvement of cortical or other higher processing in the control of the axes, rather than a hard-wired brainstem reflexive response (Wingfield et al. 1994). The importance of the mammalian amygdaloid complex for triggering HPA activation to dangerous stimuli was emphasized in chapter 1. Lesions of the homologous region produce inappropriate tameness (lack of fear) in wild birds just as they do in mammals (Butler and Hodos 1996). The medial amygdaloid nucleus and its homolog in other vertebrates has lots of sex steroid receptors, but it is unclear whether it is part of the system responsible for HPG activation to social and sexual stimuli. It tends to be assumed that information about chemical signals goes straight to brainstem structures so that only rather reflexive responses are possible, but, in fact, there are large projections from both the main and accessory olfactory systems to the amygdaloid complex and its homologs.

It is highly likely that most of the changes in steroid levels in receivers will be followed by changes in gene expression. Animals tickle each other's genomes, and the consequences include changes in the nervous system that could be lasting. The animal is not the same after it has interacted. This reinforces the importance of thinking about benefits to the recipient rather than assuming manipulation by the signaler. The extended phenotype of a hormone includes the effect of the actor's hormone-based behavior on the fitness of others, including the individual's mate and young (Ketterson and Nolan 1999), and some of these fitness consequences are hormonally mediated. This brings us to the next level of biological organization, relationships among individuals.

3

Social Relationships
and Social Organization

When multiple individuals are observed, patterns of social behavior emerge into view. The species typical social behavioral repertoire is not directed evenly or randomly toward conspecifics. Instead, biases toward or away from others reflect social relationships that vary as a function of age, sex, time, and place. The overall configuration of these relationships, or social organization, is to some extent species typical and, like other features of animal life, highly diverse. For beavers, defense of a pond-sized territory by a male and female pair is as characteristic as a flat tail and large gnawing incisors, whereas for baboons, group living with strong ties between female kin is as characteristic as dog-like snouts and large canine teeth.

To what extent are steroids mechanisms underlying this level of animal life? How do hormone–behavior relationships reflect species and population differences in social organization that have resulted from natural and sexual selection? While some of the behaviors from which social organization emerges are regulated by hormones (chapter 2), now we are dealing with a somewhat different phenomenon: not just whether and how much the animal performs the behavior, but toward which other individuals and under what circumstances. This will depend on who they are and what they are doing. Local ecological conditions will be important in addition to species typical predispositions, generating within-species population differences. We would expect permissive rather than deterministic actions of hormones and an important role for hormonal responses to other individuals and to the local environment.

Social organization varies along four behavioral dimensions that have been studied in relation to hormones. One is that species differ markedly in overall sociality. A second concerns spatial distribution and aggressive/competitive interactions. Are the animals territorial? Do they defend an area against conspecifics? If they live in groups, are there consistent predictable outcomes (winners and losers) of contests between pairs of individuals indicating dominance and hierarchies of dominance? The third concerns mating relationships and mating systems. How many opposite sex individuals does a typical member of the species mate with? Are these male–female interactions brief and limited to copulation, or more long-lasting pairbonds? The fourth concerns relationships between parents and offspring. Is there parental care after hatching or birth? By which parent? For how long? One of the challenges in mechanistic studies of social organization is to explain how this fascinating diversity in social organization can be produced by relatively conserved mechanisms.

Sociality

Some animals are found in high-density groups of conspecifics where the presence of other individuals in close proximity is tolerated (for example, caribou [*Rangifer tarandus*]), rather than as widely separated solitary individuals where interactions with others are rare and, when they occur, not very friendly (for example, komodo dragons [*Varanus komodoensis*] or polar bears [*Ursus maritimus*]). Group living animals, in turn, vary with respect to whether groups are stable in membership so that members know each other individually and are genetically related to others in the group (for example, African elephants or naked mole rats), or instead are aggregations with minimal need or opportunity for individual recognition (for example, sardines).

What makes some species "groupy"? What mechanisms determine whether individuals prefer to affiliate with others in close physical proximity? Are steroids or the peptides that they regulate involved in the overall tendency to be social? One hint that the answer might be "yes" comes from animals where sociality varies seasonally in association with changes in hormone levels (Prendergast et al. 2002; Wingfield and Silverin 2002). For example, female meadow voles (*Microtus pennsylvanicus*), like some birds and several other rodents, are territorial on long days when they are breeding but live in groups on short days. Here short-day sociality enables the animals to huddle for warmth (its function), and reproductive levels of sex steroids seem to interfere with sociality by stimulating territoriality (a mechanism). There are correlated seasonal changes in oxytocin receptor binding, with more binding in the lateral septum and two subregions of the amygdaloid complex in short-day females (Parker et al. 2001). These are brain areas known from experiments with other species to be involved in aggression. This within-species phenomenon makes it plausible to hypothesize that differences in sociality between closely related species of rodents might also be linked to steroids and peptides, a hypothesis that is examined more fully later in this chapter, in connection with mating systems, and in chapter 5.

Additional hints are that in animals that normally live in stable groups, social disruption (either separation from familiar others or contact with strangers) is a powerful stressor, and the presence of a familiar other is a major buffer against stress, reducing HPA activation (von Holst 1998; Blanchard et al. 2002). The same social manipulations have less effect, or none at all, in animals that do not normally live in groups but are housed together. The same situation has a very different meaning according to how it is interpreted by the animal. This suggests possible involvement of the amygdaloid complex. In addition to its role in triggering HPA activation, the "amygdala" seems to be where the brain decides whether strangers are to feared. Rhesus monkeys are normally terrified of strangers but those whose amygdalar regions have been removed are more affiliative, to the point of being foolishly friendly and over-confident (Emery et al. 2001).

Animals that live in stable groups for extended periods are almost always discovered to recognize each other as individuals when studied with appro-priate methods. They not only can tell others apart, but also can remember them in their absence. In house mice there is good evidence that oxytocin receptor activation in the medial amygdaloid nucleus is necessary and suffi-cient for this important kind of social memory (Ferguson et al. 2001).

Migratory locusts such as *Locusta migratoria* and *Schistocerca gregaria* show pronounced within-species variation in sociality, alternating between two distinctive forms, solitary and gregarious, that differ in appearance, physiology and behavior (Nijhout 1994). The gregarious forms can become the huge plagues of locusts dreaded by farmers. This phenomenon is strongly suspected to be hormonal or neurohormonal, with juvenile hormones as the hypothesized candidates.

Dominance

In many animal groups there are consistent dominance relationships in which winners have priority of access to critical resources such as food or mates. In a stable dominance hierarchy of familiar individuals, these relationships are often transitive, such that alpha dominates all others, beta dominates all but alpha, and so on down to poor omega, who dominates no one but is dominated by all. It is very tempting to assume that if levels of circulating hormones are measured, there will be a nice relationship between rank and hormone levels, with (in the case of a male hierarchy) alpha having the highest testosterone level and omega having the lowest. If hormonal causation of rank is assumed, then giving low ranking animals testosterone should raise their ranks and re-ducing testosterone in high ranking animals should lower their ranks. Often this is not what results, however, a finding that is not as widely known as it should be. The relationship between rank and hormones is complex, but the reasons for this are fascinating. Several principles have emerged that have helped to simplify what would otherwise be a very confusing situation.

Dominance is not a behavior but a relationship. In some species simple features of an individual's morphology, such as body size, strongly predict rank because larger size guarantees winning any aggressive interactions. Body size is a major predictor of rank in animals with indeterminate growth, such as some fishes and snakes, where adults of different ages differ tremendously in body size. Here the smallest individuals are not just lower ranking, they are food items rather than threats to rank. In these species we would predict a positive correlation between steroid levels and rank only if steroid levels increase with age. Such a correlation would reflect a causal relationship only if steroids are one of the mechanisms producing increased body size with age. Even then, hormone manipulations would probably not change rank, at least not on a short-term basis, because body size would respond only slowly.

A positive correlation between sex steroids and rank will also be expected if steroid-based morphological weaponry is the major predictor of rank. During the rutting season a male red deer without antlers has no chance of winning fights and gaining access to females. Androgen manipulations do alter dominance rank in males of this species. This happens both by direct effects on behavior and by determining presence or absence of antlers (Lincoln et al. 1972). Even in this case, however, it is not safe to assume without evidence that antler size (as opposed to presence or absence) is correlated with testosterone level. Testosterone-based badges (status signals), like weapons, will require behavioral backup to be effective. In male Harris' sparrows (*Zonotrichia querula*) black feathering on the crown and throat is correlated with rank, but experimentally increasing signal size with dye raises the bearer's rank only if testosterone is also given to provide the behavioral support (Rohwer and Rohwer 1978).

Thus, one principle is that there will be a causal relationship between testosterone and rank if rank is determined by some testosterone-dependent morphological character with a direct effect on winning ability, and if the steroid–character relationship is a graded function, not a step function (fig. 2.2). This same principle should apply to any behavior that is critical for rank. We might also expect such a relationship mainly during the breeding season, and not necessarily during the nonbreeding season, when all the males would have low testosterone (Soma and Wingfield 1999).

Does this mean that rank would be predictable from hormone level regardless of whether the individuals are strangers or are familiar with each other? Not at all, for familiarity makes all the difference in the world. Animals learn from their first few encounters with each other and winner–loser effects are strong (Pusey and Packer 1997). It has been known for many years that even in species such as chickens that drop in rank following castration, experimental elevation of testosterone may not change the dominance status of males unless they are placed into new groups (Guhl 1961). A change in rank following testosterone treatment is less likely to occur if the animals remain in their existing group where they already know each other as individuals and where each dyad has a history. This principle of "social inertia" in hormone–rank relationships seems to apply to a wide array of animals, including invertebrates (Barth et al. 1975; Dixson and Herbert 1977; Bouissou 1983; Harding 1983; Röseler 1985; Trumbo 1996; Zumpe and Michael 1996; Wiley et al. 1999). Social inertia means that testosterone is seldom sufficient to ensure high rank even when it is necessary. It makes high rank possible but guarantees nothing.

Another source of complexity is the two-way nature of hormone–behavior relationships. Hormone levels are labile. In many (not all) species, androgens such as testosterone respond rapidly to engagement in an aggressive or sexual interaction and then return to baseline (chapter 2). Thus, a male's circulating sex steroid levels depend on what has happened recently. As a result, testosterone levels are more likely to be correlated with rank during periods of instability, when ranks are being challenged through overt aggression and are subject to change, than during periods of stability (Ramenofsky 1984; Sapolsky 1992, 1993). Instability occurs if, for example, alpha falls ill or strangers join the group. The consequences, if any, of such temporary changes in testosterone are far from obvious (chapter 2).

Glucocorticoids also change during rank challenges and these changes can be even more dramatic. Aggression is a potent social stressor that activates the HPA axis at all levels in all vertebrates that have been studied. Challenges are stressful for dominants as well as subordinates, and defeat is bad regardless of the starting rank. But in the dominants (the winners of the contests), HPA activation tends to subside sooner than in the subordinates (the losers), and subordinates are more likely to experience the kind of chronic stress that is likely to be maladaptive (von Holst 1998; Øverli et al. 1999). For example, defeat is followed by immune suppression (Cooper and Faisal 1990; Jasnow et al. 2001). Thus, whether a relationship is found between circulating glucocorticoids and rank will depend enormously on when the hormones are measured relative to the onset of the challenge. In contrast to the situation with testosterone, relationships between status and glucocorticoids might actually be less pronounced, not more pronounced, during the challenge itself, which is stressful to all parties (von Holst 1998).

Even with these principles in place, there are still many species differences and discrepancies between studies of the same species in whether any significant correlation is seen between rank and either sex steroids or glucocorticoids. What else might matter? Dominance ranks are measured by human observers, but what will determine the animal's hormonal state is its own experience of that rank (Sapolsky 2002)—its own perception of how it is doing and whether that is thrilling or disturbing. In studies of testosterone responses of men to competition, it is well established that divergence in testosterone can occur solely from the psychological experience of being on the winning rather than losing side (the mood of elation rather than dejection) (for example, Bernhardt et al. 1998). A similar phenomenon could be hypothesized to occur in animals that engage in competitive interactions between coalitions or kin groups. The animal's experience will depend on the role that dominance plays in the overall social organization.

Depending on the social system, either low or high rank can be the most stressful (Creel et al. 1996; Fox et al. 1997; von Holst 1998; Sapolsky 2002). Creel (2001) summarized evidence from mammals and birds that subordinates usually have higher glucocorticoid levels except in species with cooperatively breeding groups, where subordinates are helpers and in some cases older offspring of the dominants (see section below). In hyenas the relationship between testosterone and male rank depends on whether the males are in their natal group or are immigrants. Among immigrants, rank and testosterone are positively correlated, but natal males are dominant to immigrants even though they have lower testosterone (Holekamp and Smale 1998). In a comparative meta-analysis of cortisol levels in subordinate primates, it mattered whether the subordinate had social support, for example, from kin (Abbott et al. 2003). Those with support had lower levels of cortisol. The animal's experience will also depend on what is required behaviorally to impose a subordinate status on others, how much energy that requires, how much that interferes with foraging, and whether the subordinates are willing to defer or fight back (Creel et al. 1996; Goymann and Wingfield 2004a). Alpha might be more stressed (have a more activated HPA axis) where alpha is constantly being challenged and has to spend significant time and energy keeping everyone else in line through direct physical aggression than when alpha dispenses a benevolent authority through intelligence or force of personality. Comparative tests of this and other hypotheses where the relative hormone levels of dominants and subordinates are successfully predicted in advance rather than explained after the fact would be an important advance.

Except in the case of cooperative breeders (see section below), much less is known about hormones and dominance in females, even though in some species (for example, some primates) female hierarchies are a salient feature of social organization and a good predictor of reproductive success. The same principle of social inertia discovered by Guhl (1961) in studies of testosterone and rank in roosters also applies to hens, whose ranks are increased by testosterone only when interacting with strangers. Whether this is relevant to normal hen hierarchies is unclear, because hens have lower testosterone levels than males. Because estrogens were thought to be submissive hormones, effects of estrogen on hen ranks were not well explored. They still have seldom been studied in female competition even in species where estrogens derived from circulating testosterone are known to be involved in male aggression.

Where males and females have dominance relationships, males usually dominate females whenever males are significantly larger, as in some mammals, and females can dominate males where females are larger, as in many fishes and frogs. In some mammals with more subtle size dimorphism, relative ranks of the sexes change depending on the female's hormonal condition. Hormonal organization in early development has been proposed to explain why female ring-tailed lemurs dominate males even though they have less testosterone (von Engelhardt et al. 2000) and is the currently favorite hypothesis for the "reversed" dominance of spotted hyenas (see chapter 4).

Subordinates may fail to reproduce or even to attempt mating even though they are adults. Guhl and colleagues (1945), working with chickens, were the first to investigate this phenomenon experimentally and to ask whether failure might be due to gonadal hormone inhibition caused by subordinate status, which they termed "psychological castration." They distinguished between "psychological castration" as a cause of failure to mate and cases in which the subordinate does mate if the dominant is removed briefly. They observed that subordinates given testosterone may still fail to mate, suggesting that lack of gonadal hormones is not their only problem and foreshadowing the concept of hormones as permissive, not determining (see also Collias et al. 2002). A more recent confirmation of Guhl's ideas comes from studies of male talapoin monkeys (*Miopithecus talapoin*) (Keverne 1992). Only high-ranking males mount. Dominant males have higher testosterone but only when females are around, as if it is caused by sexual activity, rather than the other way around. Giving subordinates testosterone does not raise their rank. Even when both dominants and subordinates are castrated and given testosterone (equalizing their testosterone levels), only dominants copulate. Following removal of dominants, subordinates will eventually copulate, but not for quite a while.

In some primates subordinate females fail to reproduce because of direct physical harassment by dominant females (Sapolsky 2002). Such failure could be due to HPA activation. In male cichlids (*Haplochromis [Astatotilapia] burtoni*), hypothalamic GnRH-containing neurons are significantly larger in dominant males, and the levels of GnRH gene expression (GnRH mRNA) are greater (fig. 3.1). Changes in social status in either direction result in changes in neuron size and gene expression, indicating social regulation at the highest level of the HPG axis (Francis et al. 1993; White et al. 2002). Male *Xenopus* frogs stop producing advertisement (female attraction) calls in the presence of a dominant individual but quickly begin to call again when isolated, pointing to a more rapid mechanism than HPG axis inhibition (Tobias et al. 2004).

Territoriality

Experiments on hormones and territorial establishment and maintenance have focused on males even though both sexes defend the territory with similar alacrity in a number of birds, mammals, and fishes. Territorial defense by males has been shown to be supported by steroids (either from the gonads or the brain) in a number of free-living animals (fig. 3.2). Castrated male mountain spiny lizards (*Sceloporus jarrovi*) remain on their territories but patrol them much less frequently, display less frequently, and respond less to intruders (Moore 1987). Testosterone-treated males increase their territorial behavior (Marler et al.

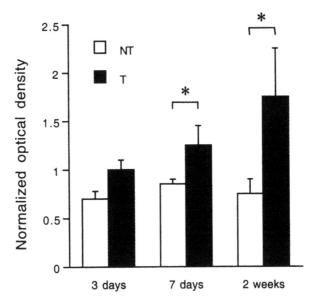

FIGURE 3.1. The effect of a change in status on GnRH gene expression in the cichlid fish *Astatoti-lapia burtoni*. Dominant males were removed, allowing the subordinates to change from a nonter-ritorial status (NT) to a territorial status (T). GnRH mRNA levels (measured as normalized optical density in an assay) increased, and by 7 days after dominant removal the increase was significant. The bars at 2 weeks are levels in males that had been in a stable dominant (T) or subordinate (NT) status for 2 weeks or more. Redrawn from White et al. (2002). © 2002 The Company of Biologists Ltd.

1995). In a number of experiments males that were castrated or given an antian-drogen subsequently had smaller territories than controls, and males given tes-tosterone had larger territories than controls (Searcy and Wingfield 1980; Sil-verin 1980; Moore 1987, 1988; Wingfield and Farner 1993; Moss et al. 1994; Chandler et al. 1994). In the most dramatic cases, males with reduced androgen action lost their territories altogether and males with enhanced testosterone not only expanded their territories (or acquired a second territory) but also acquired additional female mates. Testosterone-treated male red grouse (*Lagopus lago-pus*, also known as willow ptarmigan) enlarged their territories, and in a bad year they achieved greater reproductive success than controls (Moss et al. 1994). These kinds of positive results do not always occur, however. In a pioneering early field experiment, testosterone propionate implants increased aggression in male sharp-tailed grouse (*Tympanuchus phasianellus*) on leks but without changing territorial size or position on the lek (Trobec and Oring 1972). Why they were not able to translate aggression into territorial change is not clear. Reasonable possibilities include social inertia (these males had probably known each other for years) and the small size of lek territories.

FIGURE 3.2. A male tree lizard defending his territory with a display that shows off his blue throat and belly. Territorial defense behavior is based in part on testosterone in free-living males of several lizard species. Courtesy of Michael Moore.

Stimulating effects of a sex steroid on territoriality, when they do occur, make sense in two respects. First, the territories in most of these experiments are established and maintained during the breeding season and clearly serve reproductive functions. Second, defense of the territory is through aggressive behavior such as singing and threat displays that are known or suspected to be steroid dependent.

What is surprising is the ability of testosterone to cause an increase in territory size in intact males, who presumably already have plenty of testosterone. This could be interpreted as evidence that territorial defense behaviors are related to circulating testosterone by a graded dose–response function, not a step function (fig. 2.2). There is another plausible alternative, however. Testosterone might be altering territorial behavior not through a direct effect but indirectly, by changing some other hormone. For example, it might increase glucocorticoids, providing increased energy and locomotor activity for jacking up defense, or it might increase thyroid hormones, as methyltestosterone did in a recent study in men (Daly et al. 2003).

Territoriality is not always related to circulating hormones such as testosterone in a simple manner (Logan and Wingfield 1990; Kriner and Schwabl 1991; Gwinner et al. 1994; Wikelski et al. 1999). Chapter 2 discussed dissociations between annual cycles of testosterone and annual cycles of territoriality and the evidence that neurosteroids might support the behavior when circulating testosterone is low. In bay wrens (*Thryothorus nigricapillus*), however, neither castration nor a combination of flutamide (androgen antagonist) and ATD (estrogen synthesis inhibitor) prevent aggressive responses to simulated territorial intrusions (Levin and Wingfield 1992). Thus, activational sex steroids do not seem to be a mechanism of territorial defense in this species, although a role for earlier organizational actions of gonadal steroids cannot be ruled out.

Territorial challenges, like other challenges, can cause changes in the hormone levels of the actors. This phenomenon has been studied in free-living animals by experimentally introducing a live or stuffed intruder into the territory along with playback of territorial song or calls (in species that use such vocalizations). This method was introduced by Harding and Follett (1979), who found that male red-winged blackbirds (*Agelaius phoeniceus*) exposed to a simulated territorial intrusion quickly (in minutes) differed from controls in the relationship between LH and DHT in the blood. The same manipulation causes testosterone levels to rise in males of some other species of birds, primarily those that have relatively low testosterone at the time of the intrusion (Wingfield and Farner 1993; Hau 2001). Wingfield et al. (1990) proposed that these testosterone responses serve to successfully meet the challenge in males that cannot afford to maintain consistently high levels but nonetheless need the benefits of high testosterone during challenges. As chapter 2 explained, we need to ask what the function of such hormonal elevations might be in the short and long term and whether the testosterone itself can do anything quickly enough to have any impact during the ongoing challenge.

Winner and loser effects are pronounced in territorial behavior just as they are in dominance hierarchies. Prior residents have a clear advantage over would-be occupants. Hormones or peptides could be mechanisms behind the resident advantage. For example, the animal's own territory could become a conditioned stimulus for a favorable hormonal status.

Although females also defend territories in many territorial species, very little is known about any hormonal basis other than a correlation between seasonal changes in reproductive status and seasonal changes in territoriality. In free-living mountain spiny lizards (*Sceloporus jarrovi*) ovariectomy reduced territorial aggression between females and reduced levels of estradiol but not testosterone (Woodley and Moore 1999). Testosterone increased aggression in ovariectomized females (estradiol was not administered) but this could have been mediated by brain aromatization. The discovery that brain-produced estrogens can support territorial behavior in some males suggests testable estrogen hypotheses for females as well. Hormonal responses to territorial challenges have seldom been examined in females. There was no increase in androgens in female song sparrows following simulated territorial intrusion (Elekonich and Wingfield 2000), but whether circulating or brain estrogens increase is not known. While there might be some mechanisms that are shared by males and females, we would also expect some differences to account for sex-specific territorial defense, in which males defend against intruding males and females defend against intruding females. Here the sexes differ not in the behavior displayed, but in the sex at which it is directed.

In some animals juveniles as well as adults defend territories. Here the functions are almost certainly not reproductive, except indirectly, by promoting growth and a more fit adult size. Yet in black-headed gulls (*Larus ridibundus*), testosterone is correlated with territorial defense and testosterone treatment increases territorial aggression (Ros et al. 2002). It will be interesting to see if blocking endogenous androgen action reduces aggression in juvenile gulls.

Why are some species territorial and not others? What mechanisms could account for this dimension of diversity in social organization? Two species of zebra (*Equus burchelli* and *E. grevyi*) differ in whether the males are territorial. Chaudhuri and Ginsberg (1990) hypothesized that they would differ hormonally, but their testosterone results did not support the hypothesis. If a highly territorial species is compared with, say, a colonial species, the difference does not necessarily lie in overall aggressiveness or testosterone level, but instead in some process related to the distance threshold for tolerating the presence of others. A recent hypothesis proposes that these species differences in birds are based in the actions of vasotocin in the lateral septal region of the brain (Goodson 1998a,b; Goodson and Adkins-Regan 1999). The hypothesis is supported by a comparative study of three species of finches: zebra finches (*Taeniopygia guttata*, family Estrildidae, colonial and nonseasonal), violet-eared waxbills (*Uraeginthus granatina*, family Estrildidae, territorial and nonseasonal), and field sparrows (*Spizella pusilla*, family Emberizidae, territorial and seasonal). The territorial status of the species predicts the effects of lateral septum vasotocin manipulations on species-typical aggressive behavior but phylogeny and seasonality do not.

Mating Systems

Social mating systems are patterns of reproductive relationships: who mates with whom, with how many, how often and when. Animal mating systems are highly diverse and sometimes quite puzzling. It is now widely appreciated that underneath the behavioral level of social mating systems are genetic mating systems as revealed by offspring parentage: who has produced how many offspring, with whom, and with how many others. For example, some pairs of socially monogamous biparental birds have young in their nest that have a genetic parent (or two!) that is not a pair member. This happens either because one of the female's eggs has been fertilized by another male through an extrapair copulation or because another female has dumped an egg in the nest. Many more female mammals and other animals produce offspring with multiple fathers than was previously suspected (Zeh and Zeh 2003). Identification of genetic parentage has led to many such surprises, has revealed interesting "slippage" between mating and fertilization, and has challenged received wisdom about mating systems. Many female mammals are solitary, are receptive only very briefly, and can't store viable sperm, whereas female birds have no sharp window of receptivity, can fly, and can store sperm. The facts challenge the stereotype of birds as more monogamous than mammals (Burley and Parker 1997). At the same time, because there are many truly monogamous birds (Stutchbury and Morton 2001), they throw into sharper relief the puzzle of why some birds are so monogamous genetically as well as socially and others aren't.

Social mating systems emerge from choices and decisions by individuals implementing reproductive strategies and tactics. Such strategies and tactics are taxonomically variable. They are also often sexually dimorphic because of differences between the sexes in the anatomy, physiology, and energetics of reproduction. It is exciting to ask if hormones or hormonally regulated peptides might be proximate mechanisms creating some of the species diversity or sex differences.

One of the most famous links to a reproductive mechanism is the association between testis size and mating system. In four different groups of animals (mammals, birds, frogs, and insects), males of species with mating systems producing strong sperm competition have relatively larger testes (Harcourt et al. 1981; Birkhead and Møller 1992; Møller and Briskie 1995; Gomendio et al. 1998). But does this story have anything to do with hormones, or suggest any hormonal hypothesis for mating systems? Spermatogenesis is androgen dependent, but among species of apes, relative testis size does not predict circulating testosterone level (Short 1979). If this is true for other animal groups then testis size would be a red herring in a search for steroid mechanisms underlying mating systems.

Comparative studies of rodents, birds, and amphibians have found different circulating hormone levels or brain peptide receptor distributions in species with different mating systems (Wingfield and Farner 1993; Houck and Woodley 1995; Emerson and Hess 1996; Young et al. 1998). This comparative evidence provides a foundation for the manipulation experiments necessary to establish causation. Such experiments have been carried out in birds and rodents, generating exciting new ideas about mechanisms of mating systems.

Many temperate zone songbirds are socially monogamous; each male is paired with one female and vice versa. It is also common among socially monogamous temperate zone songbirds for the males to assist with some part of the parental care (nest building, incubation, or feeding of young) or even to share these duties more or less equally with the female. Nonetheless, exceptions occur, for example, in pied flycatchers (*Ficedula hypoleuca*), males have two or more females (are polygynous) yet visit all their nests. Wingfield (1984) proposed that hormones are a proximate cause of this aspect of songbird mating systems and carried out hormonal manipulation experiments in free-living birds that supported this hypothesis. In socially monogamous song sparrows and white-crowned sparrows, males given testosterone implants were more likely to subsequently acquire a second female or even a third than the controls. The controls continued to have one female each, as expected.

The fact that these induced polygyny experiments worked at all is amazing but also puzzling. If having two females (which would surely increase reproductive success) is so easy—just a matter of having a bit more testosterone—then why don't more males have more testosterone so they can have two females? Why hasn't polygyny evolved in these sparrows or in more birds generally? Both within and between species, polygynous males tend to sustain high testosterone levels for weeks, whereas monogamous males have a shorter period of elevated testosterone limited to the early breeding phase (Beletsky et al. 1995). By giving monogamous males the testosterone profiles of a polygynous species, they were made behaviorally polygynous. But why do monogamous males' testosterone levels drop so early on in the season, and why are there more socially monogamous than polygynous temperate zone songbirds? House sparrow pairs can produce multiple broods in a season, and the testosterone profiles of the males seesaw up and down as they go from nest building and fertilizing eggs (testosterone up) to chick feeding (testosterone down) and back to fertilizing eggs again (up again) (Hegner and Wingfield 1986). Why don't they just leave the testosterone up all the time?

These questions are addressed by a framework for thinking about hormones and mating systems in temperate zone songbirds that combines the challenge hypothesis with the concept of hormonally mediated trade-offs (Wingfield et al. 1990; Ketterson and Nolan 1994). According to this set of ideas high testosterone levels are not required for the basics of reproduction such as spermatogenesis or copulation (for which more modest levels suffice) but are required mainly for competition with other males (responses to challenges, for example, territorial intrusions) and attracting females, that is, the kinds of sexually selected activities for which more testosterone is supposed to be better. Because testosterone is costly, it will be elevated only when it is needed, for example, to face challenges (a challenge effect). The costs of testosterone that have prevented many males from having continuously high testosterone are those discussed in chapters 1 and 2, including energetic costs, reduced fat stores, interference with immune function, increased risk of predation, interference with pairbonds, wounding during aggression, and increased mortality (Wingfield et al. 2001). Of particular interest are costs due to other behavioral actions of testosterone, especially reduced paternal care, that might vary depending on the species and thus help explain species diversity in mating systems. Yes, some testosterone-treated male sparrows acquired two females, but was their short-term reproductive success actually increased? Male house sparrows and pied flycatchers given testosterone implants produced fewer surviving young, not more, because they failed to feed their young adequately and the young starved (Silverin 1980; Hegner and Wingfield 1987). This would be an effective brake on the evolution of higher testosterone and polygyny-promoting behavior in any case where significant male parental care is obligatory.

Thus, the breeding season testosterone profile and mating system that have evolved in each species reflect a workable solution to this cost–benefit calculus. Not all males are polygynous because for some the costs are too great or the benefits too small or both. Where males must contribute substantially to offspring survival, there will be limits to their ability to be socially polygynous. If the solution is that testosterone levels should be low once the eggs are laid because the male's contribution to incubation and chick care is essential, any need for higher testosterone to meet challenges from other males during this period will be achieved through short-term elevations in testosterone, not by raising the baseline. (Note that once again this assumes short-term consequences of testosterone spikes). A more sustained period of high testosterone would be more beneficial for a male's reproductive success, and more likely to evolve, in species such as red grouse (willow ptarmigan) where highly precocial young need only one parent to survive, or those where females can compensate for reduced male care without a cost to themselves or the young. Female dark-eyed juncos paired with testosterone-treated males fed their offspring more frequently (Ketterson et al. 1992), but whether they suffered a cost afterward is not known.

The discovery that many socially monogamous birds are not genetically monogamous has added a new twist to this framework. Testosterone levels might determine how a male allocates himself to reproduction with his partner (through parental behavior) versus other females (through extrapair copulations). This hypothesis is supported by a heroic experiment with free-living dark-eyed juncos combining hormonal manipulation of a set of males with determination of genetic reproductive success in the entire population (fig. 3.3) (Raouf et al. 1997). Testosterone-treated males produced more offspring than controls through extrapair matings but had lower reproductive success in their own nest, because of poorer paternal care. Their net reproductive success was about the same as controls. What differed was the proportion due to extrapair matings. All told, the framework provides an explanation (not the only one, but an insightful one) into why few temperate zone songbirds pursue a strategy of regularly breeding polygynously and why so many engage in extrapair copulations during their testosterone peak instead (Wingfield and Silverin 2002).

As this conclusion implies, both the challenge hypothesis (challenge effects) and mating–parenting trade-off ideas are intended to account for species diversity as well as for the solution arrived at and currently maintained by indi-

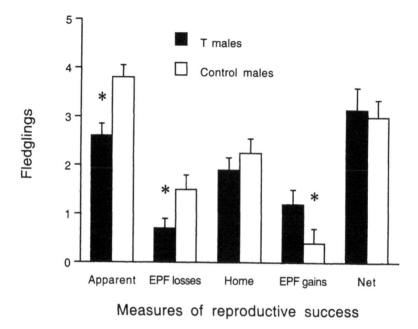

FIGURE 3.3. Experimentally elevated testosterone (T) causes male dark-eyed juncos to shift their reproductive tactics toward extrapair fertilization (EPF). Testosterone-implanted males fathered more chicks in other nests than control males did (EPF gains). They had fewer fledglings in their own nests (apparent reproductive success), but fewer of the chicks in their own nests had been fathered by other males (EPF losses). Overall, they genetically fathered the same number of fledglings as the control males (net reproductive success). Redrawn from Raouf et al. (1997). © 1997 The Royal Society.

viduals of any particular species. The comparative aspects of these hypotheses are very relevant to issues that are discussed in chapters 5 and 6. There it will become clearer why hypotheses about evolution within a species, which occur over relatively short time periods when certain kinds of brakes on change are strong, can't safely be generalized to comparisons between species or clades, which occur over longer time periods that allow these brakes to be lifted. For now, let's focus on the empirical evidence of testosterone profiles and responses and ask whether hormonal responses to challenge actually are associated with social monogamy and male parental care, as claimed. The original comparative evidence (Wingfield et al. 1990) was not "corrected" for phylogeny (for nonindependence of data points due to phylogenetic relatedness). When a recent data set was subjected to a phylogenetic analysis, androgen responsiveness was indeed related to mating system in the predicted manner, with monogamous males having a greater difference between peak and baseline levels, but the relationship to paternal care did not emerge except in passerines, where incubation was more predictive than chick feeding (Hirschenhauser et al. 2003). Outside of birds, there is little association between social mating system, paternal care, and hormonal responses to challenges. For example, many male mammals with no paternal behavior show the phenomenon of increased testosterone in response to a successful challenge. Nor is high testosterone incompatible with parenting (chapter 6). The hypotheses seem to work mainly for birds (and possibly for 11-ketotestosterone in some teleosts [Hirschenhauser et al. 2004]), perhaps because the costs of relatively high testosterone are greater for them. Finding that aggressive interactions cause sex steroid elevations is not by itself evidence for the challenge hypothesis (which is about relationships between those responses and systems of mating and parental care) but instead is evidence for the more taxonomically widespread principle discussed earlier that rank is reflected in hormone level mainly during periods of instability (during ongoing challenges).

Where sexual selection is acting on females, we can ask if female hormonal profiles differ according to mating system. The relevant profiles wouldn't necessarily be those of androgens. Higher testosterone levels might be favored for female–female competition but interfere too much with ovulation to evolve. Several species of birds, including some shorebirds, are polyandrous and also "sex-role reversed," that is, females aggressively compete for male mates and males assume all parental duties. The hypothesis that this role reversal is caused by a reversal in sex steroid levels has not been supported (Fivizzani et al. 1990; Eens and Pinxten 2000). Yes, the sex difference in prolactin is reversed, but based on the research in other birds this is just as likely to be because sitting on eggs causes increased prolactin as because males had higher prolactin to begin with that caused them to sit on eggs. The mechanisms responsible for this kind of sex-role reversal are still a mystery. One possibility is a reversed sex difference in steroid-related enzymes, receptors, or other brain mechanisms downstream of the receptors, but thus far nothing unusual has been reported (Schlinger et al. 1989). Another possibility is some kind of organizational hormone effect. Whatever it is would have to "reverse" brain mechanisms for mating and parenting strategies without interfering with basic reproductive physiology and anatomy. Ligon (1997) reported that males of the sex-role reversed coucals (*Centropus* spp.) have a single functional testis. This is indeed unusual for a bird, but it is not obvious how this could produce sex-role reversal, especially given that males still produce more testosterone than females (Goymann and Wingfield 2004). An activational effect seems unlikely (if you removed one testis from a non-role-reversed species, would the male's role change?), but a single testis could conceivably be a sign of something unusual during embryonic organization, such as an ovary (feminized testis) that disappeared.

Most mammals are not socially monogamous. The exceptions are interesting material for testing hypotheses about hormone regulated mechanisms promoting monogamy, especially when they have congeners that are not monogamous, allowing comparative tests with close relatives. The most successful work of this kind has focused on voles (*Microtus*) and deermice (*Peromyscus*). Prairie voles (*M. ochrogaster*) are socially monogamous and tend to affiliate with other individuals rather than live solitarily, whereas mountain and meadow voles (*M. montanus* and *M. pennsylvanicus*) are polygynous in their mating (and probably polyandrous as well) but not very social otherwise. These social systems are correlated with species differences in the distribution and density of oxytocin and vasopressin receptors (that is, in expression of genes for these proteins) in regions of the brain important for social behavior such as the lateral septum and extended amygdala (Carter et al. 1995; de Vries 1995; Young et al. 1998). Distributions of other kinds of receptors do not differ between congeners. Are these brain differences causally related to the mating systems themselves? Experiments in *Microtus* support the hypothesis that peptide actions are critical for pairbond formation (see section below). In *Peromyscus*, however, the correlation between lateral septum and extended amygdala vasopressin receptor density and mating system is in the reverse direction from *Microtus*, supporting an alternative hypothesis that on a wider taxonomic scale aggressive behavior is a better predictor of species differences in these brain characters than mating system (Bester-Meredith et al. 1999). There has also been some interest in possible organizational influences on monogamous vs. polygynous tendencies. For example, Carter et al. (1995) proposed that male prairie voles are in a developmental sense feminized males compared to polygynous male *Microtus*.

There remains a lot to be learned about hormonal involvement in mating systems. The territory is still largely uncharted, which makes the achievements thus far even more impressive. There are many fascinating questions to ask about hormonal and other mechanisms of mating strategies and tactics. For example, in socially monogamous species that engage in extrapair copulations, do these mechanisms differ between courting and mating with the pairmate, where there is lots of time to get ready because courtship is extended, and mating with an extrapair individual, where the opportunity may appear suddenly, courtship is brief or nonexistent, and there is little time to prepare? Are hormones or hormone-regulated mechanisms involved in adaptive ejaculate allocation, for example, increased ejaculate size in the face of sperm competition (Nichols et al. 2001)? Olsson (2001) reported a "voyeur" effect in the dragon lizard (*Ctenophorus fordi*). A male's copulation was prolonged, increasing ejaculate volume, if he saw another male copulate with that female. Could this be a rapid effect of a sudden elevation in a peptide or steroid?

Mate Choice

In most mating systems, animals make some choices about which other individuals to try to pair and mate with. Animals have distinct mating preferences based on the species, sex, and individual characteristics of others. Even in what appear at first glance to be wildly promiscuous animals, closer examination usually reveals some selectivity by one or both sexes. Complete promiscuity (everyone mates with as many others as possible) and random mating are seldom either a realistic possibility or an adaptive approach to mating. If there are exceptions, they would probably not be vertebrates but rather sessile marine invertebrates that release clouds of gametes into the water. Mate choice is a two-way decision process, for each individual is not only a chooser but also a chosen. What are the hormonal mechanisms of these choices?

A preference for mating partners of one's own species is obviously highly adaptive given the inevitable costs of mating and the likelihood of infertile offspring in hybrid matings. Studies of the development of species preferences have revealed that sexual imprinting is the process in many birds and some mammals. During sensitive periods early in life, animals learn the features of their family members and then use these as a guide to their own species identity and species mating preference (Immelmann 1985). It is sometimes assumed that hormones are involved in this process, for example, that changing hormone levels during development might open or close the sensitive period "window," but thus far this hypothesis has received little support (Hutchison and Bateson 1982; Balthazart 1983; Bolhuis 1991). Several recent experiments with zebra finches suggest a second stage of sexual imprinting occurring later, in young adulthood, that stabilizes (consolidates) what was learned from the family earlier in life (Bischof 1994). This second stage coincides with the onset of adult reproductive behavior and fertility. Hormonal changes are an attractive possibility for a mechanistic basis of consolidation.

A preference for the other sex as mating partners is again obviously adaptive but until recently little was known about the mechanisms behind this preference (Adkins-Regan 1988, 1998). Preference based on the sex of the chosen (sexual partner preference) is highly sexually dimorphic and constitutes a glaring exception to a tendency of animals to mate assortatively (to mate with individuals having a similar phenotype). Here we can ask about hormonal mechanisms in the same way we would for other strongly dimorphic social behavior. We can begin by posing an activational hormone hypothesis. Do males choose females, whereas females choose males, because the choosers have different levels of sex steroids circulating in adulthood? Does testosterone make males interested in females whereas estradiol makes females interested in males? Some role for activational hormones seems particularly likely in species where preferences change depending on current hormonal status. For example, in choice tests female rats show a bigger preference for males over females during estrus than during diestrus; the preference for a male is lower during pregnancy and higher during lactation (Eliasson and Meyerson 1975). Female rhesus monkeys and meadow voles prefer the company of other females to the company of males except during the breeding season. At the onset of the breeding season they switch to preferring the company of males. Male voles prefer female odors only if they have breeding testosterone levels. In both species these seasonal preference switches have been shown experimentally to be based on seasonal changes in gonadal sex steroids (Ferkin and Zucker 1991; Ferkin and Gorman 1992; Michael and Zumpe 1993). Treatment of female rough-skinned newts with testosterone plus AVT causes them to be attracted to female odors, unlike control females (Thompson and Moore 2003), suggesting steroid and peptide activation of males' attraction to female stimuli. Involvement of activational hormones seems less likely in species with a constant unchanging preference independent of current reproductive status. Thus, in zebra finches, which are continuously paired even when conditions are not conducive to breeding, there seems to be little or no effect of activational hormone manipulations on sexual partner preference (Adkins-Regan and Ascenzi 1987; Adkins-Regan 1999b).

More often than not, activational hormone manipulations simply alter the level of interest in sexual activity per se (sexual motivation) rather than changing sexual partner preference. Taken together with the highly dimorphic nature of sexual partner preference, this suggests that organizational hormones might be the primary source. Experiments in several species of mammals (including ferrets, rats, mice, hamsters, dogs, and pigs) and in Japanese quail have confirmed this hypothesis (Adkins-Regan 1988; Baum et al. 1990; Vega Matuszczyk and Larsson 1995; Balthazart et al. 1997; Bakker 2002a,b). Manipulations of sex steroids early in development can actually sex reverse preference, producing females that choose other females in choice tests or males that choose other males. None of these mammals, however, forms any particular relationship with mating partners beyond copulation itself; these are neither pairbonding nor socially monogamous species. Japanese quail show strong sexual dimorphism in "peeping Tom" tests, with males but not females liking to spend time looking through a high window to see a female in the next room. Females hatched from eggs treated with an estrogen synthesis inhibitor act like males in these tests, but the mating system of this species is unknown.

Thus far the best evidence for a contribution of organizational hormones to sexual partner preference in a pairbonding animal comes from experiments with zebra finches. Females treated with estradiol as nestlings, who grow up to have a masculinized song system, also prefer other females for pairing (Mansukhani et al. 1996). This effect is seen only if the females have spent their juvenile life in an all-female social group. An all-female environment by itself does not alter female pairing preference, suggesting some reinforcement (consolidation?) of organizational masculinization through social influence. Genetic females hatched from eggs injected with fadrozole, who have testes instead of ovaries, strongly prefer to pair with other females (fig. 3.4) (Adkins-Regan and Wade 2001), further evidence for an organizational hormone hypothesis.

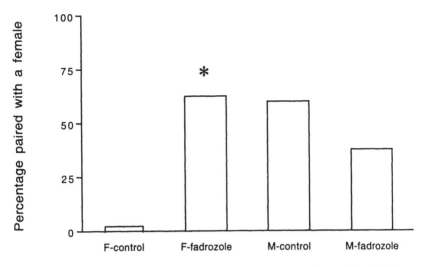

FIGURE 3.4. Masculinization of sexual partner preference caused by an early steroid manipulation. Female zebra finches hatched from eggs injected with fadrozole, an estrogen synthesis inhibitor, preferred to pair with other females instead of males, whereas no control female preferred to pair with another female. The fadrozole treated females had testes or ovotestes instead of ovaries. Males hatched from eggs injected with fadrozole were not different from control males. Reprinted from Adkins-Regan, E. and Wade, J. 2001. Masculinized sexual partner preference in female zebra finches with sex-reversed gonads. *Horm. Behav.* 39: 22–28. © 2001 Academic Press (Elsevier).

Animals have distinct preferences for particular individuals from among the pool of opposite-sex animals of the appropriate species. This is the dimension of mate choice that has been a major focus in behavioral ecology. There seem to be three kinds of potentially interesting relationships between hormones and individual mate preferences. First, the traits that seem to make individuals attractive desirable choices are often steroid based. Here the hormonal action is in the chosen, not the choosers, and hormones are mechanisms of the traits being assessed. Second, social interactions leading up to mate choice can cause hormonal changes in both the assessors and the assessed. Third, there might be hormonal influences on the preferences of the choosers.

Both of the first two relationships were examined in chapter 2 but it is worth speculating about their potential implications for mate choice a bit further here. If acute changes in circulating or brain hormones really do have rapidly occurring effects (directly or via peptides) that influence the interaction while it is still going on, then one can imagine the following scenario. A female approaches several males one at a time and pays attention to her internal responses to them as well as their behavioral responses to her. As she approaches a male his behavior changes (augments) according to how much he raises his hormone levels, which is, in turn, a function of both his ability to do so (his quality) and his interest in that particular female (her value on the market).

She also has some sense of her quality because she is monitoring her internal hormone levels. She selects the male that causes the greatest or most rewarding hormonal response in her. This kind of interactive complexity is probably biologically realistic. Certainly behavior during mate choice can be highly interactive, as when female bowerbirds (*Chlamydera maculata*) select a male (Borgia and Presgraves 1998). Pair formation is a negotiated process (Todd and Miller 1999). It would be exciting to know the mechanisms of those negotiations. It is also important to find out if there are hormonal or other physiological mechanisms common to both mate attraction (by the signaler) and mate preference (by the receiver) that help explain how they co-evolved (Fisher et al. 2002). Vasotocin stimulates not only advertisement calling by male bullfrogs (*Rana catesbeiana*) but also responses of females to call playback (Boyd 1994), providing a common genetic basis for both parties, the genes for vasotocin and its receptors.

The third potential relationship between hormones and individual mate choice, hormonal effects on the chooser's preferences, has been largely ignored. Mate preferences for individuals are usually sexually dimorphic. Females are interested in different traits in males than males are in females. For example, female peacocks choose mates on the basis of elaborate tails with "eyes"; males either don't choose at all, or, if they do, the choice can't be based on elaborate tails with eyes because females don't have them (Petrie et al. 1991). Animals that sexually imprint acquire some within-species as well as between-species mate preferences through family example but males and females do this differently. Female but not male zebra finches imprinted on an experimentally produced novel trait (a red feather on the parents' heads) and later preferred opposite sex partners that had this "trait" (Witte and Sawka 2003). What are the sources of these sex differences? A good starting point might be the same activational and organizational hypotheses that have helped in understanding sexual partner preference.

Pairbonding

In socially monogamous mating systems, both choosers and chosen form and maintain an exclusive and distinctive social relationship with an individual of the other sex. These relationships are expressed (and measured) as a preference for being with the partner rather than another individual of the same sex when given a choice, combined with engagement in specific behaviors such as mutual preening that are not directed at others. The hormonal mechanisms of pair formation and maintenance are likely to vary according to (1) the duration of the relationship, (2) whether initial formation is limited to a particular reproductive stage or is stimulated by engaging a particular reproductive activity such as mating, and (3) phylogeny, that is, the taxonomic group, because socially monogamous pairbonds have a scattered distribution among animal species.

A pattern seen in some north temperate birds is for pairs to form in the spring and dissolve in the fall. Both sexes then pair with new partners the following spring. The seasonal nature of the pairbonds and the correlation with changes in sex steroid and other hormone levels suggest an activational hormone hypothesis for both their formation and their dissolution, with rising or high levels promoting formation of bonds and low or falling levels promoting their breakup. In a pioneering early field experiment, California quail (*Callipepla californica*) implanted with testosterone (males) or the synthetic estrogen stilbestrol (females) during the nonbreeding season, when the birds live in groups, formed pairs but without showing any courtship or copulatory behavior (Emlen and Lorenz 1942). That is, pairing was advanced in time by artificially advancing the "spring" increase in sex steroids. Somewhat the opposite occurred in canvasback ducks (*Aythya valisineria*), however. Steroid treatment (testosterone in males, estradiol in females) increased courtship by males but without stimulating receptivity in females or pair formation (Bluhm et al. 1984). Females require more than hormones to be interested in a courting male.

In mating systems with long-term social monogamy, pairbonds can last for several years or several breeding episodes, in some cases for the life of the individuals, with a new pairbond occurring only if the partner dies. This mating system is probably most common in birds (Mock and Fujioka 1990; Black 1996; Ligon 1999), and because it is common in tropical species, could conceivably be the most common avian social mating system (Kunkel 1974; Freed 1987; Stutchbury and Morton 2001). Long-term monogamy raises interesting questions about hormonal mechanisms underlying pair formation and maintenance but so far little research has been done to try to answer them. What keeps the birds together from one breeding episode to the next? Are hormones (for example, low testosterone) the mechanistic basis for the association between long-term monogamy and opportunistic or tropical breeding? In long-term monogamy with mate fidelity the initial choice of a partner is critical for lifetime reproductive success, yet is often made without benefit of prior breeding experience. For example, zebra finches form their permanent pairs at a very young age (50–90 days), passing rapidly from a preference for the parents and siblings to a preference for partners and (after a brief period of "dating") the formation of their permanent pairbond (Zann 1996). What causes this transfer of preference? Does some kind of late organizational effect occur at this time that ensures the permanence of the new relationship?

Both steroids and the peptides they regulate have been discovered to be key mechanisms for the pairbond-related behaviors of socially monogamous voles and deermice (Young et al. 1998). The prairie vole has been the most thoroughly investigated (Carter et al. 1995). In this species, pairbonding is expressed as a preference for close proximity to the partner and aggressiveness toward other individuals. It requires an extended interaction. For the female, either mating or cohabitation without mating is sufficient, but for the male an extended bout of mating is required in addition to cohabitation. This suggests that something about the animals' physiology that is different for males than females (possibly because of early organizational effects of sex steroids [Cushing et al. 2003]) changes in response to the interaction that then stimulates pair formation.

Sex steroids do not seem to be facilitators. No ovarian hormones are required for pair bonding by females; male testosterone levels drop with pairing (De Vries 2002). With respect to corticosteroids, interacting with an initially unfamiliar individual might alter HPA activity, and, indeed, glucocorticoid levels drop with pairing. Adrenalectomy has the opposite effect on males and females; it promotes partner preference in females and interferes with it in males, whereas stress or corticosterone treatment promotes it in males and inhibits it in females (DeVries 2002).

Oxytocin is a candidate mechanism for females. The vagino-cervical stimulation of mating or parturition is known to cause release of oxytocin in the brain in several female mammals, and mating is sufficient (though not necessary) for female prairie voles to pairbond. Oxytocin administered into the brains of females increases partner preference in the absence of mating, whereas an oxytocin receptor antagonist prevents mating-induced partner preference (Williams et al. 1994). The corresponding candidate for males is arginine vasopressin (AVP). Pairing stimulates changes in the lateral septum indicating synthesis and synaptic release of AVP; experimental manipulations have confirmed the hypothesis that these peptides are necessary and sufficient for males to pair (de Vries 1995; Young et al. 1998). The administration of AVP into the lateral septum of males has a similar effect as oxytocin in the females, increasing partner preference and affiliation and postmating aggression toward nonpartners even in males that have not mated (Winslow et al. 1993). These peptides do not affect social behavior in congeneric montane voles, which are polygynous. Also, oxytocin does not work in males and AVP does not work in females; their effects on pairbonding are sex-specific (Young et al. 1998).

Evidence points to a particular AVP receptor subtype, the V1a receptor, as an important site of action for effects on pairbonding in males. The V1a receptor is not steroid regulated in prairie voles. Infusion of a V1a receptor antagonist into the brain blocks pairbonding induced by mating (Liu et al. 2001). The gene for this receptor is expressed in areas of the brain regarded as part of a "reward" circuit, including the medial amygdaloid nucleus, nucleus accumbens, and the ventral pallidum, as if the receptor is a mechanism regulating whether a social interaction will be rewarding (Bester-Meredith et al. 1999; Young et al. 2001; Aragona et al. 2003). Pitkow et al. (2001) used a viral vector to transfer the prairie vole *V1a* receptor gene into other prairie voles to increase the number

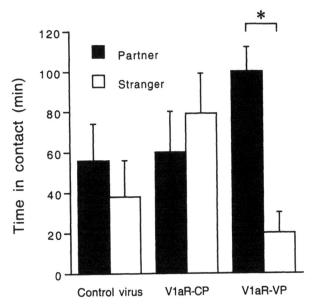

FIGURE 3.5. The effect of transfer of the gene for the vasopressin type 1a receptor (V1aR) on partner preference formation in male prairie voles. Such gene transfer causes increased expression of the receptor. Males with increased *V1aR* expression in the ventral pallidal area (VP) showed a significant preference for the familiar female partner over a female stranger after a 17-hour cohabitation (as indicated by time in contact), whereas males with increased expression in the caudate putamen (CP) and control males (those injected with a control virus instead of the gene transfer viral vector) did not. This study tested the hypothesis that the V1aR might underlie species differences in mating systems and pairbonding through an experimental, rather than correlational, approach. Redrawn from Pitkow, L. J., Sharer, C. A., Ren, X., Insel, T. R., Terwilliger, E. F. and Young, L. J. 2001. Facilitation of affiliation and pair-bond formation by vasopressin receptor gene transfer into the ventral forebrain of a monogamous vole. *J. Neurosci.* 21: 7392–7396. © 2001 by the Society for Neuroscience.

of receptors in the ventral pallidum, to see if this would cause males to bond with the female without having to mate first (fig. 3.5). After transfer they indeed showed a preference for a familiar female without mating with her.

It is exciting to see how this research with voles connects interesting social behavior to steroids and peptides and from there to receptor proteins and gene expression. It is not yet known whether peptides contribute to pairbonding in animals other than mammals, in animals that do not have to mate to pair (are there any?), or in animals where mating does not cause peptide release in the brain (are there any?), but the research with voles should inspire a look at the physiology of pairbonding in other kinds of animals.

Relationships between males and females in established pairs contain elements of both cooperation and conflict, which is to be expected given that the partners have some overlapping interests but are not genetically related. The

FIGURE 3.6. A titi monkey family (female on the right, male carrying young on the left). Measurement of glucocorticoid levels following social separation shows that the male and female are more attached to each other than to the offspring, and the offspring is more attached to the father than to the mother. Courtesy of Katherine Floyd and Sally Mendoza.

hormone and peptide levels of the animals are likely to change as the pair goes from harmony to strife and back again. Do individuals monitor their stress (glucocorticoid) levels in deciding when to give in? In spite of the occasional conflict, however, from a psychological mechanism viewpoint pairs in some species can be considered to be attached to one another. Attachment is a concept applied to relationships where there is individual recognition and close contact over an extended time period, and where separation is followed by signs of distress and behavior designed to reestablish contact (Zeifman and Hazan 1997; Insel and Young 2001). In mammals, separation of attached individuals activates the HPA axis so reliably that increased glucocorticoid levels in response to separation are now sometimes used as a measure of how attached the animals were.

Attachment is not the same as affiliation, for animals in affiliative relationships do not necessarily show distress or HPA activation when experimentally separated. This point is beautifully illustrated by a study comparing socially monogamous titi monkeys (*Callicebus moloch*) with nonmonogamous squirrel monkeys (*Saimiri sciureus*) (Mendoza and Mason 1986). Paired titi monkeys spend quite a bit of time in intimate physical contact, bodies pressed against each other and tails twined around each other (fig. 3.6). Squirrel monkeys are highly social and interactive, but do not form pairs within the group. Experimental separation of males and females housed together caused increased cortisol in titi monkeys (where separation meant loss of the pairbond partner and attachment object) but did not have this effect in squirrel monkeys (who although housed together were neither pairbonded nor attached).

When reunited with the familiar pairbond partner, glucocorticoid levels return to baseline. This "calming" effect of being with the partner is also seen as pairs first form and glucocorticoids (and in males, testosterone as well) decline (Reburn and Wynne-Edwards 1999; Ginther et al. 2001). These hormonal changes may be important for setting in motion the cooperative biparental and food-sharing behavior of the pair (French 1997). Harmonious male–female pairs of tree shrews had lower levels of circulating glucocorticoids than incompatible pairs (von Holst 1998). Pairbond partners can also act as a buffer against external sources of stress. In pairbonded marmosets, guinea pigs, and tree shrews, individuals show less HPA activation to a novel physical environment if the partner is also present (Sachser et al. 1998; von Holst 1998; French and Schaffner 2000).

Much less is known about hormonal consequences of pair formation and separation in nonmammals. In zebra finches separation of a pair is followed by changes in behavior suggesting that the birds are distressed and searching for the mate (Butterfield 1970). Pairs that were separated showed elevated circulating corticosterone, which returned to baseline upon reunion (Remage-Healey et al. 2003). A corticosteroid response following separation might be beneficial by rapidly stimulating the kind of locomotor activity that helps the bird locate the partner or find a new one. But HPA axis activation accompanying separation and the energy required to re-pair might also contribute to the costs of abandoning the first mate and switching to a second that have prevented the evolution of divorce in this species.

Parent–Offspring and Sibling Relationships

These relationships, like those between mating partners, involve genetically different actors and a complex mixture of cooperation and conflict. The parent's interests are to leave behind surviving offspring over the course of the lifespan, not to invest maximally in any particular offspring. Yet from an individual offspring's perspective, maximal investment is exactly what is desired, and siblings be damned.

The concept of attachment figures importantly in research on hormones and the cooperative side of parent–offspring relationships in mammals. Infant mammals are often very attached to the mother and vice versa, as revealed by constant close physical contact, obvious distress (such as loud calls) if separated, and vigorous behavior directed at relocating each other. Separation of the two parties often results in dramatic HPA activation in one or both (Hennessy 1997). This makes sense given what the function of HPA activation is supposed to be in the lives of animals. In the wild separation would be a very dangerous situation for the young. From the mother's perspective it threatens to drastically reduce her short-term reproductive success, especially if the litter size is one. HPA activation directs energy and behavior toward resolving the crisis (finding the other party) and increases fitness for both.

In species with biparental care, measurement of blood glucocorticoid levels in response to separation provides an intriguing window into who really matters, as shown in a study with titi monkeys (Hoffman et al. 1995). In this species both of the closely pairbonded parents care for the infant, but the infant is carried more often by the father. Infants show no HPA activation if the mother alone is removed, nor do mothers show any activation if the infant is removed briefly. But infants show HPA activation if the father alone is removed, and (not surprisingly) show marked activation if both parents are removed. Most surprising is that the adults show HPA activation to separation from each other but not to brief removal of the infant! Perhaps this is because the animals are territorial and so timid that wild infants would never actually move away from the parents, in contrast to group-living macaques where infants are curious and exploratory and other adults try to grab them.

Brain peptides have been shown to be mechanisms underlying the separation distress shown by very young animals (Nelson and Panksepp 1998). Oxytocin treatment blocks separation distress in baby rats and chickens (that is, oxytocin is comforting), and AVP has a similar effect in infant rats. This suggests that the attachment concept and its associated mechanisms might be quite conserved, at least in birds and mammals.

Parent–offspring imprinting, on the other hand, is a special kind of attachment that is probably less widespread and certainly not universal. It has been selected for only under particular developmental and ecological conditions that require an unusually rapid attachment process. In parent–offspring imprinting (called filial imprinting when it occurs in the young and parental or maternal imprinting when it occurs in the parent) an attachment is formed during a limited critical or sensitive period shortly after birth or hatching. The attachment must be formed quickly because the young are precocial, the mother is going to have to keep moving around to forage or avoid predators, the mother may be mixing with many other mothers and young, all the young need to keep up with the mother to be safe and get fed, and the mothers need to know who their own offspring are so that they don't waste energy caring for individuals that are not genetically theirs. The classic examples of parent–offspring imprinting come from ground-nesting birds with precocial young, such as ducks, where the female leaves the nest site once all the ducklings have hatched, and sheep, where all the births in the herd occur during a short time period and the herd is on the move immediately after the lambs can walk. It could be hypothesized to occur in colonial bats such as the evening bat (*Nycticeius humeralis*) as well. In their colonies, mothers and young are packed in densely, the mothers forage at a distance, yet in early lactation mothers nurse only their own young, as if they learned fast which one was theirs (Wilkinson 1992).

The critical period property of imprinting directs attention to the physiological mechanisms responsible for the opening and closing of the critical period "window." Hess (1959) proposed that the developing fear/avoidance system (neophobia) closed the critical period window for filial imprinting by ducklings, suggesting a possible involvement of the HPA axis. Manipulations of corticosterone or ACTH do alter behavior in imprinting tests, although more by nonspecifically affecting performance of the behaviors used to measure imprinting than by affecting the imprinting process itself (Martin 1981). The attachment to the parent (or other imprinting object) then ends as the bird matures. In chickens this coincides with an increase in HPG activity, and experimentally administered sex steroids induce premature detachment from the imprinting object (Gvaryahu et al. 1986).

Maternal imprinting (also called maternal "bonding") by sheep occurs during a roughly 12-hour period after birth. During this "window" the mother labels her lamb with her own scent (by licking it), accepts it as an object to provide with food and protection, and then rejects all lambs that smell different. An elegant program of experimental research involving several labs has made impressive progress discovering the mechanisms that produce these interesting time-limited behavioral changes (Lévy et al. 1996). Prolactin is unimportant for maternal imprinting. A combination of priming by steroids (estradiol and progesterone, which would normally increase during gestation) and vagino-cervical stimulation (produced by the birth process) opens the "window." The latter works because it causes an increase in oxytocin in the PVN. These mechanisms in turn act on the olfactory bulb via norepinephrinergic neurons to produce an olfactory memory of the lamb.

With litter or clutch sizes greater than one, the early social environment includes siblings as well as parents. Their relationships are not always harmonious, and can even result in death, as when female spotted hyenas kill their male littermates. Brood reduction in nestling birds is a distinctive and extreme form of size-based dominance. Many birds have asynchronous hatching; chicks hatch on different days and because of the rapid growth rates typical of baby birds a brood contains chicks of different sizes corresponding to hatching order (large ones first hatched, small ones last hatched). In some of these species later-hatched chicks are less likely to survive, and the cause is aggression by the early hatched. In the nazca booby (*Sula granti*), a seabird with two-egg clutches, the first-hatched chick (alpha) always kills the second-hatched (beta). While it lives, beta has higher corticosterone levels than alpha (Tarlow et al. 2001). Is this good or bad for beta? While costs of chronic HPA activation could hasten beta's demise in this species, in the blue-footed booby (*Sula nebouxii*), dominant and subordinate chicks end up having remarkably similar survival rates and reproductive success, as if subordinates are at a disadvantage prior to fledging but dominants are at a disadvantage thereafter (Drummond et al. 2003).

The postfledging survival of some birds depends on their condition and size at fledging, which is in part a function of how they fared in competition with their siblings for parental feeding. Parents have some control over this as well through their food allocation to the young in the nest, and (in the case of mothers) through their investment in each egg. This provides an interesting way to think about the sex steroids of maternal origin that have been found in the yolks of freshly laid eggs. In the majority of birds that have been studied, testosterone and other androgens increase with laying order; that is, later-laid eggs in a clutch have more androgens per milligram of yolk. Schwabl (1993, 1996a, 1997a), who discovered this pattern, proposed that it was a mechanism whereby the mother could mitigate the effect of hatching asynchrony, leveling the playing field to ensure that later hatched young would have as favorable a chance of survival as earlier-hatched young (see also Winkler 1993). Allocation of androgens to yolks could help the female's postfledging offspring compete with those of other females as well (Schwabl 1997a). In support of this competition hypothesis, testosterone treatment of eggs increases begging vigor in canaries, increases the weight of the muscle used to hold the head upright during begging in red-winged blackbirds (*Agelaius phoeniceus*), and enhances the growth and begging behavior of black-headed gull chicks (*Larus ridibundus*) (Schwabl 1996b; Lipar and Ketterson 2000; Eising et al. 2001; Eising and Groothuis 2003).

There is an alternative hypothesis, the maternal investment hypothesis, for the function of yolk steroids (Gil et al. 1999). Female zebra finches that were paired with males with red leg bands (more attractive males) laid eggs with more testosterone in the yolks (Gil et al. 1999). Eggs laid by female canaries had more yolk testosterone when those females had been exposed to canary songs known from prior work to stimulate female displays than when those same females were exposed to less stimulating canary songs (Gil et al. 2004). These results are consistent with the idea that females should invest more if the young have a high-quality father (Burley 1988). Male eggs laid by dominant female chickens had more yolk androgen than their female eggs, as if these high-quality females were shifting their investment toward their sons, consistent with Trivers and Willard's (1973) model of parental manipulation of sex ratio (Müller et al. 2002). Higher-quality female starlings (older females and those producing early and large clutches) produced eggs with more yolk androgens (Pilz et al. 2003). (See also chapter 4 for the hypothesis that yolk steroids are a sex ratio adjustment mechanism.)

Both hypotheses contain two claims. One is that yolk androgens are subject to regulation by the female, according to laying order, condition, breeding density, or social interactions. There is some evidence that yolk steroid levels vary as a function of such influences (Schwabl 1997b; Gil et al. 1999, 2004; Reed and Vleck 2001; Whittingham and Schwabl 2002; Pilz and Smith 2004). This claim seems plausible, because yolk steroid levels parallel blood steroid levels (Adkins-Regan et al. 1995) and many kinds of internal and external conditions affect blood steroid levels of animals. However, little is known about effects of condition and social interactions on testosterone and other androgen levels of females. The other claim is that yolk androgens have fitness enhancing effects on the chicks hatched from the eggs. So far there is not enough empirical work to know how solid or species general this claim is, or whether the benefits of yolk testosterone outweigh the potential costs. Elevating yolk androgens can decrease rather than enhance survival (Sockman and Schwabl 2000). The eventual survival and reproductive success of chicks that show increased growth and vigor with elevated yolk androgen are unknown.

When experimentally elevated yolk androgen does increase growth or vigor, are these effects activational or organizational? Yolk androgens are measured when eggs are freshly laid, because only then is it safe to assume that steroids are maternal in origin. But the embryo feeds off the yolk for an extended period, even beyond hatching in some species. So any effect of yolk androgen on behavior during the nestling period could be activational, organizational, or both, depending on when the maternal yolk androgens have been used up. Any effects that lasted well beyond the first few posthatching days or that emerged later in life more likely would be organizational, at least in principle. Thus far there are very few established organizational effects of early androgen administration in birds aside from feminization of males due to testosterone aromatization; however, it has usually been adult, not juvenile, phenotypes that have been studied (Balthazart and Adkins-Regan 2002).

Are yolk steroids an evolved adaptive mechanism on the part of the mother? This is certainly what is implied when it is said that females are "depositing" or "allocating" androgens to yolks. Maternal steroids have been found in the yolks of eggs of other vertebrates as well (Bern 1990; Dickhoff et al. 1990). Steroids readily diffuse into lipids such as yolk. While yolk itself is expensive stuff the steroid might be in it simply through passive diffusion and not because of any special mechanism or need for getting it there. Yes, yolk androgen increases with laying order in most birds, but this is what would be expected from the mother's testosterone cycles during laying (Schwabl 1996a). Thus far there is little evidence in birds that yolk levels are adjusted independently from blood or follicular levels (Hackl et al. 2003; Williams et al. 2004). This does not rule out an adaptive function, but it does shift the likely target for the selection to the chick rather than the mother. If steroids are inevitably going to show up in the yolk (that is, if it would be too expensive for the mother to prevent that) then chicks that capitalized on their presence to benefit their own fitness would be at an advantage.

FIGURE 3.7. A Florida scrub jay brooding chicks. Alloparents feed the brooding female and the chicks. Courtesy of Stephen Schoech.

As with many other interesting adaptive function hypotheses, it is challenging to determine whether, how, why, and for whom the feature is adaptive. It is especially difficult when the phenomenon (here maternal yolk steroids) is found in a wide array of vertebrates. Does that mean it is so wonderful that it arose early in the vertebrate lineage and has remained useful ever since, or that it is neutral baggage from the past that is not worth getting rid of? Among birds, the exceptions to the general pattern of increases in androgens with laying order might be easier tests of adaptive mechanism hypotheses. Cattle egrets have two-egg clutches, but androgens are higher in the first-laid egg, suggesting that they might contribute to the obligatory brood reduction that occurs (Schwabl et al. 1997). It will be important to see if this hypothesis is confirmed through hormonal manipulation experiments, and if in a comparative analysis species with obligatory brood reduction have a different yolk testosterone pattern than other birds. Furthermore, it will be important to consider all the other steroids in yolks and not just androgens.

Cooperative Breeding and Alloparenting

In cooperatively breeding social systems animals participate in the care of eggs or young that are not their own (fig. 3.7). To a human observer reproduction seems to be a group effort. Familiar examples of cooperative breeders include gray wolves, which live in packs consisting of a socially monogamous pair, who are the genetic parents of the pups, along with other wolves. The pups are fed by all the pack members, both males and females (Asa 1997). Over three percent of mammals and birds have some form of cooperative breeding (Emlen 1997). In the most puzzling cases, the genetic parents appear to be aided by adult nonreproductive helpers (sometimes their older offspring) that are socially subordinate to them, generating a very lopsided distribution of reproduction (high reproductive skew) between dominants and subordinates. The biggest evolutionary puzzle is why the subordinates put up with this arrangement. Progress has been made in understanding how cooperative breeding could evolve through theoretical and empirical work addressing the costs and benefits for both the genetic parents and the alloparents in different ecological contexts, taking into account genetic relatedness and indirect fitness in calculating benefits. (The term "alloparents" is now preferred over "helpers" because the latter implies that the alloparents actually increase the reproductive success of the genetic parents, which is a hypothesis rather than a safe assumption.) In reproductive skew theories, cooperative breeding is treated as part of a larger category of social systems in which reproduction in animals living in a group is partitioned between same-sex adults in dominance hierarchies (Vehrencamp 1983; Reeve et al. 1998; Cant and Johnstone 2000).

Cooperative breeding is a complex mixture of several kinds of behavior and social relationships with known or suspected connections to hormones, including mating systems and pairbonds, dominance, parental behavior, and delayed dispersal. Furthermore, consideration of hormones is essential for understanding the costs and benefits to alloparents and dominants. This has been appreciated ever since it was realized that some alloparents are not prereproductive juveniles but rather adults that are not reproducing in the presence of their parents or other dominants. Guhl's concept of "psychologically castrated" subordinate males has led a number of researchers to wonder if avian helpers are suffering the same fate. This raises the important issue of who is in control here. Chapter 2 discussed hormonal responses caused by others as signaling acts and why we should not assume that they benefit only the sender. Similarly, there might be benefits to the subordinates, or, as Reyer et al. (1986) pointed out, their status may in some sense be chosen rather than imposed. "Reproductively inactive" will be used here rather than "reproductively suppressed" because suppression implies control by the dominants with subordinates having no say in the matter. "Inactive" is more neutral, indicating simply that subordinates are not reproducing without implying why. "Inactive" is also more neutral than "socially induced infertility," which implies failure to produce gametes, which is not always the case. Similarly, phrases like "physiological castration" or "psychological castration" can be confusing unless they are clearly defined with respect to exactly what aspects of physiology or behavior are absent and distinguish between the hormonal and gametic functions of the gonads.

The most extreme case of lack of reproduction by alloparents occurs in the highly eusocial insects. In honeybees the alloparents are female worker bees that are a different morphological and behavioral caste from the queen and outnumber her hugely. Under normal circumstances workers have rudimentary ovaries and never reproduce. Removal of the queen, instead of causing workers to become reproductives, causes workers to produce a replacement queen by feeding a larva to become a future queen. Only if that fails and the hive is queenless for an extended period will some workers slowly become reproductives (Seeley 1995). For the vast majority of alloparent bees reproductive inactivity is complete and lasting.

The closest vertebrate analog of the eusocial insects is the colonially breeding naked mole-rat (*Heterocephalus glaber*, a bathyergid rodent) (Faulkes and Abbott 1997; Lacey and Sherman 1997). There is one breeding female in a colony, who is larger than the other females, and one to three breeding males. These individuals are the dominant members of the colony. Very few of the hundreds of other colony members ever reproduce. If the queen dies or is removed, however, a contest for succession ensues among some other colony females. Circulating testosterone is a strong predictor of which female will win, and the winner goes on to reproduce (Faulkes and Abbott 1997). Unlike the honeybee workers, alloparent mole-rats have reversible reproductive inactivity.

At the other vertebrate extreme (from the standpoint of reproductive physiology) are golden lion tamarins (*Leontopithecus rosalia*) (French 1997), in which the reproductive systems of alloparenting subordinates seem to be just as active as those of dominants with respect to hormone production, ovulation, and spermatogenesis. In species where alloparents have fully functional gonads, differences in reproductive success between dominants and subordinates are produced by behavioral methods such as absence of mating behavior by subordinates (to avoid incest?), active prevention of subordinate mating by dominants, or killing of any subordinate offspring by dominants (Kleiman 1980; Wasser and Barash 1983; French 1997). The entire physiological range is seen in the cooperatively breeding African mole-rats, from strong inactivity of gonads in both sexes of alloparents in the naked mole-rats to functional gonads with behavioral brakes on reproduction in several members of the genus *Cryptomys* (Faulkes and Bennett 2001).

When alloparents have smaller gonads, a set of hormone-related questions is raised. Are they actually failing to produce gametes? (Small gonads can still be functional.) Are their sex steroid levels lower? At what level is the HPG axis down or off? At the level of the gonads only? Are pituitary gonadotropins lower? Are hypothalamic GnRH cells less active? In seasonally breeding species, does the HPG state of subordinates resemble that of nonbreeding dominants or is it different (third) state? What are the social and environmental cues that impact the HPG axis of the subordinates (and the dominants!) and what are the neural and hormonal mechanisms connecting those cues to the HPG axis? Are the dominants actually the cause of inactivity in subordinates? If so, is there an identifiable cue emanating from a dominant that is detected by one of the subordinate's sensory systems that then results in that inactivity? Or does it result from repeated interactions with the dominant rather than any particular identifiable cue? Is it better to think of the subordinate as making a decision about whether to reproduce based on external (including social) and internal conditions and then ask about hormonal mechanisms that enter into the decision? How can diversity in whether alloparents are gonadally inactive be explained? And, because of the dominance and offspring care components of cooperative breeding, we can ask if the principles that emerged from the study of other kinds of dominance and parental behavior seem to apply here as well, and whether any new ones are needed.

Only in a few cases do we have some answers to any of these questions (Faulkes and Bennett 2001). The most complete story with respect to the cues involved comes from experiments with laboratory housed groups of marmosets, especially common marmosets (*Callithrix jacchus*) of the New World family Callitrichidae (Abbott et al. 1989, 1998; Carlson et al. 1997; French 1997). Callitrichid pairs produce twins that are very large compared to the size of the parents. In addition, females have a functional postpartum estrus and can be simultaneously pregnant and lactating. No wonder they need help not only from the father but from others as well! In some family groups daughters that remain with the parents fail to ovulate, as confirmed by measurement of progesterone levels. Downregulation appears to be at the level of the hypothalamus or pituitary. If daughters are removed and placed in their own room, they begin to ovulate, confirmining that they are mature enough to breed; if returned to the family ovulation ceases. If daughters are in their own room but urine from the mother (the breeding adult female) is placed in their cage, they fail to ovulate for a while and then eventually start ovulating. This confirms that it is cues from other individuals that are responsible for failure to ovulate and also shows that chemosensory cues alone can produce reproductive inactivity, at least initially and in laboratory experiments. (Chemical cues from dominants are also known to mediate effects of dominants or queens on subordinate reproduction in *Bombus* bees and in the ant *Camponotus floridanus* [Trumbo 1996; Endler et al. 2004].) Urine does not prevent ovulation if it is from an unfamiliar female who is dominant in another group—individual recognition is required. This means that the cue is learned, not "hard-wired," and that there is no need to posit specially evolved neuroendocrine mechanisms (Abbott et al. 1998). In addition to cues from the mother, which in normal situations would not be limited to those in urine, cues from unfamiliar unrelated males can stimulate ovulation.

In naked mole-rats nonbreeding subordinates have both lower basal LH and a smaller LH response to a GnRH challenge; in some individuals the entire HPG axis is "off" (Faulkes et al. 1990; van der Westhuizen et al. 2002). Chemosensory cues do not seem to be the mechanism whereby the dominant breeding female leads to reproductive inactivity, but instead patterns of behavioral interactions, especially those of an aggressive nature. In Damaraland mole-rats (*Cryptomys damarensis*), on the other hand, the HPG axis is downregulated but female subordinates do not begin to breed when removed from the colony, indicating that something other than the dominant breeder is the stimulus, possibly the absence of an unrelated male to breed with (Clarke et al. 2001).

In most other cases of reproductive inactivity in mammals and in birds, it is not known what it is about the group and the relationships between the subordinate and the rest of the group that is responsible. Dominant removal experiments with appropriate control manipulations have seldom been done, and so the dominant cannot be assumed to be the source, as opposed to, say, the absence of an unrelated individual of the other sex. The latency for the subordinate to breed would provide clues to what is inactive. If the subordinates begin mating immediately upon dominant removal, then inactivity was probably behavioral (and perhaps peptide related) rather than reflecting a turned-off HPG axis, which would take longer (at least a few hours) to ramp up.

It is usually assumed that any differences between dominant breeders and subordinate alloparents will take the form of greater gonadal activity in dominants. Yet in groove-billed anis (*Crotophaga sulcirostris*) the dominant males' testes are quite regressed but the subordinate male alloparents' testes are not (Vehrencamp 1982). This actually makes sense in the context of avian behavioral endocrinology, because the dominant male does much more of the incubation and incubation is often associated with regressed gonads in birds. This study serves to remind us that the relationship between sex steroids and dominance status in any kind of social organization is never a simple one that generalizes well across species, and also that cooperative breeding is not just about dominance but also about parental behavior (see below).

One popular hormonal hypothesis for reproductive inactivity in subordinates is the social stress of conflict-ridden behavioral interactions with dominants—that subordinates are experiencing high enough levels of stress to produce chronic HPA activation, which in turn inhibits the HPG axis (Reyer et al. 1986). This hypothesis predicts that subordinates should have higher levels of glucocorticoids than dominants. Subsequent research has found little evidence for this in birds (Schoech et al. 2004). In reviewing the literature Creel (2001) found the opposite in cooperatively breeding groups of mammals. Dominants were more likely to have higher glucocorticoid levels whenever a difference was found. Some of the hormone levels had been measured in only one sex, the nonalloparenting sex, however. An additional difficulty is that when comparing animals that differ in physiological reproductive status (as dominants and subordinates in cooperatively breeding groups often do), it is risky to interpret differences in glucocorticoid levels as necessarily reflecting differences in stress. For example, dominant female mongooses have higher, not lower, glucocorticoids. Creel et al. (1996) interpret this to mean that the dominant is experiencing more social stress because she has to keep all the subordinates in line. An equally plausible alternative is that her higher levels are due to the energetic demands of mammalian reproduction itself and would happen even if she were not part of a cooperatively breeding group.

Can a functional perspective generate some unifying principles about hormone levels? There is good reason to think that the theoretical advances in understanding the evolution of cooperatively breeding systems might also be a source of hypotheses to account for the species differences in the sex steroid and glucocorticoid status of subordinate alloparents (Abbott et al. 1998; Creel 2001). At the heart of most theories is that the costs and benefits to the subordinates of different reproductive strategies will depend in part on their relatedness to the dominant(s) and to the dominants' offspring (Emlen 1997). When the parties are closely related, the benefits of helping are greater (all other things being equal) because they include inclusive fitness gains. The close genetic relationship between worker honeybees and their younger nestmates has long been thought to be why forgoing direct reproduction altogether and instead raising the queen's offspring has been an evolutionarily successful path for honeybees. Once the benefit to cost ratio is great enough, the subordinates should forgo reproducing themselves. A physiologically active reproductive system is costly (and a complete waste, if the dominant is just going to kill any subordinate offspring anyway), and shutting it down to save energy is a good idea. Avoidance of inbreeding with the same-sex parent also could be an important selective pressure for shutting down the reproductive system when relatedness is high (Mumme 1997; Faulkes and Bennett 2001). With high relatedness, high reproductive skew would not necessarily require chronic overt conflict, and neither party would have to be highly stressed simply because of their roles.

Where subordinates are unrelated to dominants, we would not expect them to be so accommodating as to shut off their own reproductive systems in response to cues from dominants, at least not without putting up a struggle. The conditions under which this could evolve (could be beneficial for subordinates and not just for dominants) would be more limited and it should be rare for an unrelated individual to have the entire HPG axis shut down. Instead, we would expect chronic overt conflict (or at least extended negotiation) over who gets to reproduce and prevention of reproduction in subordinates through dominant behavior (enforcement of the rules), with the occasional concession to prevent subordinates from leaving altogether (Vehrencamp 1983; Reeve et al. 1998; Johnstone 2000). These are the more likely cases for conflict or stress to be elevated and to play a role in any turning off of the reproductive system that does occur.

Does relatedness predict the hormonal patterns in the literature? Naked mole-rats, the eusocial vertebrates, have high within-colony relatedness and also extreme inactivity, with the HPG axis shut down in both sexes (Faulkes and Abbott 1997). In marmosets and tamarins daughters often don't cycle or ovulate but unrelated female alloparents do ovulate (French 1997). In dwarf mongoose (*Helogale undulata*) groups, the less related females (those related only to the dominant female) sometimes reproduce in addition to alloparenting, whereas the more related females (those related to the dominant male as well) just alloparent (Creel and Waser 1997). In red-cockaded woodpeckers (*Picoides borealis*) male breeders and male alloparents have similar testosterone levels but alloparents related to the breeding female have lower testosterone levels than alloparents that are unrelated to her (Khan et al. 2001). On the other hand, male Florida scrub jays that are alloparenting in their mother's territory are not hormonally different from those in territories where the breeding females are not their mothers (Schoech et al. 1996). The cases with related and unrelated alloparents in the same group that do differ hormonally are informative with respect to likely cues for reproductive inactivity. If the related individuals are hormonally downregulated but the unrelated ones aren't, then either the inactivity emerges from patterns of behavioral interactions that differ by relatedness, or any more specific cues for it are learned. Whether dominants and subordinates differ in glucocorticoids and in what direction doesn't seem to be related to the degree of reproductive skew in the group (Creel 2001).

What other predictions about steroids can be made from a functional perspective? What kinds of costs would bias the likelihood that an alloparent would evolve to be reproductively "on" vs. "off"? All other things being equal, greater energetic costs of reproduction should result in greater physiological inactivity in subordinates, a hypothesis supported by a comparative analysis of carnivores and studies of callitrichid primates (Creel and Creel 1991; French 1997). Looking within a species, the condition of the individual might bias decisions about whether to try to invest energy in reproducing. That could be one reason why daughters of golden lion tamarins housed in captivity (with abundant food) have ovulatory cycles, whereas wild daughters don't (French 1997).

On energetic grounds we would also expect sex differences in the hormonal status of subordinate alloparents (Wasser and Barash 1983; Creel and Waser 1997; Faulkes and Bennett 2001). Current or future energetic costs of reproduction (of being physiologically "on") are usually higher for females than males in birds and mammals. If all other costs and benefits of alloparenting are equal, a shutdown of the HPG axis or of the gonads should be more likely to accompany alloparenting in females than in males. This hypothesis receives some support from mammals (Creel and Waser 1997; Faulkes and Bennett 2001). Even in naked mole-rats, where both sexes have relatively turned off HPG axes, females tend to be more turned off than males (Faulkes et al. 1990; van der Westhuizen et al. 2002).

Thus, there have been several interesting proposals for how to integrate function and mechanism to better account for the degrees and kinds of reproductive inactivity in alloparents. These different hypotheses are not easy to test. The "gold standard" of acceptance would be if (1) some successful predictions could be made in advance, and (2) some of the tests were manipulation experiments rather than correlational studies. Good tests also require knowing the age of the individuals. Subordinate alloparents are often younger than the dominant breeders. It is common to find that steroid levels and even gonad size change with adult age and so when dominants and subordinates differ in age, hormonal differences between them are completely confounded by age and can't be assumed to be related to breeding roles (Schoech et al. 2004). Looking on the bright side, where there are significant increases in gonad size and sex steroid levels with age in ancestral members of a lineage, that might provide a path toward the evolution of alloparenting by younger adults provided other conditions produce a favorable benefit/cost ratio.

Let's turn to the parental behavior aspect of the alloparenting role. Mammal mothers lactate and it is hard for other individuals to duplicate their role as a vital source of food for infants. Hard but not impossible. Female meerkat (*Suricata suricatta*) alloparents lactate! For a mammal this is the physiological ultimate in alloparental commitment, and not surprisingly these lactating females are the full sibs of the pups they nurse (Creel et al. 1991). How is it possible for these females to lactate if they are not pregnant and have not given birth any time recently? Creel's hypothesis is that they are pseudopregnant, and alloparental lactation is an adaptive consequence of pseudopregnancy. In the case of the meerkats, it has not been confirmed that the females are in fact pseudopregnant. Female wolf alloparents definitely are, however, and don't differ hormonally from the dominant breeding/lactating female (Asa 1997).

Pseudopregnancy is a common phenomenon in mammals, not just in cooperatively breeding species, in which the stimuli of mating induce the hormonal state of pregnancy, complete with cessation of ovulation and increased prolactin and progesterone levels (Erskine 1995; Allen and Adler 1985). In these mammals, much of the endocrinology of the pregnant state is due not to the fetus, but to the mating that produced the fetus. If the meerkats really are pseudopregnant, then they must have mated, but without getting pregnant. How? With whom?

It has been suggested that the hormonal state of pseudopregnancy, especially with respect to prolactin, might be favorable for the behavior of alloparenting and not just for lactation (Asa 1997). Both sexes of alloparental wolves have elevated prolactin at the time of the birth of the pups. Pseudopregnancy could be an ancestral mammalian state that incidentally makes it a little easier for alloparental behavior to evolve in females once there are benefits to it (Scantlebury et al. 2002). The state of "real" pregnancy does facilitate maternal behavior, but it is estrogen more than any other hormone that is responsible. As chapter 2 pointed out, we can't always assume that prolactin is the parenting hormone or, given the ease with which parental behavior is induced in adults by mere exposure to young, that any hormonal mechanism is required at all.

Thus far the only experimental evidence for prolactin or other hormonal involvement in alloparental behavior in a mammal comes from common marmosets (Roberts et al. 2001).

In birds as in mammals, alloparental behavior is often assumed to have a similar hormonal basis as parental behavior by the breeders, and prolactin is usually the candidate hormone hypothesized (Schoech et al. 1996; Brown and Vleck 1998). This hypothesis is largely untested, however. The best evidence for a causal role of prolactin in chick care comes from a species (ring dove) that does not have alloparenting (chapter 2). In cooperatively breeding jay alloparents, prolactin does rise before the chicks hatch, and is higher in helpers that bring food than in those that do not (Schoech et al. 1996; Brown and Vleck 1998). In red-cockaded woodpeckers breeders and male alloparents have similar prolactin levels (Khan et al. 2001). However, there are no experimental manipulations of prolactin in these species to show that prolactin or any other hormone is responsible for the parental behavior, and no basis yet for concluding that anything special of a hormonal nature has evolved in species with alloparents that is not present in related nonalloparental species. Reduced testosterone levels in alloparents have been predicted to occur in birds (Vleck and Brown 1999), but when azure-winged magpies (*Cyanopica cyanus*) were given testosterone implants, the helpers showed an increase, not a decrease, in parental behavior (de la Cruz et al. 2003). Also unknown is whether biases in the sex of alloparents (when both sexes of young remain in the area) have any hormonal basis, either activational or organizational.

Conclusions

In this chapter, social behavior has been viewed at the level of social organization and there has been a greater emphasis on considering function together with mechanisms. This adds another layer of complexity but forges closer connections with some of the ecologically relevant social phenomena central to evolutionary behavioral ecology. The last 15 years or so have seen bold and often successful attempts to uncover the principles that will predict an animal's hormonal status from its social role and interactions and vice versa. These have led to new insights about both steroids and social organization, for example, between testosterone and mating systems of birds. Future studies of hormonal mechanisms may help to discriminate between competing models for alloparenting and other forms of seemingly cooperative behavior. Social conflict is a big stressor and if not resolved quickly it exerts serious costs. Such costs would weigh heavily in a cost/benefit calculus and could be an important incentive for cooperative behavior (von Holst 1998).

4

Development of Sexes and Types

Social behavior can be sexually dimorphic as a result of natural or sexual selection. This chapter examines how these sex differences develop, that is, how dimorphic adult behavioral phenotypes are created during ontogeny. Steroids are central to understanding this process of sexual differentiation. Important themes include the distinction between organizational and activational effects of hormones and between gonad directed and brain directed differentiation. These concepts are then applied to the fascinating phenomena of adult sex change and discrete behavioral types of the same gonadal sex.

Sex Determination and Morphological Sexual Differentiation

Most vertebrates and many invertebrates come in two sexes that often differ behaviorally as well as in internal and external anatomy. What happens when fertilized eggs go down two alternative developmental paths, one producing females and the other producing males? How are the two phenotypes created? This is the problem of sexual differentiation. A distinction is often made between gonadal sex determination and sexual differentiation of the rest of the phenotype, or between determination of the sex of the fertilized egg and sexual differentiation of the gonads.

The discovery of some of the mechanisms underlying sexual differentiation of anatomical reproductive systems has been a huge success story in embryology and developmental biology. The underlying concepts have also guided research on sexual differentiation of behavior. Sexual differentiation of reproductive systems in an embryo is a chronological sequence of events: first recognizably ovarian or testicular tissue forms, then the internal reproductive organs take on a male or female form, and lastly any external reproductive organs become male or female. What establishes the sex of each organ? What is the causal connection, if any, between one step and the next?

In animals that have sex-determining genes or sex chromosomes the fertilized egg has a genetic sex. Does this then explain sexual differentiation? Anyone who thinks of the genome as a blueprint for an organism would say "yes." But the genome is merely a starting point and in no way explains how the organism gets from start to finish (Waddington 1939). All of development is epigenetic, consisting of cascades and regulatory networks of contingently related events in which molecules resulting from gene expression interact with the internal and external environment. The task is to figure out those responsible for sexual differentiation.

The first question is, how do the sex genes and sex chromosomes determine the sex of the gonads? This is a very active research topic and answers are arriving fast for a few mammalian species. In mice and humans development of testes requires a particular chunk of DNA normally found on the Y chromosome to be present in the genome, called "sex-determining region of the Y chromosome" or *Sry* (Wachtel 1994; Vaiman and Pailhoux 2000; Chadwick and Goode 2002). *Sry* is dominant and will ensure testis development (through cascades involving other chunks of DNA, of course) no matter how many X chromosomes are also present. The putative protein coded by *Sry* is called "testis-determining factor" or TDF. It is not yet known, however, what *Sry* expresses, how it works, or what its targets are. Without that knowledge, it seems premature to assume that its product actually determines a testis, rather than inhibiting ovarian differentiation to permit testis development to occur. At any rate, *Sry* is expressed briefly, upregulating (through an unknown route) *Sox9* and leading to differentiation of Sertoli cells, the sperm-producing cells (Lovell-Badge et al. 2002). *Dax1* is thought to antagonize this process and instead promote ovarian development, although recent evidence points to a testis promoting function instead (Meeks et al. 2003). *Sf1* is expressed more in future testes, reflecting the greater steroidogenic activity of mammalian embryonic testes.

With one exception so far, nothing analogous to *Sry* has been discovered in other vertebrates with sex chromosomes, nor do models positing simple dominance by a bit of DNA on the sex determining chromosome seem to work as well. The exception is the medaka (*Oryzias latipes*), a teleost fish, which was discovered to have a testis-determining chunk of DNA (*Dmy*) on one of its sex chromosomes (Matsuda et al. 2002). In contrast to *Sry*, which is mammal specific, several of the other genes or members of the gene families whose expression products are also required for gonadogenesis in mammals have been discovered in other vertebrates, including *Sf1*, *Dax1*, and *Sox9*. These appear to be relatively conserved, as are some of their functions related to creating steroid-producing testes or ovaries out of initially bipotential tissue in vertebrates. It is as if the initial switch has changed in evolution, not the rest of the ovarian and testicular cascades (see also chapter 7) (Chadwick and Goode 2002).

In some (not all) teleosts, turtles, lizards, crocodiles, birds, and marsupials there is interesting evidence that sex steroids themselves might be part of the gonadogenesis cascades, as first proposed by Yamamoto (1969). This might seem odd (wait—aren't gonads supposed to produce steroids, not the other way around?) but any tissue can produce sex steroids if the genes for the steroidogenic enzymes are expressed there. It has been known for many decades that treatment of embryos or fish fry with sex steroids can sex reverse the gonads of some species, including species with genetic sex determination. Estradiol, in particular, is quite effective at producing ovaries in genetic males, and androgens sometimes feminize gonads (via aromatization) or masculinize them. Even more compelling is that exposure of embryos to nonsteroidal, nontoxic estrogen synthesis inhibitors such as fadrozole causes substantial gonadal sex reversal, resulting in genetic females with gonads that are more testicular than ovarian (Elbrecht and Smith 1992; Wibbels and Crews 1994; Wade and Arnold 1996; Kitano et al. 1999; Afonso et al. 2001). Fish turned into males by this treatment can produce sperm and fertilize eggs (Piferrer et al. 1994). Evidently the P450arom gene is a switch in the cascade for some nonmammalian vertebrates.

With chromosomal sex determination as in birds, mammals, and a number of other vertebrates, sex ratios are expected to be (and often are) close to 1:1. What mechanisms are responsible when biased sex ratios occur, and are steroids any part of them? No clear answers have emerged yet but there might be different mechanisms in birds and mammals. In birds, which have a ZZ/ZW sex chromosome system, the chromosomal sex of the future embryo is determined in the female just prior to ovulation (Z or W bearing ovum) before the egg is fertilized, whereas in mammals, which have an XX/XY sex chromosome system, it is the male sperm (X or Y bearing) that fertilizes the egg that determines the genetic sex. Both mammalian and avian embryos are exposed to steroids of maternal origin during the critical period for gonadal differentiation, mammals through the placenta and birds through maternal steroids in the egg yolk. Female rodents that are between two males in the uterus are slightly masculinized because of exposure to sex steroids from their male neighbors and also produce male-biased litters as adults (Vandenbergh and Huggett 1994). Because mammalian gonads are rather refractory to sex-reversing effects of steroids, the mechanism for this and other cases of biased sex ratios in mammals is unlikely to be steroid effects on gonadal sex. Such a mechanism would be more plausible in vertebrates whose differentiating gonads are sensitive to steroids, but there is no evidence yet that maternal yolk steroids can determine the sex of an embryo (Roosenburg and Niewiarowski 1998; Eising et al. 2003). In the lizard *Anolis carolinensis*, a species with XX/XY chromosomal sex determination, eggs that when incubated produced male embryos had more yolk testosterone than those producing female embryos, even when the yolk was sampled from the eggs taken directly from the oviduct before embryonic gonads had developed (Lovern and Wade 2003). Because treatment of lizard embryos with androgens, even high dosages, causes only partial, not complete, gonadal sex reversal (Adkins-Regan 1987), a more likely possibility for this result is that the testosterone concentration in the yolk influenced the sex of the fertilizing sperm.

To continue with the sexual differentiation sequence, what is the link between the sex of the newly formed gonads and the development of other sex organs such as phalluses? Vertebrate embryonic gonadal tissue manufactures steroidal and nonsteroidal hormones. Classic experiments by Wolff (1959), Jost (1985) and many others showed that these hormones establish the sex of the internal and external reproductive anatomy (Adkins-Regan 1981; Adkins-Regan 1987; Wilson et al. 1995; Mittwoch 1998). That is, the sex-determining genes such as *Sry* don't determine the sex of the nongonadal reproductive organs directly, but instead act through the gonads and through hormonal, not local, signals. In mammals, testicular hormones lead to cascades for building the male anatomical phenotype. Genetic male mammals whose gonads are destroyed very early or whose gonadal hormones are blocked lack a penis and prostate and have a uterus. In those birds that have been studied, ovarian estrogens lead to cascades for building the female anatomy by eliminating male typical structures. Genetic female ducks whose gonads are destroyed early develop a male phallus. With respect to external anatomy, in mammals, the testes organize maleness and in those birds studied the ovaries organize femaleness.

These organizing effects are limited to the time period in development when these anatomical structures are forming—a classic example of critical periods in development. Once these organs have assumed a clear sexual phenotype, however, such experimental sex reversal is no longer possible. Sex differences in steroid levels are consistent with these scenarios. During the critical period mammalian testes produce more steroids (including testosterone) than ovaries, whereas embryonic chicken, duck, and quail ovaries produce more estrogen and a higher estrogen to androgen ratio than testes (Tanabe et al. 1983, 1986; Woods 1987; Ottinger et al. 2001). This is gonad-directed sexual differentiation, organizational style. Once a bird or a mammal reaches sexual maturity, the organs it has may grow (be activated), but they don't change their sex.

It is very tempting to take these important principles about how reproductive anatomy is established and generalize them to other sexually dimorphic morphological characters. And, indeed, this works for some of them, for example, the duck syrinx, which is larger in males because of the absence of embryonic ovaries (Wolff 1959). But in the absence of the necessary developmental experiments, such generalizations are risky. Sexually dimorphic plumage in birds is a good example. Because males are often fancier (for example, more colorful), it is widely assumed that male-typical plumage is either organized or activated by testosterone. While there are a few examples of male color ornaments that are indeed activated by adult testosterone (for example, Evans et al. 2000), there are even more that require no gonadal steroids at all (Owens and Short 1995), and in no case thus far is the plumage dimorphism known to be hormonally organized in earlier development. Another risky assumption, which the plumage example also illustrates, is to use the mammalian developmental pathway, in which the testis has the organizing role, to account for dimorphic characters of birds or other vertebrates where the ovary has an important organizing role. The mammalian method is not a standard vertebrate method. Instead, there is a diversity of alternative methods, including but not limited to the duck method.

In some vertebrates such as crocodiles, most turtles, and some teleost fish, sex is determined not by sex chromosomes or sex genes (or not solely by them), but by embryonic or larval temperature, a type of environmental sex determination. While this might seem to be a radically different way of developing a sexual phenotype, there is evidence that much of the developmental sequence determining anatomy may be similar to what transpires in vertebrates with sex genes, with the major difference lying in the initial triggers that set off the cascades leading to ovaries vs. testes (Wibbels and Crews 1994; Pieau et al. 1999; Morrish and Sinclair 2002). In genetic sex determination the expression of the gene(s) producing the product(s) that set off an ovarian or testicular cascade is genetically regulated, whereas in temperature sex determination this same gene expression (for example, P450arom expression, for an ovarian cascade) is temperature sensitive. Treatment of embryos of newts (*Pleurodeles waltl*) or fry of flounder (*Paralichthys olivaceus*) with an estrogen synthesis inhibitor produces testes at temperatures that would normally produce ovaries (Chardard and Dournon 1999; Kitano et al. 2000). The biggest mystery is the ultimate causes of temperature sex determination. Because the mother crocodile or turtle selects the nest site and therefore the incubation temperature, temperature sex determination seems like a method for females to manipulate offspring sex ratios so as to increase fitness. But how would this increase fitness, especially in very long-lived animals, and why is this a method used by some animals and not others? Sex determination by environmental temperature remains a tantalizing evolutionary puzzle (Janzen 1996; Shine 1999).

Sex Differences in Behavior and Brains

Sex differences in social behavior are apparent in many animals. They reflect the outcome of both natural selection (reproductive mode, foraging strategies, etc.) and sexual selection (mate competition and attraction). Some of them go hand in hand with sexually dimorphic morphology, as when male ungulates with horns compete for access to females through ritualized head-butting contests. Sex differences in social behavior, like other sex differences, also vary tremendously in magnitude. At one extreme are species with males that court in groups, for example, lekking birds. During the breeding season males spend most of their days on the lek, interacting frequently with their male neighbors and springing into male-specific courtship displays whenever females are visiting, whereas after visiting females go off to quietly incubate eggs, with fewer adult interactions. At the other extreme are animals such as albatrosses and anchovies. Here sex differences in social behavior (or in anything) are so slight that humans are hard-pressed to tell the sexes apart. In between are the many animals with sexes sharing a common behavioral repertoire but where specific behaviors are performed more frequently by one sex or in different social contexts.

A vocabulary limited to "sex differences" and "dimorphic behaviors" doesn't capture this diversity very well. Strictly speaking, "dimorphic" refers to morphological, not behavioral, traits, and refers only to sex differences that are genuinely dichotomous (have nonoverlapping distributions). In practice, however, the term is also used to refer to behaviors with substantial overlap between the sexes but significant differences between the means of the female and male distributions (to allomorphic behavior [Fox et al. 1999]). If a researcher cares to look hard enough at enough individuals, there will usually be some small but statistically significant sex difference. This does not automatically mean that the difference is biologically significant or that focusing on finding a sex difference is the most exciting thing to do in studying the behavior.

It has been known for several decades that the nervous system, although largely monomorphic, has some sexually dimorphic features (Raisman and Field 1971; Jacobs 1996; Woodson and Gorski 1999; Forger 2001; Simerly 2002). At first this came as a surprise to scientists and even now such discoveries are news. Some dimorphisms have shown up in brain regions with known links to social behavior such as the preoptic area and anterior hypothalamus. Others are in areas such as the hippocampus that function in memory consolidation and spatial cognition. Some sex differences lie in fine structural details such as the number of synapses of a particular type onto neurons or the number of dendritic spines on neurons in a particular part of the cortex, and require careful measurement of many neurons to be detected. Some are biochemical, such as sex differences in the production of steroid synthesizing enzymes, steroid receptors, or peptides by neurons. Some lie at the level of entire circuits or the size of projections from one nucleus to another. The most astonishing show up at the level of the gross structure of the brain, as when a particular nucleus is so much larger in one sex than the other that stained brain slices through the nucleus can be accurately sexed by eye without magnification. That is what Nottebohm and Arnold (1976) discovered in the telencephalon of the canary. These song system nuclei of canaries and other oscine passerines remain the most sexually dimorphic brain structures yet discovered in any vertebrate.

Many of the sex differences in nervous systems that have been discovered take the form of characters that are larger or more abundant in males. This seems odd, because females also perform behaviors not seen in males. A likely solution to this puzzle is the realization that these male character states, like the male behaviors they support, have probably resulted from sexual selection, which often acts more strongly on males (Kelley 1988; Jacobs 1996). The telencephalic song nuclei of the male orange bishop (*Euplectes franciscanus*) are as extravagant as his bright orange and black plumage and his odd courtship display, more so (to human senses) than his "song," an unpleasant screech (Arai et al. 1989).

A more difficult problem is pinning down the causal connection between sexual dimorphism in some aspect of brain structure or chemistry and sexual dimorphism in behavior (de Vries 1995; Forger 2001). How can we know if the former is the cause of the latter? In the case of the songbird song system, the causal assumption seems plausible because this system is dedicated to singing and song perception. But most brain regions serve multiple behavioral functions, so that the causal connection often rests on shakier correlational logic. For example, the medial preoptic area supports mating behavior in virtually all male vertebrates that have been studied. In Japanese quail, a well-placed lesion in the medial preoptic nucleus (a subregion of the overall medial preoptic area) eliminates male mating behavior. This medial preoptic nucleus is larger in male than female quail, and female quail do not show the male mating pattern (mounting) (Adkins-Regan and Watson 1990; Balthazart et al. 1996; Panzica et al. 1996b). But can it be concluded that the reason females don't mount is because they have a smaller medial preoptic nucleus? Not without going into the brain and removing the extra that males have, without disrupting anything else, and seeing if the males start behaving like females. Such an experiment is technically very challenging.

Where do sex differences in behavior come from developmentally? Many possibilities have been hypothesized and confirmed for some particular behavior, not all of which involve hormones. In the case of animals with big brains and complex social lives, certain sex differences could be situational, arising largely from social context in which they are measured. For example, the magnitude of sex differences in rhesus monkey behavior such as aggression, submission, rough and tumble play, and foot-clasp mounting varies markedly, from no sex difference to a large one, depending on the social context in which it is assessed and the ranks of the individuals in the group (Goldfoot and Neff 1985; Wallen 1996; see Itzkowitz et al. [2003] for a teleost example). Some sex differences are secondary consequences of local demography. For example, where females but not males disperse, any alloparenting by older siblings will be limited to males. That doesn't mean that females lack the capacity for alloparenting. Some sex differences are learned, as when both sexes sing but females copy songs of other females, whereas males copy songs of other males, producing a sex difference in song type in the population (Price 1998; Enggist-Dueblin and Pfister 2002). Some could be products of socialization and social learning, as when juvenile female and male rats diverge in their play behavior in part because they grow up in different social groups (Meaney et al. 1985).

A highly interesting emerging idea is that some sex differences in mate choice that have been viewed as "hard-wired" sex-specific strategies might actually be flexible tactical decisions based on an assessment of the current social context. For example, a change in the operational sex ratio of a mating group of syngnathid pipefish produces sex reversal in choosy vs. indiscriminate mating tactics (Berglund and Rosenqvist 2003). In the case of animals with highly dimorphic body size, it is fun to imagine what would happen if you could experimentally sex reverse body size in adult animals without changing anything else. Would males and females continue to behave in the species-typical, sex-specific manner, or would the behavior change quickly as the animals realized that those tactics no longer make sense with their new body size?

How might steroids and other hormones be mechanisms producing sex differences? A very simplistic hypothesis is that males and females produce different hormones, which in turn "cause" the sex differences. This seems unlikely. Notwithstanding the common phrases "male sex hormones" and "female sex hormones," and as stated in previous chapters, no sex-specific steroids or other hormones related to social behavior have been discovered in the circulation of any vertebrate. Levels of a steroid might be too low in one sex to be detectable in an assay, but this is as likely to mean that the assay isn't sensitive enough as to mean that the steroid is absent and biologically irrelevant. The brain produces its own steroids, either from "scratch" or from other steroids that reach it from the circulation, but none that are sex specific have been discovered so far. Nor has a sex-specific brain sex steroid receptor been found. To be sure, there are some brain nuclei containing intracellular steroid receptors that are involved in dimorphic behavior. It is uncommon, however, to find sex differences in the number or location of those receptors, with some notable exceptions such as the greater number of androgen receptors in the telencephalic song system nuclei of songbirds (Arnold and Saltiel 1979).

Instead, most evidence supports three other hypothesized relationships between hormones and the creation of sex differences in social behavior. First, sex differences can be due to activational consequences of sex differences in the levels of circulating gonadal hormones in adults (the activational hormone hypothesis). Second, they can be due to organizational effects of gonadal hormones occuring earlier in development (the organizational hormone hypothesis). Third, they might be due to a sex difference during development in a brain produced steroid or gene product factor independent of the gonads (the direct differentiation hypothesis).

Sex Differences Due to Activational Hormone Effects

Both sexes produce the same hormones, but not always in the same amount. For example, most male mammals have more testosterone in the circulation (and reaching the brain) than females do, whereas most female mammals have more estradiol in the circulation than males do. Although the hormone is always present, its level might be too low in one sex to activate behavior shown by the other sex. This could account for some sex differences in behavior. In the popular media this is widely touted as the source of human sex differences. As always, its truth value rests on a proper experimental test.

An activational hormone hypothesis is tested by equating (insofar as possible) the hormone levels of adult males and females, by, for example, giving both sexes estrogens or androgens, gonadectomizing all of them first if possible to ensure that all their hormone exposure is coming from the controlled dosage of the treatment. This assumes that gonadectomy eliminates most of the circulating sex steroids, which is not always the case. The females and males are then tested in an identical manner, using the same social stimuli, along with controls with normal sex-typical hormone levels. If the sex difference in behavior is absent in the hormonally manipulated animals but present in the controls, then an activational hypothesis is confirmed. The sex difference can disappear either because both sexes converge toward some in-between state, or because the hormone treatment sex reverses the behavior of one sex, either raising or lowering it to the level of the other sex.

An activational hypothesis has been supported for a wide array of social behaviors in all kinds of vertebrates. For example, female rats and green anoles that are ovariectomized and treated with testosterone show the complete male copulatory sequence, including the motor patterns of intromission and ejaculation, even though they have no intromittent organ (Barfield and Krieger 1977; Adkins and Schlesinger 1979). Human transsexuals given a cross-sex hormonal milieu undergo sex reversal of some cognitive abilities (Slabbekoorn et al. 1999). Female canaries given testosterone start to sing like males, female goldfish given 11-ketotestosterone court and spawn like males, and female weakly electric fish given androgens begin to give off signals with a male-typical waveform or pulse rate (Leonard 1939; Stacey and Kobayashi 1996; Zakon 1998). Male Japanese quail given estradiol are completely receptive to the copulatory attempts of other males, and begin to emit the female mate attraction "cricket" call (Adkins and Adler 1972). It is very impressive to see these dramatic behavioral sex reversals taking place in adult animals.

The Organization of Behavioral Sex Differences in Mammals

For reasons that will become clear as the story unfolds, the organizational hypothesis will be discussed first for mammals, then for birds, and finally for other animals. This hypothesis is a logical next step when an activational hypothesis is tested and rejected, that is, when the sex difference in behavior persists even when adult sex steroid levels have been equated, or when no sex difference in circulating hormone levels is detected. When that happens, it is as if the "soma" itself (the neural and morphological substrates upon which hormones act to stimulate behavior) is sexually differentiated because of something that happened earlier in development. The organizational hypothesis has been the subject of hundreds of experiments in a wide variety of animals and is very much worth examining in some detail. Although in its original form it does not seem to account for sexual differentiation of behavior in all vertebrates, nonetheless it has been spectacularly successful for some animals and quite rightly continues to be the dominant conceptual framework.

Phoenix, Goy, Gerall, and Young (1959) were the first to apply the concept of hormonal organization to behavior (to the copulatory behavior of guinea pigs) and their version is the classic original (see also Young et al. 1964; Goy and McEwen 1980). The essence of it is that (1) gonadal hormones produced during early development (in mammals, fetal or early postnatal development); (2) act on the soma during a critical period; (3) to permanently organize the sex of the soma; (4) such that in the presence of testicular hormones development will be predominantly masculine; (5) and in the absence of testicular hormones development will be predominantly feminine. Notice the same asymmetry here that Jost (1985) recognized for anatomical differentiation: behavioral males are organized by the testis, but behavioral females are not organized by the ovary. Instead, they are organized by the absence of a testis. A common (but semantically loaded) way this is sometimes expressed is to say that femaleness is the neutral or default developmental path and maleness is imposed by the testis. In the language of molecular developmental biology, a testis switches the embryo from one set of cascades to an alternative set. Once the male-typical and female-typical behavioral substrates (somas) have been organized, then adult activational hormones are required for their expression. No matter how well masculinized the male is, he won't show normal levels of copulatory behavior without adult testicular hormones, nor will a female show receptive behavior without ovarian hormones. A combination of organization and activation is required for the normal adult sex differences in behavior, and the phrase "organizational theory" (or hypothesis or concept) typically subsumes hormonal activation as well. But these adult activational hormones do not change the fundamental behavioral sex of the animal. Rather, they allow what has been laid down earlier to be expressed.

Experiments with mammalian species in more than four orders have lent strong support to all components of this theory (fig. 4.1) (Goy and McEwen 1980; D'Occhio and Ford 1988; Gerall et al. 1992; Wallen and Baum 2002). For example, genetic female rats given a single shot of testosterone shortly after birth (when they are still quite undeveloped) are permanently masculinized and defeminized. They are almost (not quite) turned into behavioral males. They mount other females when given testosterone as adults (to activate their masculine substrate) but are unreceptive to males when given estrogen plus progesterone. (Note that these adult hormone treatments are absolutely essen-

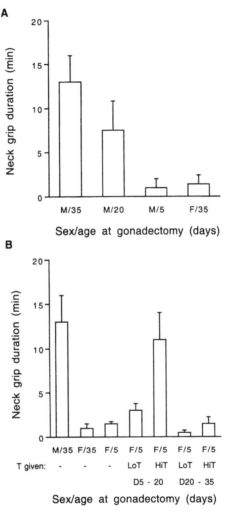

FIGURE 4.1. Organization of masculine sexual behavior in ferrets. A mating male mounts a female, grips her neck, and shows pelvic thrusting. Subjects were tested as adults following implantation with testosterone to activate masculine sexual behavior. Shown here are the results for the duration of neck gripping in 20-minute tests; the results for pelvic thrusting and mounting were very similar. (A) Males and females with intact gonads throughout early development showed very different levels of neck gripping (M/35 vs. F/35). Males castrated early (M/5) failed to masculinize, whereas those castrated later (M/20) had already passed through the critical period for masculinization and neck gripped. (B) Females gonadectomized early (F/5, no testosterone) were not different from control females (F/35), whereas females gonadectomized early and given a high dose of testosterone within the critical period (D5-20) were fully masculinized. A lower dose of testosterone given early had a similar but smaller effect. Testosterone given on days 20–35 came too late and did not masculinize females. Results similar to these have been obtained in a number of mammalian species, confirming the Phoenix et al. (1959) organizational hormone hypothesis. Redrawn from Baum, M. J. and Erskine, M. S. 1984. Effect of neonatal gonadectomy and administration of testosterone on coital masculinization in the ferret. *Endocrinol.* 115: 2440–2444. © 1984 The Endocrine Society.

tial to distinguish between organizational and activational effects. Without them there is no way to know if the animal's behavior is due to the early treatment as opposed to an activational hormone deficiency/excess.) This effect does not occur if the shot of testosterone is given more than 10 days after birth; it is limited to a critical period. Genetic males whose testes are removed shortly after birth are permanently feminized. They are receptive to mounting attempts of other males when given estrogen plus progesterone but are less likely to mount females even when given testosterone. Again, this degree of sex reversal is not possible if the manipulation is done more than a few days after birth. Measurement of sex steroid levels during this period confirms that the baby males are producing more androgens than the baby females.

In these rats and in other mammals, especially those with short gestation periods, some of the behavioral defeminization produced by testicular androgens is actually being done by estrogens produced in the brain by aromatization from testosterone, as predicted by the aromatization hypothesis. A single shot of estradiol will mimic some of the effects of testosterone in those species (Feder 1981; Naftolin and MacLusky 1984; Wallen and Baum 2002). The timing of the critical period varies relative to birth according to whether the animal is precocial or altricial. With a few exceptions the critical period occurs before birth in precocial species and during the neonatal period in altricial mammals.

Over four decades later, it is clear that the organizational hypothesis is a remarkably successful theory of behavioral development. The sheer fact that a single shot of testosterone on one day of early life causes a genetic female rat to be a behavioral male forevermore is an astounding experimental result and a testimony to the power of critical periods. As advances are made in understanding how sex genes lead to the development of a testis, and how steroids act on the cells of the developing brain, we can see the possibility of a complete picture emerging of the cascade linking genes to behavior via steroids (McCarthy 1994).

Furthermore, in mammals this theory has very wide behavioral as well as taxonomic generality. Sex differences in virtually every category of behavior that has been studied have some organizational basis, including aggressive behavior, sexual partner preference (preference for males or females as mating partners), parental behavior, vocalization, spatial learning ability, home range size, dispersal, and juvenile play behavior (Beatty 1992; Wallen and Baum 2002). In the case of juvenile behavior and some adult behavior as well, the sex differences are organized by sex steroids but don't require any activation to be expressed. There's still lots to learn, however. Notice that the list does not yet include some of the sex differences of interest in behavioral ecology, such as the tendency of many ungulates to segregate by sex socially and spatially (Ruckstuhl and Neuhaus 2002) or the sex specificity of targets for sexually selected aggression, as when males defend territories against male intruders, whereas females defend them against other females. These share with sexual partner preference (a dimension of mate choice whose organizational and activational basis was discussed in chapter 3) the interesting property that the behavior itself is not what is sexually differentiated, but rather the sex of the target for the behavior. It will be interesting to see if there is a common process (cognitive, neural, hormonal) underlying all of them.

There are still big gaps in the taxonomic coverage, however. To pick a non-random example, nothing is known from apes and the jury is still out on the theory's importance for human behavior. There are plenty of documented sex differences in humans, most of which are rather small but a few of which are moderate in magnitude (childhood play preferences) or even quite sizable (gender identity, sexual partner preference, violent aggressiveness). Girls exposed to excess androgens prenatally (through "experiments of nature") are slightly masculinized in their childhood play preferences compared to their sisters (the controls), but this statistically significant result has a rather small effect size compared to the outcomes in the nonhuman animal experiments, even for the behaviors that show the biggest sex differences, such as sexual partner preference (Meyer-Bahlburg and Ehrhardt 1982; Collaer and Hines 1995; Zucker 1999). Human parents and societies socialize and in other ways direct the development of dimorphic behavior to a degree that probably has no match in the nonhuman world.

The nervous system is now known to be an essential component of the "soma" that is responsible for organizational (as well as activational) hormone effects on behavior. There are important parallels between masculinizing and feminizing effects of early hormone treatments on behavior and on the anatomy and neurochemistry of the brain and spinal cord (Arnold and Gorski 1984; Cooke et al. 1998; Matsumoto 1999; Woodson and Gorski 1999; Forger 2001). For example, if a nucleus such as the medial preoptic nucleus is larger in males, then females given testosterone (or estradiol) during the behavioral critical period will have a substantially masculinized nucleus size as adults, and males castrated early will have a feminized nucleus size. The neural mechanisms for establishing the size of a nucleus during development in mammalian brains are primarily those involved in the regulation of neuron death (apoptosis), rather than birth (neurogenesis). That is, a larger nucleus results when steroids reduce cell death rates, preserving more neurons (see, for example, Sengelaub and Arnold 1989). Early steroid effects on neurochemical sex differences also parallel behavior. For example, the male rat BNST expresses higher levels of vasopressin mRNA than the female BNST, and early androgen or estrogen treatment masculinizes the expression levels of females (Han and de Vries 2003). The biochemical mechanisms downstream of steroid receptors that make the preoptic area and hypothalamus male typical rather than female typical are beginning to be discovered (Auger et al. 2000, 2001; Perrot-Sinal et al. 2003, Amateau and McCarthy 2004).

In addition to this reductionistic approach, it is important to ask about the behavioral and functional processes that are organized. The possibilities are very similar to those discussed in chapter 2 for activational steroid effects. What is different is the time period for the action. Gonadal steroids could organize sex differences in sensory processes that could then lead to a sex difference in behavior. Good candidates for this hypothesis include mate preferences, which in many mammals are based on chemical signals. Both the vomeronasal organ (the "other" olfactory organ found in many mammals) and the accessory olfactory bulb to which its receptors project are subject to effects of early steroid manipulations (Segovia and Guillamón 1996). Studies of aromatase knockout mice suggest that sexual differentiation of aspects of the main olfactory system required for recognition of the sex of partners depends on estrogens (Bakker 2003). Steroids could organize sex differences in motor systems. The best understood effect of this kind is seen in the spinal cord and penile muscles of rats (Breedlove 1984). Both the bulbocavernosus muscles and the spinal motor nucleus innervating the bulbocavernosus muscles are much larger in male mammals. Early exposure to testosterone (either naturally occurring, in baby males, or experimentally administered, to baby females) results in the male-typical phenotype; early treatment with flutamide causes genetic males to have a female-typical system. Gonadal steroids could organize motivational, attentional, or other cognitive processes. In rhesus monkeys, prenatal androgen affects predispositions to find certain kinds of stimuli attractive and engaging in certain behaviors rewarding (Wallen 2002). In rats males have better spatial memory in maze tasks. Females given steroids early are masculinized and males castrated early are feminized both in their performance and in the morphology of the hippocampal cells that have been linked to spatial performance (Williams et al. 1990; Isgor and Sengelaub 1998).

When animals develop in a social environment, as in group-living species or those produced in litters, there are interesting possibilities for organizational hormone effects to be subject to social influences—to be enhanced or reduced by them or mediated in part by them. The influences of others can even begin before birth, as revealed by intrauterine position effects. In several rodents and in pigs it has been found that female fetuses packed between two males in the uterus are slightly masculinized in external anatomy and behavior compared to females situated between two females, an effect that is caused by exposure to testosterone from the male neighbors' testes (vom Saal 1981; Rohde Parfet et al. 1990; Clark and Galef 1998). Female rhesus monkeys exposed to elevated testosterone prenatally show elevated mounting frequency later on (compared to controls, who also mount), but the magnitude of the masculinization depends quite a bit on their social environment and rank, with dominant females mounting more often (Goy and Roy 1991; Wallen 1996). The behavior of mother rats contributes to hormonal organization of adult sex differences in behavior and in the nervous system (Moore 1995). Mothers lick the perineal region of male pups more. The stimulus for this differential treatment of male pups is chemosensory. Females given a shot of testosterone are licked at male-typical rates, and as adults they are slightly masculinized in their copulatory behavior and in the spinal nucleus of the bulbocavernosus. Sex differences in play in juvenile rats that originate in neonatal hormone exposure are amplified by subsequent social experience (Meaney et al. 1985).

In the classic organizational theory no ovarian hormones are needed for the female behavioral phenotype to develop; sex differences arise from the presence versus absence of a testis. This principle works well for mammals with long gestation periods, but the ovaries do make a minor contribution to female development in rats (Wallen 1998). Regardless of whether the ovaries are involved, one can ask if steroids, including estrogens, produced outside the ovaries contribute anything. Female mice with the aromatase gene knocked out are more affected than females whose ovaries are removed early (Bakker et al. 2002a). This could reflect a nonspecific role for neurotrophic actions of locally produced estrogen during development in both sexes.

Another fundamental component of the classic organizational theory is that the critical period for organization occurs early in development, corresponding to a period of elevated testicular steroids in males. How well has this claim held up over the subsequent decades? In some species fetal and perinatal life is not the only time in the prepubertal period that steroids are elevated in males. Primate males and females have elevated sex steroids at and shortly after birth, well after the classic organizational period (primates are more precocial than altricial in their general development). Male pigs have elevated circulating testosterone and estradiol throughout the period between birth and puberty, and although they are very precocial anatomically, their critical period for behavioral organization by steroids extends well after birth, up to four months of age (Berry and Signoret 1984; Ford and Christenson 1987; Adkins-Regan et al. 1989). Puberty itself is not only the beginning of the activational hormone period for adult behavior, but a time when rising steroids could conceivably have permanent effects on later behavior, an idea that is also explored in chapter 6 in connection with research on tree shrews. In seasonally breeding mammals such as sheep, the onset of each breeding season involves a puberty-like rise in circulating steroids. It seems unlikely that only in pigs and tree shrews do juvenile or pubertal steroids have long-term or permanent effects on behavior. Until recently, there had been more hints and hypotheses than solid evidence for late organization (Dixson et al. 1998). It has long been known that castration of male mammals before puberty can have a more drastic effect on sexual behavior than castration after puberty. There is now evidence that steroid manipulations at the time of puberty can have long-lasting effects on the mating behavior of hamsters and rats (Farrell and McGinnis 2004; Schulz et al. 2004).

One well-known behavioral example of pubertal organization comes from humans, oddly enough. Rising pubertal androgens permanently deepen the pitch of the male voice through an effect on the larynx. Men who lose their testes as adults retain their male-typical voice pitch, and if an adult woman happens to be exposed to higher than normal androgens, she too will experience a permanent lowering of voice pitch. But whether there are any such permanent effects in humans that occur through hormone actions on the brain is not yet clear.

Arnold and Breedlove (1985) proposed that all structural (morphological) changes in neurons occurring in adulthood in response to hormones were organizational effects, but such effects are not permanent, instead disappearing when the hormone levels are lowered. That's one hallmark of an activational hormone effect, and so such neural changes in no way violate the organizational theory. What such structural change does show, which is remarkable in its own right, is the previously unsuspected plasticity of the adult nervous system.

The Direct Genetic Differentiation Hypothesis

What if the sex difference in behavior is still there when sex steroid levels are equalized early in development and in adulthood? For example, what if genetic male mammals whose testes are removed very early still develop some male-specific behavioral and neural traits? A direct genetic differentiation hypothesis has been proposed for such outcomes (Arnold 2002). Genes do nothing directly, of course, but instead operate through their protein products and the cascades and networks that those switch on or enter into. "Direct" in the context of sexual differentiation refers to developmental pathways for brain differentiation that don't go through the gonads. In the classic organizational hormone theory, sex-determining genes on the sex chromosomes create the sexual phenotype indirectly, via the gonads, and it is the gonadal hormones, not the sex gene proteins, that act on the soma to stimulate the expression of other genes (not the sex genes), which then produce the proteins that create male and female nervous systems. The expression of the sex genes has no role in this process beyond the early development of the gonads themselves. The alternative captured by the direct genetic differentiation hypothesis is that Y-chromosome genes such as *Sry*, which are part of the gonadal sex determination cascade, are also expressed in the cells of the brain; these brain expressed gene products act locally to permanently establish some aspects of the neural and behavioral sexual phenotype. The differentiation signal is tissue or cell autonomous. The discovery that several Y-chromosome genes, including *Sry*, are indeed expressed in the brain makes this a plausible complement to the organizational hormone theory (Lahr et al. 1995; Mayer et al. 1998). "Complement" is a better word than "alternative" here, because we know that a lot of mammalian sexual differentiation happens through the gonads.

What might be some candidates for this kind of sexual differentiation? There are some promising morphological, cellular, or biochemical features of the nervous system that could be behaviorally relevant (Forger 2001; de Vries et al. 2002; Reisert and Pilgrim 1991). In some New World monkey species all males have dichromatic vision (two cone types and spectral distributions) but some females have trichromatic vision because the polymorphic opsin locus is X-linked (Jacobs et al. 1993). Body size is another candidate. In humans, XY men are taller than XX men (XX people with testes and male anatomy), as if part of the Y contributes to height directly (Ogata and Matsuo (1993). XY mammalian embryos grow faster than XX embryos even before any gonads develop (Mittwoch 2000).

Some of the anatomy of the Australian macropod lineage of marsupials clearly fits a direct genetic hypothesis instead of a classic organizational theory. Marsupials have a male-determining Y chromosome with a structurally *Sry* piece of DNA, although it is not clear if *Sry* is a gonad determiner (Graves 2002). The scrotum (which is not homologous to the eutherian scrotum), pouch (female homolog of the scrotum), and mammary glands of the tammar wallaby (*Macropus eugenii*) begin to take on a male or female form before the gonads have formed. Early testosterone treatment does not masculinize females, and in individuals with abnormal sex chromosome constitutions, external anatomy is better predicted by whether there are one versus two X chromosomes present than by whether a testis versus ovary is present (Renfree et al. 1995). The rest of the reproductive anatomy follows the Jost model that works for eutherian mammals.

There are very few behavioral contenders for direct genetic differentiation thus far. Behavioral experiments with a New World marsupial (*Monodelphis domestica*) have shown the kinds of early hormone treatment effects seen in other mammals, not the negative results just described for some features of marsupial anatomy (Fadem 2000). There are certainly many cases where early plus adult hormone manipulations do not completely sex reverse behavior, but there is always the possibility that this is because the experiment has not adequately mimicked the hormonal profile of the entire critical period. Advances in genetic manipulation technology have enabled a more powerful approach to testing the direct genetic-differentiation hypothesis. In this new approach, mice with different combinations of sex chromosome status (XX or XY), *Sry* status (*Sry* present or absent), and gonadal status (ovaries or testes) are created so that the behavior of animals with the same gonads but different sex chromosome genes can be compared. For example, XX males (with *Sry* attached to an autosome to produce testes) can be compared with XY males with respect to aggressive behavior (mice are good for studying aggression) to examine the contribution of Y chromosome genes other than *Sry*. Thus far, behavioral measures correspond better to the type of gonad, in agreement with the organizational hormone theory (de Vries et al. 2002; Reisert et al. 2002). The agreement is not complete, however, leaving the door open for a contribution according to the direct genetic hypothesis. No matter how many minor direct genetic effects are unearthed in mammals, however, the fact remains that gonadal hormones, both organizational and activational, are responsible for a substantial chunk of the male phenotype, because their early removal causes a fair degree of sex reversal in males that cannot be compensated for by activational hormones.

The Development of Sex Differences in Birds: Progress and Puzzles

The original organizational hormone theory referred only to mammals, although one of the authors later speculated that it might shed light on other vertebrates (Young 1965). How well does it apply to birds? There are plenty of cases where sex differences in avian behavior persist even when adult steroid levels are similar. Most experiments to test an organizational hypothesis have used either Japanese quail (and occasionally chickens, which are in the same family, the Phasianidae) or zebra finches (family Estrildidae).

Japanese quail are excellent for studying the development of sex differences because of their unusually rapid development (six weeks from hatching to sexual maturity), their high egg production, and their well-defined, easy-to-measure behavior. Male Japanese quail crow and strut and stand tall. They head-grab, mount, and try to achieve cloacal contact with females (and other males too). They readily approach females and if a female is behind an opaque barrier they are very motivated to peek at her through a narrow window by standing on tip-toe (Domjan and Hall 1986). Females don't do any of these things. Instead, they are receptive to mating attempts by males and when not mating are usually focused on the ground and on foraging or dust bathing. All the sex typical behaviors are highly hormone dependent and some of the sex differences are due largely to activational hormones (Balthazart and Adkins-Regan 2002). The biggest exception is male-typical copulatory behavior and interest in females. No matter how much testosterone an adult female is given, she will not show this behavior.

A series of experiments has confirmed that early steroid exposure is the mechanism for sexual differentiation of this behavior, providing strong support for all but one of the components of the original organizational hormone theory (Adkins 1975; Balthazart and Adkins-Regan 2002). Males hatched from eggs treated with estradiol fail to head-grab and mount as adults even when given testosterone to stimulate the behavior; they, like normal females, are demasculinized birds (fig. 4.2A). The time period for this effect in this highly precocial species is the prehatching embryonic period, especially before day 13 of the 18-day incubation period. A single injection of a very low dose of estradiol is quite effective. Testosterone also "deletes" male-typical copulatory behavior, via aromatization to estrogen. Females hatched from eggs treated with an estrogen receptor antagonist or an estrogen synthesis inhibitor show the complete male-typical copulatory sequence as adults and will look through a peephole at other females (fig. 4.2A). Although less research has been done with chickens, it too demonstrates the power of estrogens to demasculinize males exposed as embryos (Wilson and Glick 1970; Sayag et al. 1989).

Thus, sex steroids permanently organize the sexual phenotype of the copulatory and sexual interest behavior of quail during an early brief critical period. The big contrast with mammals, and the violation of the original organizational theory, lies in which gonad has the organizational role. In mammals, a testis imposes maleness; without it a female develops. In quail all the evidence indicates that an ovary imposes femaleness; without its actions a male develops. This is the same mirror-image contrast that occurred in earlier research on differentiation of the reproductive anatomy in mammals and birds. The sexual dimorphism in early sex steroid levels is also reversed in quail compared to mammals. Quail embryos of both sexes produce both estrogens and androgens, but by day 10 of incubation female embryos have higher estradiol levels and higher estradiol to testosterone ratios (Tanabe et al. 1979; Schumacher et al. 1988; Ottinger et al. 2001). The most important thing that is missing from this account of behavioral sexual differentiation in quail is direct evidence that it is gonadal sex steroids, not sex steroids from somewhere else, that organize behavior. Because of technical limitations imposed by a prehatching critical period, there are no experiments of the sort that have been accomplished many times in mammals, in which the gonads were removed prior to the critical period and behavior was later observed.

The neural mechanisms responsible for these dramatic behavioral effects of early hormone treatment are still somewhat obscure (Balthazart et al. 1996; Balthazart and Adkins-Regan 2002). Although there is a nucleus in the preoptic area that is 40% larger in males, its size seems to be determined more by activational than organizational hormone effects. The same conclusion was reached for the origin of the sex difference in aromatase activity in the brain. What looks promising is a vasotocinergic system hypothesis (Panzica et al. 2001). In both quail and chickens, there is marked sexual dimorphism in vasotocinergic innervation in brain regions known to be involved in sexual and vocal behavior. In the medial preoptic nucleus and the bed nucleus of the stria terminalis small vasotocin immunoreactive neurons are seen only in males. There is a close parallel between effects of early steroid manipulations on this innervation and on the birds' ability to show the male-typical copulatory sequence (fig. 4.2). For example, the same egg treatment with estradiol that deletes male mounting also produces a female-typical vasotocinergic innerva-

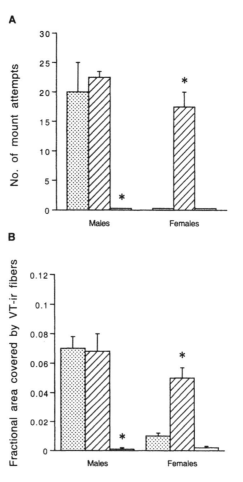

FIGURE 4.2. Organization of mating behavior and vasotocinergic innervation in Japanese quail. Birds hatched from eggs injected on day 9 of incubation with vehicle (control, dotted bars), estradiol benzoate (EB, clear bars) or R76713 (racemic Vorozole, an estrogen synthesis inhibitor, diagonally striped bars). As adults they were gonadectomized, given testosterone, and tested for mating behavior. Following immunocytochemistry for vasotocin, the fractional area of three brain regions covered by immunoreactive vasotocinergic fibers was determined. Embryonic exposure to EB strongly demasculinized both mating behavior in males (as indicated by mounting attempts, A, with similar effects on other mating measures not shown here) and also vasotocinergic innervation in the bed nucleus of the stria terminalis (B, with similarly large effects in the medial preoptic nucleus and the lateral septum not shown here). Embryonic exposure to R76713 had no effect on males but strongly masculinized females both behaviorally and neurally. The close parallel between the behavioral and neural effects suggests a causal relationship, one in which the normal behavioral sex differences result from actions of ovarian estrogens during embryonic development on vasotocin innervation that lead to female-typical behavior (absence of mounting) in genetic females. Redrawn from Panzica, G. C., Castagna, C., Viglietti-Panzica, C., Russo, C., Tlemçani, O., and Balthazart, J. 1998. Organizational effects of estrogens on brain vasotocin and sexual behavior in quail. *J. Neurobiol.* 37: 684–699. © 1998 John Wiley & Sons. Reprinted by permission of Wiley-Liss Inc., a subsidiary of John Wiley & Sons, Inc.

tion pattern. Now it will be important to determine how having more vasoto-cinergic cells and projections makes mounting occur. All in all, this is a valuable model for understanding avian sexual differentiation. Once the chicken genome is sequenced, a molecular approach to the development of the related Japanese quail can proceed.

Songbirds, as their name implies, are famous for singing and for learning songs. There are marked species differences in whether and how singing is sexually dimorphic. Diversity ranges from species in which only males sing through those in which males sing more or sing more complex songs but females sing too, to those where both sexes sing identical songs and duet (sing antiphonally) (Brenowitz and Kroodsma 1996; MacDougall-Shackleton and Ball, 1999). This diversity can be seen even within a single lineage. For example, among estrildid finches, only males ever sing in zebra finches (genus *Taeniopygia*), whereas both sexes sing, but sing different songs, in the genus *Uraeginthus* (Gahr and Güttinger 1986). In songbirds the degree of sexual dimorphism in the neural song system is correlated (although not perfectly) with the degree of song behavior dimorphism (Brenowitz and Kroodsma 1996; MacDougall-Shackleton and Ball 1999).

Where there are sex differences in singing and in the song system, are these activational, organizational, or both? In canaries and a number of other species activational hormones are doing most of the work. Adult treatment of females with testosterone produces pretty good singing by females and an impressive amount of neural masculinization as well, effects that while dramatic, are readily reversible (Leonard 1939; Nottebohm 1980; DeVoogd and Nottebohm 1981; Brown and Bottjer 1993).

Zebra finches have been the favorite passerine subjects for early hormone manipulation experiments because of their strongly dimorphic singing and neural song system, their willingness to produce lots of chicks in a lab, and their rapid development. Females never sing even if given testosterone implants as adults (Arnold and Saltiel 1979), suggesting organization earlier in life. There are pronounced sex differences in the overall volumes of song system nuclei HVC, RA, area X, LMAN, DM, and nXIIts (fig. 2.11), as well as sex differences in the number, size, connectivity, and neurochemistry of neurons in some of these nuclei (see Table 3 in Balthazart and Adkins-Regan 2002). In addition, the male's HVC and LMAN have many more androgen receptors (Arnold and Saltiel 1979).

In the first organizational hormone experiment in a songbird, Gurney and Konishi (1980) exposed nestling female zebra finches to estradiol or dihydrotestosterone and when they were adults gave them testosterone, dihydrotestosterone, or estradiol. Females treated with estradiol and then androgens sang and their song system nuclei HVC and RA were greatly enlarged, reaching the male-typical size. This was an amazing result, a testimony to the power of early hormones to alter brain development. They interpreted their results according to the mammalian organizational hormone theory, which made sense

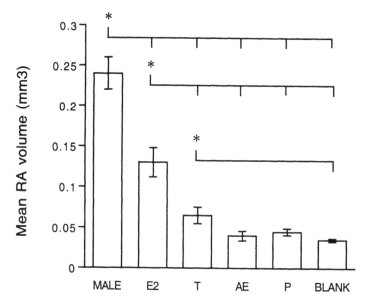

FIGURE 4.3. Effectiveness of different steroids in masculinizing the song system of zebra finches. The volume of the song system nucleus RA was greatly increased in females given estradiol (E₂) implants at hatching. Other steroids were much less effective. MALE, untreated males; T, testosterone; AE, androstenedione; P, progesterone; BLANK, empty implants (control females). Redrawn from Grisham, W. and Arnold, A. P. 1995. A direct comparison of the masculinizing effects of testosterone, androstenedione, estrogen, and progesterone on the development of the zebra finch song system. *J. Neurobiol.* 26: 163–170. © 1995 John Wiley & Sons. Reprinted by permission of Wiley-Liss Inc., a subsidiary of John Wiley & Sons, Inc.

because female mammals given estrogen early and then androgens as adults are also masculinized. Several other labs subsequently reported similar effects of early estrogen treatment in which some females sang as well as males and song system nuclei were substantially masculinized (fig. 4.3) (reviewed in Balthazart and Adkins-Regan 2002). The critical period for good singing was the nestling period, especially the first week after hatching, which is not unexpected in this highly altricial species (Pohl-Apel and Sossinka 1984; Simpson and Vicario 1991a; Adkins-Regan et al. 1994).

This important discovery established the zebra finch as an excellent species for teasing apart the mechanisms whereby steroids regulate brain development. Using the dimorphic features of the song system as the outcomes, it was found that treatment of nestling females with estradiol increases the incorporation of new neurons into HVC and area X (by influencing where the neurons migrate), increases the number of androgen concentrating cells in HVC, saves some neurons from an early death in RA and LMAN, and preserves connections between HVC and RA and between LMAN and RA that are essential for singing (Gurney 1981; Nordeen and Nordeen 1989, 1994; Simpson and Vicario 1991b; see also Table 4 in Balthazart and Adkins-Regan [2002]). As in the mammalian research, steroids seem to be regulating not neurogenesis, but instead the other processes that determine how many and what kind of neurons there will be in a nucleus.

In the meantime it has become apparent that the mammalian version of the organizational hormone theory simply won't work for the zebra finch. There are many reasons, but the most compelling are these. First, no replicable sexual dimorphism in circulating levels of steroids or steroid-synthesizing enzymes in the gonads has been detected. Zebra finches have high aromatase levels in the telencephalon and so brain-produced estrogens could be important but brain steroids and related enzymes and receptors aren't dimorphic either (for example, Schlinger and Arnold 1992; Perlman et al. 2003). That is, no dimorphic signal to send males down a different path than females has been found. Holloway and Clayton (2001) did find a sex difference in estradiol in the culture media containing brain chunks that included HVC and RA, and presented evidence that this locally produced estradiol is the signal for establishing a projection from HVC to RA, but this dimorphic estradiol was coming from brain chunks 25 to 32 posthatching days old, whereas the critical period for the estradiol effect on nucleus enlargement and singing is already over long before then. Second, no manipulations have ever substantially feminized males (have prevented males from becoming males), whereas in mammals this is easily accomplished by early castration or by giving steroid receptor antagonists. Third and most damaging, females treated with fadrozole as embryos, so that they have testes or ovotestes, are not masculinized either in singing or in the brain (do not differ from controls), and males treated with estrogen as embryos, who have some ovarian tissue, also don't differ from controls (Wade and Arnold 1994; Balthazart et al. 1994; Adkins-Regan et al. 1996; Wade et al. 1997). The histological state of the gonads doesn't seem to have anything to do with the phenotype of the song system. Instead, the system always corresponds to the genetic sex of the bird, just like the plumage does (Wade et al. 1999).

This dissociation between the state of the gonads and the phenotype of the song system makes this species a tantalizing candidate for a direct genetic differentiation hypothesis (Arnold 2002). No tissue or cell autonomous mechanisms leading from the sex chromosomes to the neurons have been discovered yet, however. What on earth is estradiol doing when it produces masculinization? Interfering with a signal that normally inhibits female typical brain development? It is ironic indeed that a vertebrate with such whopping brain dimorphism should be so hard to figure out!

Early steroid effects on sexual partner preferences of zebra finches are also a mixture of progress and puzzles. The same early estradiol treatment that masculinizes singing in females also masculinizes sexual partner preference, producing a shift toward preferring other females as pairbond partners (Mansukhani et al. 1996). Treatment of nestlings with fadrozole, however, does not feminize the partner preference of males, making a mammalian organizational theory untenable (Adkins-Regan et al. 1996). Treatment of female embryos with fadrozole so that they have testes nearly sex reverses partner preference (Adkins-Regan and Wade 2001). How can estradiol and fadrozole both masculinize the same behavior? Evidently one or both of them are acting indirectly, by enhancing or suppressing some other signal, possibly a signal from the ovary.

The only zebra finch behavior that is even remotely consistent with the quail version of the organizational hormone theory is mounting (copulation). Females seldom mount other birds, and males treated with estradiol as nestlings mount less than controls (Adkins-Regan and Ascenzi 1987). However, fadrozole does not increase mounting in females, so even this behavior doesn't fit the quail model (Adkins-Regan et al. 1996; Wade et al. 1999). Very little else is known about sexual differentiation of behavior or brains in songbirds or any other birds. It is not even clear that the dramatic masculinizing effect of early estradiol exposure on the zebra finch song system generalizes to nonestrildid songbirds (Casto and Ball 1996).

This is an unfortunate situation, because understanding sexual differentiation in birds is very important for several problems in behavioral ecology. Three examples will illustrate this point. First, there is much interest in the possibility that maternal yolk steroids could have long-term consequences for the behavior of the offspring (chapter 3). Winkler (1993), for example, proposed the intriguing hypothesis that in cooperatively breeding species, yolk steroids could be a mechanism whereby a female biases some offspring toward being future alloparents. Yet nothing is known about organizational hormone effects on parental behavior in any birds. Furthermore, the focus thus far has been on yolk androgens, yet when steroids are administered to avian embryos or nestlings, androgens have relatively weak effects on behavior, if any, when compared to estrogens. Second, there has been interest in possible hormonal mechanisms underlying sex role reversal in birds such as phalaropes (*Phalaropus*) in which males perform all the parental care and females compete for males. Adult sex steroid levels are not reversed and in fact are entirely unexceptional, casting doubt on a simple activational hormone hypothesis of role reversal (Fivizzani et al. 1990). But there is still no test of an organizational hormone hypothesis. It's not even obvious what such a hypothesis would look like, because the male and female still have to produce sperm and eggs (they can't be reversed all across the board). Third, there is an interesting difference between birds and mammals in which sex is more likely to disperse from the natal area as juveniles (Pusey 1987; see also chapter 6). Females are more likely to disperse in many birds and males are more likely to disperse in many mammals. This phenomenon is interesting from both a functional and a mechanistic perspective. There is evidence that organizational effects of testosterone are generating the sex difference in dispersal probability in Belding's ground squirrels (*Spermophilus beldingi*) (Holekamp et al. 1984), but nothing is known about hormonal mechanisms leading to the sex difference in birds.

Sexual Differentiation of Behavior in Other Vertebrates

It is fairly common to find that adult sex steroid treatments largely sex reverse behaviors in lizards, frogs, and fishes, supporting an activational hormone hypothesis and eliminating the need to posit any earlier developmental process such as organizational hormone actions. When sex differences persist, there have been few attempts to test an organizational hypothesis of their origin. Sometimes there are good reasons for this. For example, the animals may mature so slowly or be so big that developmental and behavioral experiments are impractical (imagine crocodiles or sharks).

Three species of lizards, green anoles (*Anolis carolinensis*), striped whiptails (*Cnemidophorus inornatus*) and leopard geckos (*Eublepharis macularius*), have proven to be practical experimental subjects for hormone manipulations and hold out promise for an understanding of the development of sexually dimorphic behavior. These animals exhibit clear sex differences in courtship, sexual, and aggressive behavior and in some of the neural mechanisms producing the behaviors.

In the green anole some of the behavior, such as masculine courtship and copulatory behavior, is substantially although not completely sex reversed following adult hormonal manipulation, indicating a sex difference produced by activational hormone actions, whereas others, such as sexual receptivity, are not, as if they might be hormonally organized (Mason and Adkins 1976; Wade 1999). The young are precocial at hatching and it is not known if gonadal steroid production is sexually dimorphic in the embryos. Juvenile males have higher testosterone levels than females by two weeks posthatching, but whether steroid treatment then has a permanent and specific effect on adult reproductive behavior is not yet known (Lovern et al. 2001).

Gonadectomized female striped whiptails respond to estrogen by becoming receptive to mating, but males don't. There is a corresponding sex difference in the ventromedial nucleus of the hypothalamus, an essential nucleus for female-typical receptive behavior. Estrogen increases both estrogen and progesterone receptor mRNA in females but not males. With greater time since castration, males become feminized, pointing to a likely activational basis for this sex difference (Wennstrom and Crews 1998).

In leopard geckos gonadal sex is determined by incubation temperature, providing an excellent opportunity to ask whether the organizational hormone theory works for the behavior of species with temperature sex determination as well as those with genetic sex determination. When adult males and females were gonadectomized and given the same hormone treatment (estradiol to stimulate female behavior or the androgens dihydrotestosterone or testosterone to stimulate male behavior) there were still big sex differences in behavior (Rhen and Crews 2000). Males, unlike females, were almost never sexually receptive even after estradiol, and males were more likely to attack another male than females were following testosterone. An organizational hypothesis for these sex differences has not yet been tested directly through hormone manipulations in early development. Animals of the same gonadal sex that result from eggs incubated at slightly different temperatures differ behaviorally, but so far these within-sex individual differences seem to have an activational basis (Rhen and Crews 2000).

The African clawed frog (*Xenopus laevis*) has been the focus of the most systematic work with any amphibian (Kelley and Tobias 1999). Only males produce advertisement calls. Male calling is dependent on gonadal hormones; castrated males do not call but they start calling if given testosterone or dihydrotestosterone. Females given these same hormones never produce advertisement calls; the sex difference is not activational. As in zebra finches, the vocal motor control system is sexually differentiated in multiple ways (biochemical, cellular, neurophysiological) and at multiple levels from brain to vocal organ. An organizational hypothesis, mammalian version, for the development of these sex differences has been thoroughly tested (Watson and Kelley 1992; Watson et al. 1993). Ovariectomized females given testis transplants start to call but only after many months, as if time is required to reconfigure the system. The earlier they receive the transplants, the better the quality of their calling. Administration of androgenic steroids does not produce the same effect as testis transplants, pointing to a different category of testicular hormones as critical for masculinization. Masculinization by testes that occurs more readily earlier in life is consistent with the mammalian version of the organizational hypothesis. But there is no critical period. Instead, there is a steadily decreasing likelihood of full masculinization as the females get older, with significant masculinization still possible in adulthood. In addition the time period required for the masculinization to occur is much longer than in mammals or zebra finches.

Many teleost fishes have stable adult sex of the sort typical of other vertebrates, and some (such as the medaka, *Oryzias latipes*) have heteromorphic sex chromosomes that determine gonadal sex. In fact, with over 23,000 species, there are usually plenty of teleosts to illustrate almost any aspect of reproduction! Some are very monomorphic and seem to have little sexual differentiation to account for beyond the gonads. In several behaviorally dimorphic species (for example, *Betta splendens* and *Carassius auratus*) the sex-specific behavior of adults is readily sex reversed through steroid manipulations, as if the brain is very bipotential and any sex differences are created largely through dimorphism in adult circulating hormone levels (Badura and Friedman 1988; Francis, 1992; Stacey and Kobayashi 1996).

In several commercially important farmed teleosts, there are strong economic incentives for manipulating sex ratios. To do this, very young fish are treated by putting steroids in their food or water. Such treatment can be very effective, sex reversing the gonads and (apparently) everything else in half the individuals to give all male or all female populations. Depending on the species, genetic females can be turned into gonadal males with androgens or with estrogen synthesis inhibitors like fadrozole, genetic males can be turned into gonadal females with estrogens, and occasionally both directions of sex reversal are possible in the same species (Adkins-Regan 1987; Borg 1994; Kavumpurath and Pandian 1994; Piferrer et al. 1994; Kwon et al. 2000). In the Nile tilapia (*Oreochromis niloticus*) a marked sex difference in aromatase gene expression in the gonads appears just a few days after the eggs are fertilized. Feeding larvae food containing the aromatase inhibitor fadrozole results in 80–100% males, half of which are XX males (Afonso et al. 2001; Kwon et al. 2001). Estrogens appear to be a signal for ovarian development, as in chickens and zebra finches.

These sex-reversed fish are evidently behaviorally sex reversed as well, for they can attract and mate with untreated individuals of their own genetic sex to produce offspring. Almost no systematic comparisons with controls of the behavior of sex-reversed fish have been carried out, however. Because the gonads themselves are sex reversed, behavioral sex reversal could be due to either organizational or activational effects or a combination of both. Because of the evidence from other species that adult teleost brains are remarkably sexually bipotential, the phenomenon could be entirely activational. In other words, there is little evidence for any hormonal organization of teleost fish behavior (Francis 1992).

Thus far the best evidence in a teleost for hormonal organization of something other than the gonads themselves comes from the mosquitofish *Gambusia affinis* (Rosa-Molinar et al. 1996). As in other poeciliid fish, males have a gonopodium, a specialized anal fin. During mating it is swung out and forward to guide sperm to the female's genital pore. This unusual fin movement is possible because of a modification of the axial skeletal formula accompanied by an alteration in the innervation of the fin. If females are exposed to 17α-methyltestosterone during the late embryonic period, they develop a fully functional gonopodium. This effect of androgen on the development of the gonopodium is hypothesized to be mediated by a cascade involving regulation of *hox* genes (conserved genes for the development of segmental body plans) by retinoic acid (Rosa-Molinar et al. 1997). So far this fascinating example of sexual dimorphism in a teleost seems to develop according to a mammalian organizational hormone theory, although it is not yet known whether treatment of males with androgen receptor antagonists early in development prevents a gonopodium from forming.

Thus, with the possible exception of *Gambusia*, none of these "other" verte-brates (lizards, frogs, fish) follow either the mammalian or the quail/chicken versions of the organizational hormone theory of behavior (Crews 1993). Even when sex differences in behavior do not disappear following adult steroid ma-nipulations, there is no evidence yet that the sexual phenotype was perma-nently established by hormones during an early critical period. Very few spe-cies have been studied developmentally, however, and the vast majority of sexually dimorphic vertebrate lineages are completely absent from research on sexual differentiation of behavior.

Do Invertebrates Have Hormonally Organized Sex Differences in Behavior?

Invertebrates use different kinds of molecules for reproduction than vertebrates do. Their reproductive hormones aren't usually steroids, and they aren't neces-sarily produced by the gonads. Nonetheless, those that occur as sexes undergo sexual differentiation. It has been known for many decades that hormones are a mechanism of sexual differentiation in some invertebrates. The proof that the differentiation signal was hormonal came from parabiosis experiments in which joining a male and a female together caused one of them to change sex. In the coelenterate *Hydra fusca*, for example, a female parabiosed to a male is fully masculinized and capable of producing sperm, suggesting a dominant role for the testis in sexual differentiation (Brien 1962). In experiments with the coleopteran glow-worm *Lampyris noctiluca*, the testis organizes masculinity during early development, and larval females implanted with a testis become adults with testes and male sex structures, male behavior, and the capacity to fertilize females (Naisse 1966). As in the teleost fish experiments, the complete gonadal sex reversal is an amazing phenomenon as well as a sign that gonads are hormonally organized, but it also makes it impossible to tell whether the behavioral sex reversal is activational or organizational. To date *Lampyris* seems to be the only insect for which hormonally based sexual differentiation of any kind has been shown (Nijhout 1994; Ringo 2002). Other kinds of perma-nent effects of hormones occurring during early development are common in insects, but the sexes aren't usually made this way.

In daphnid crustaceans males are produced and sexual reproduction occurs under poor environmental conditions. Methyl farnesoate, a member of the juvenile hormone category, has been proposed to be the signal that transduces a poor environment into male development (Olmstead and Leblanc 2002). In isopod, decapod, and amphipod crustaceans peptide hormones from the androgenic gland (a separate gland from the testis) are thought to be responsi-ble for male differentiation, including testicular differentiation (Charniaux-Cotton 1965; Sagi and Khalaila 2001). The androgenic gland hormone both stimulates male characters and inhibits female characters. Behavior is altered as well. In crayfish *Cherax quadricarinatus* females with androgenic gland transplants directed some male typical courtship and copulatory behavior toward other females (Barki et al. 2003). Male crustaceans that lose the androgenic gland are feminized. This has been demonstrated experimentally (Carlisle 1960) but also happens under free-living conditions when parasites destroy the gland. A particularly interesting story has unfolded in the isopod *Armadillidium vulgare*, with *Wolbachia* bacteria as the parasitic culprits. *Wolbachia*-infested "neo-females" are sufficiently female-typical behaviorally and anatomically to be mated and inseminated by males, although males prefer real females when given a choice (Moreau et al. 2001). Thus, there is clear evidence for behavioral consequences of the presence versus absence of an androgenic gland. It is not clear, however, that the organizational concept applies to any of these results. The phenotype is feminized by loss of the androgenic gland even in adult males. An activational hypothesis of sexual differentiation seems to be sufficient.

The tobacco hornworm moth *Manduca sexta* has a sexually dimorphic olfactory system and olfactory behavior. Males but not females have specialized olfactory receptor cells on the antennae that project into a macroglomerular complex in the antennal lobe of the brain, and only males fly toward female sex pheromone. When antennae are grafted onto opposite sex larvae, they develop the characteristics of the donor's sex, not the host's sex, indicating that the dimorphism does not arise from hormonal influences (Schneiderman et al. 1982).

When it comes to developmental research subjects, *Drosophila melanogaster* and *Caenorhabditis elegans* lead the pack. In these animals sex is determined by the ratio of sex chromosomes to autosomes, but neither has hormonally induced sexual differentiation, either activational or organizational. Instead, differentiation of the brain and body is cell autonomous, occurring through direct genetic sexual differentiation. The molecular genetic cascades that are reponsible for male- and female-specific behavior, brain, and bodies are complex, but excellent progress has been made at sorting them out (Baker 1989; Hodgkin 2002; Wilkins 2002). Given how conserved some of the molecular mechanisms of development are in animals, it is not impossible that some of these same gene families might be involved in direct genetic sexual differentiation in vertebrates.

There are many highly sexually dimorphic invertebrates whose sex is determined during development by the environment, not simply the genes (Korpelainen 1990). For example, male sea worms of the genus *Bonellia* live as little parasites inside the females. Sex is determined in large part by whether a larva lands on a female, in which case it becomes a male. Hormones from the female "host" are the mechanism for this fascinating phenomenon (Jaccarini et al. 1983).

Sex-Changing Fish

Some teleost fish undergo complete functional sex reversal as a normal part of their adult life. Such species are mainly found in the order Perciformes. This mode of reproduction is called sex change or sequential hermaphroditism. Behavior, external dimorphic characters such as coloration, and gonads change from male to female (protandry) or, more commonly, from female to male (protogyny). Gobies are among the smallest vertebrates, and some can change sex in both directions (Nakashima et al. 1995).

Adult sex change is an amazing fact of nature. Nearly all other vertebrates, and many teleosts as well, have a fixed adult sex. Sex-changing teleosts stand out from this crowd. Sex change in these fish occurs not at a particular age but in response to external cues. It is sometimes regarded as a form of environmental sex determination. Most importantly with respect to social behavior, it is social interactions that provide the environmental stimuli triggering sex change in perciform sex changers (Robertson 1972; Godwin et al. 2003). In *Anthias squamipinnis* groups consist of one male and several females. Removal or disappearance of the male causes the dominant female to turn into a male (Shapiro 1983). It is the dominant female's interactions with the other females, together with the removal of inhibitory influences from her interactions with the male, that provide the stimuli for sex change, as if females assess the potential costs and benefits of continuing as a female versus male and make a decision about whether to change.

The coral reef-dwelling bluehead wrasse (*Thalassoma bifasciatum*) has a complicated but fascinating social system that includes both sex change and morph change (fig. 4.4). There are three morphological sexual phenotypes: initial-phase females (animals that began reproductive life as females and have not yet changed sex), initial-phase males (animals that began reproductive life as males and have not yet become terminal-phase males), and terminal-phase males (animals that used to be initial-phase females or males) (Warner 2002). Terminal-phase males are larger, have a different color pattern (including a blue head), and are the least numerous of the three. Although there are two male morphological types, there are three male mating types, one for initial-phase males and two for terminal-phase males. Initial-phase males are nonterritorial and spawn in large aggregations or by sneak fertilization in the territory of a terminal male. Nonterritorial terminal-phase males behave like initial-phase males. Territorial terminal-phase males defend a territory used as a spawning site. If there is no territorial terminal-phase male in the group, the largest initial-phase female will become a terminal-phase male. Her behavior changes relatively rapidly; the ovaries are then replaced by testes over a 7- to 10-day period.

FIGURE 4.4. A mating swarm of bluehead wrasses. This species exhibits both sex change and morph change. The mechanisms underlying both have been studied experimentally. Courtesy of John Godwin.

Understanding adult sex change means finding answers to both functional (why) and mechanistic (how) questions. Some progress has been made in theorizing why sex change might increase lifetime fitness and in predicting when it should occur in the animal's life (Ghiselin 1969; Charnov and Bull 1977; Warner 1988). Teleosts often have indeterminate growth and pronounced sex differences in the relationship between size and reproductive success. For example, where males compete for matings and larger males win, reproductive success may increase with size for males more than it increases with size for females, making it advantageous to reproduce as a female when small and then become a male at a larger size (protogyny). Such size–sex relationships are central to most models of why sex change has evolved as a reproductive mode. Recent versions also incorporate intense sperm competition between males and the concept of reproductive skew (for example, Muñoz and Warner 2003). The role of social interactions in the actual initiation of sex change means that each fish is making a decision based on who else is in the group rather than simply changing sex at a particular absolute body size. How are these social interactions "transduced" into gonadal sex reversal? What is the role of steroids and peptides in this process? Why are teleost fishes so plastic in their sexuality? These are fascinating mechanistic questions. The search for answers is having a very stimulating effect on the field of teleost neuroendocrinology.

A classic organizational hormone hypothesis (not just for behavior, but for gonads as well) is obviously severely violated by adult sex change. Activational hormone actions alone could account for the behavioral part of sex change given adult teleost bipotentiality (Francis 1992). But should we rule out altogether a role for any kind of organizational hormone effect? Perhaps not, for two reasons. First, some sex-changing species such as *Thalassoma* (see above) have initial sexual differentiation in which the animals start out as either males and females, with some individuals then changing sex later in life, two different stages of sex determination separated in time. Nothing seems to be known about how this initial-phase sex determination occurs, but it would not have to be different than in other teleosts. Second, most sex-changing teleosts change sex only once (the gobies mentioned earlier are exceptions [Grober and Sunobe 1996]). Is this because there is no reason (advantage, stimulus) to going back even though the mechanisms are there to permit it (reversible activation), or because the change involves a permanent (irreversible) mechanism? Perhaps there is a second period of organization in adulthood that, unlike the first one, has a temporally flexible "window of opportunity" (Adkins-Regan 1985). An experimental manipulation is the best way to find out. The angelfish *Centropyge ferrugata* is protogynous but when the neomales were made to lose contests with other males, they turned back into females, gonads and all (Sakai et al. 2003). It will be interesting to see which species can do this and which can't.

Behavioral and morphological sex change is correlated with changes in levels of circulating sex steroids and in levels of peptides such as GnRH and vasotocin in the preoptic area of the brain. As individuals change from female to male, levels of androgens, especially 11-ketotestosterone, rise and levels of estrogens fall, and the number and/or size of GnRH- and vasotocin-producing cells in the preoptic area changes from female typical to male typical (Bass and Grober 2001). As individuals change from male to female, levels of androgens fall and levels of estrogens rise. These correlations have inspired several different hypotheses for the mechanisms of sex change: steroid hypotheses, a GnRH hypothesis, and peptide hypotheses.

These are not mutually exclusive. Which one is correct might depend on the type of sex change (how dimorphic the species is), the rapidity of sex change, and perhaps even the direction of sex change. A set of elaborate mechanisms unique to sex changers might not be necessary for the behavioral part of sex change, especially when behavior changes rapidly, as it sometimes does. Suppose all individuals have the neural circuitry at hand for both female- and male-typical behavior. Then behavioral sex change should be conceptualized not as the creation of a new behavioral repertoire, but as the inhibition of an old one (female behavior in the case of protogyny) as the new one is released from inhibition. Where spawning behavior is very similar in males and females (the circuits are the same), all that needs to change is the sex with which the animal spawns, not the behavior itself. Expressed in the language of behavioral ecology, the animals might be changing tactics, not strategies (Oliveira et al. 2001a,b,c). The same peptide-based mechanisms that are suspected to regulate rapid shifts between tactics as a function of social circumstances in non-sex-changing vertebrates should suffice.

What is truly unique in sex changers is the adult gonadal change and the fact that it is driven by the social environment and therefore by the brain. In sex changers sex determination is top down, not bottom up as in genetic sex determination. Thus, most recent hypotheses incorporate the concept of brain directed sexual differentiation first stated explicitly by Francis (1992).

Back to the hypotheses. The hypothesis that the changes in sex steroids cause the changes in behavior and coloration is the oldest of the three (Chan and Yeung 1983). In non-sex-changing teleosts, male-typical behavior and morphology are often dependent on 11-ketotestosterone and female-typical behavior is dependent on estrogens (chapter 2). This hypothesis assumes that the gonads start changing sex first, and as they change the internal sex steroid milieu changes, which in turn causes changes in behavior and appearance. If this is true, then treatment of animals with steroids prior to naturally occurring sex change should reverse the sex of behavior and appearance regardless of the social environment. In some protandrous species treatment with estrogens causes earlier than normal sex reversal. In some protogynous species treatment with 11-ketotestosterone or other androgens induces sex change and in some it doesn't (Cardwell and Liley 1991; Borg 1994; Devlin and Nagahama 2002). In *Thalassoma* females treated with 11-ketotestosterone acquire terminal-phase male coloration (Semsar and Godwin 2004). Another test is to remove the gonads of animals and then expose them to the right social environment for sex change to see if behavioral and morphological sex change are prevented. When this experiment was done with female *Thalassoma*, their behavior changed sex anyway (Godwin et al. 1996). So much for a causal role for the gonads in this species! A "gonads first" hypothesis also doesn't look promising whenever the earliest signs of sex change in behavior and coloration begin very rapidly, within hours after the right change in the social environment, but the gonads reverse over a period of days or weeks.

One steroid-based alternative to the "gonads first" hypothesis is that extragonadal steroids are responsible, a "steroids first" version of a top-down hypothesis. A change in the pattern of aggressive interactions could alter glucocorticoid or brain neurosteroid levels, which then might act rapidly through membrane steroid receptors to reverse behavior and through other pathways (including the HPG axis) to eventually reverse the gonads (Grober and Bass 2002; Perry and Grober 2003). Teleosts do undergo changes in circulating steroid levels in response to aggressive interactions (Bonga 1997) and teleost brains produce a lot of estrogen (Callard 1984). In non-sex-changing teleosts sex steroids cause gonadal sex reversal in very young animals. Whether brain steroid production changes in response to aggressive interactions in ways that impact the gonads is not yet known.

A GnRH hypothesis posits that the change in social interactions causes a change in GnRH, which in turn alters behavior directly and also begins a cascade down the HPG axis to result in gonadal sex change (Bass and Grober 2001). In several sex changers that have been studied, there are sex differences in the number of GnRH cells in the preoptic area and sometimes in levels of LH as well, with males having more and larger GnRH cells. In protogynous species where 11-ketotestosterone doesn't produce sex change, GnRH usually does (for example, in *Thalassoma bifasciatum*), and it sometimes changes the sex of the gonads as well as the sex of the behavior (Devlin and Nagahama 2002).

According to the vasotocin hypothesis, the change in social interactions causes a change in vasotocin and behavior, which in turn causes gonadal sex change (Grober and Bass 2002). Here the GnRH alterations are a consequence, not a cause, of behavioral sex change. There is correlational evidence to support this hypothesis. In *Thalassoma* the number of cells producing AVT mRNA, and the amount of vasotocin mRNA in those cells, is low in initial-phase females and high in terminal-phase territorial males, who behave aggressively toward initial-phase fish (Godwin et al. 2000). In the marine goby (*Trimma okinawae*), which has very fast sex change, forebrain vasotocin cell size is larger in females and changes very rapidly coincidentally with sex change (Grober and Sunobe 1996). What is needed to pin down a causal hypothesis is to show that brain vasotocin manipulations alter the probability of sex change. In one of the few studies of this kind, females treated with AVT in a setting conducive to sex change acquired territories (underwent behavioral sex change) (Semsar and Godwin 2004). Neuropeptide Y, another peptide candidate, is part of the regulatory mechanisms for GnRH and when given to female *Thalassoma* caused ovarian degeneration (Kramer and Imbriano 1997). Because it was given systemically, this could have been a peripheral effect instead of evidence for a top-down role for a brain peptide.

A new hypothesis, the monoamine hypothesis, has recently been added to this list. The brain mechanisms that regulate the HPG axis include the monoamine neurotransmitters, and the neurotransmitter dopamine is also a hypothalamic releasing factor. In support of this hypothesis, drugs that alter monoamine activity stimulate or retard gonadal sex reversal in *Thalassoma duperrey* (Larson et al. 2003).

An exciting possibility for the future is a deeper integration between function and mechanism in understanding sex-changing teleosts. As with other complex phenomena, there may be different functions for sex change depending on the species or environment, and different mechanisms supporting those functions. Different mechanisms might not exact the same costs, and those costs could help explain when sex change occurs and when it doesn't.

Within-Sex Types (Within-Sex Dimorphism)

Adult bluehead wrasses not only can change sex, but come in three forms: females plus two different kinds of males. Scattered around the vertebrate and invertebrate worlds are other species (most of which are not sex changers) with adults of the same gonadal sex that fall into two or more discrete phenotypic categories. This phenomenon goes by several names, including alternative types, alternative morphs, polymorphisms, and polyphenisms. "Alternative" implies that these are mixed evolutionarily stable reproductive strategies. The true reproductive success of the types is seldom known, however (Sinervo and Zamudio [2001] is an exception). The more neutral phrases "within-sex types" or "multiple types" will be used here. Like sex change, within-sex types raise important questions about the mechanisms producing the types as well as the evolution of these systems that seem so odd to human sensibilities (Rhen and Crews 2002). Many explanations for why they evolved point to extreme male competition as the selective force (Gadgil 1972).

The most widely known examples of multiple types are the castes of eusocial insects. In some ants there are multiple types of males, females, or both that carry out different behavioral roles. Among honeybees, a drone is a drone but a female can be either a queen or a worker. Only a queen lays eggs, and only workers care for young, forage, and waggle dance to communicate food source locations to other workers. These are radically different lives. The life course of the female bee (queen or worker) is determined in part by what she was fed as a larva. A larva fed highly nutritious food becomes a queen. The long-lasting influence of the early diet suggests some kind of organizational hormone effect. In support of this idea, experiments have shown that eating highly nutritious food increases juvenile hormone levels, and that actions of these juvenile hormones during a critical period plus inhibitory pheromones produced by adults determine the two female castes of honeybees. Queens are made, not laid. Similar effects of juvenile hormones are thought to be responsible for the castes in some ants and termites, although there is recent evidence for some genetic contribution to caste in a few species (Nijhout 1994; Elekonich and Robinson 2000; Volny and Gordon 2002; Cahan and Keller 2003).

In vertebrates, most examples of multiple types that have been studied are teleost fish with two male types. The male types differ in reproductive and other social behavior as well as in size and appearance. What is special about types is that they are discrete categories, not just points on a continuum. So are sexes. By analogy with sexual differentiation we can ask whether steroids are part of the developmental mechanisms producing differences between types of the same gonadal sex. Moore (1991) was the first to explicitly apply

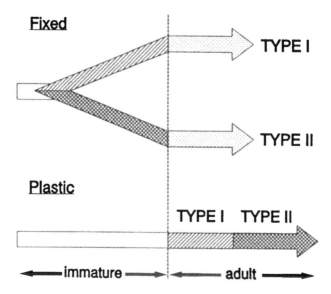

FIGURE 4.5. Moore's (1991) schematic representation of his relative plasticity hypothesis, showing the predicted relationship between the ontogenetic pattern of phenotypic change in morph
type (fixed in adulthood vs. plastic in adulthood) and the period of life when hormonal differences between types are expected to occur. Areas that are shaded with different patterns
represent stages of development when hormone levels would differ. Fixed types are predicted
to be mainly organizational, and plastic types are predicted to be mainly activational (Moore et
al. 1998). Reprinted from Moore, M. C. 1991. Application of organization–activation theory to
alternative male reproductive strategies: a review. *Horm. Behav.* 25: 154–179. © 1991 Academic
Press (Elsevier).

the organization/activation theory to male types (fig. 4.5). His proposal was as
follows. First, when the types are determined prior to adulthood and are stable
(fixed, not plastic) thereafter, they have been created by organizational hormone actions, not by adult hormones. The source of the types that is mediated
by early hormones can be either genes or environment, but once the course is
set, it can't be changed. As adults the types will not differ in hormone levels
except during a short-term response to a social interaction such as a challenge.
Second, when the types are determined in adulthood according to environmental or social conditions and the animals can change from one to the other (are
plastic, not fixed) in response to those conditions, the types are produced by
activational hormone actions, and therefore as adults the types will differ in
circulating hormone levels. The mechanism for the adult plasticity is the sensitivity of the HPG and HPA axes to the external world. Moore's framework
corresponds to the distinction made in behavioral ecology between fixed alternative strategies and condition-dependent alternative tactics or roles.

The tree lizard (*Urosaurus ornatus*) was the animal chosen to test this relative plasticity model (Moore et al. 1998). There are three kinds of males. "Orange-blue" males have orange-blue-colored dewlaps (throat fans extended during displaying), are resident, hold territories, and pursue a mating strategy based on territory possession. "Orange satellites" have orange dewlaps, are resident, and pursue a satellite mating strategy, positioning themselves near the territory of an orange-blue male. "Orange nomads" have orange dewlaps and are nomadic rather than resident. "Orange satellite" and "orange nomad" appear to be alternative condition-dependent tactics available to males with the orange dewlap phenotype; orange males are satellites in environmentally good years and nomads in environmentally harsh years.

The difference between orange-blue and orange males is hypothesized to be due to an organizational steroid effect. There are three kinds of evidence for this. First, the proportions of the two dewlap color types are relatively stable in any given wild population, and no change in either direction in the dewlap color of marked males has been observed. Second, the two morphs don't differ in adult hormone levels (Thompson and Moore 1992). Third, implanting hatchlings with testosterone during the first month posthatching increased the proportion of animals that acquired orange-blue dewlaps, whereas implants on day 60 were ineffective (Hews et al. 1994; Hews and Moore 1996). An injection of progesterone on the day of hatching produced the same effect as testosterone (Moore et al. 1998).

The difference between the two orange dewlap tactics is hypothesized to be due to activational effects of steroids. Specifically, environmental conditions such as drought and poor food supply cause hormonal changes that trigger a shift to the nomadic tactic that is reversed if conditions improve. Both orange and orange-blue males have higher corticosterone when conditions are poor, but orange males have lower testosterone than orange-blues when conditions are poor, and only orange males are sensitive to the testosterone depressing effects of experimentally elevated corticosterone, which is probably why only they become nomadic when things get bad (Knapp and Moore 1997; Knapp et al. 2003).

So far the model looks promising for this species, although not all the pieces are in place. Most importantly, it has not yet been shown that early treatment can alter the behavioral type of the male, rather than just the dewlap color. In the side-blotched lizard (*Uta stansburiana*), orange-throated males have higher testosterone levels than the two other male morphs and defend larger territories. Young animals given testosterone assume the behavior of the orange type, as if orange behavior is being activated and is plastic after early life (Sinervo et al. 2000).

Moore's model is potentially valuable for thinking about any species with male types, not just lizards. For example, the shorebird known as the ruff (*Philomachus pugnax*) has a lek mating system with two kinds of males, inde-

FIGURE 4.6. Male ruffs on a lek. The independent morph male has the dark ruff and the satellite morph male has the white ruff. Females injected with testosterone show these same displays. Courtesy of David Lank and Oene Moedt.

pendents and satellites (fig. 4.6). The two types differ in plumage coloration and also in their courtship and territorial behavior on the lek. Adult females given testosterone implants display the behavioral repertoires of the two male morphs in the proportions that would be predicted by a model in which male type is determined by autosomal genes (Lank et al. 1999). This suggests that the two male types require activational testosterone for their expression. But because all females received the same amount of testosterone, the results also suggest that the individual's type that will be expressed has been determined earlier (is in the "soma") through a process that has occurred in the females as well as in the males. The process itself would not have to involve hormones, but Moore's model hypothesizes that it does.

A number of teleosts have male types. The midshipman fish (*Porichthys notatus*) has been the subject of the most sustained and focused research to date on the mechanisms producing these male types (Bass 1992). Type I male midshipmen are larger, and large ones build and defend nests under rocks and produce a long, sustained courtship vocalization (hum) by means of large androgen-dependent sonic muscles. They have testes that are relatively small for their body size. Type II males are smaller (female size), always pursue a sneaker mating strategy, do not hum, have small (female-size) sonic muscles, and have relatively large testes. How do the hormone levels of these types differ? As is typical of other teleosts with male types, the types don't differ much in testosterone, but type I males have high 11-ketotestosterone levels, whereas type II males have low to undetectable 11-ketotestosterone levels (Brantley et al. 1993; Borg 1994). When either 11-ketotestosterone or testosterone was administered to juveniles, their sonic muscles enlarged to type I size but the effect on their vocalization or other behavior is not yet known (Brantley et al. 1993). The two types also differ in hindbrain aromatase, with type I's having the lowest levels and type II's resembling females, which have higher levels (Schlinger et al. 1999). The actions of oxytocin family peptides on vocalization are also different. The vasotocin cells in the preoptic area of type II males resemble those of juveniles instead of type I males (Foran and Bass 1999). Vasotocin alters the neural output of the brain's vocalization system in type I males but isotocin is the effective peptide in type II males and females (Goodson and Bass 2000). The two categories of males begin to diverge morphologically well before sexual maturity and no spontaneous changes from one type to the other have been observed.

How would these facts about *Porichthys* be interpreted according to Moore's model? The early type establishment and absence of spontaneous change point to fixed, not plastic types, with gonads (specifically, high expression of the gene for the enzyme that produces 11-ketotestosterone) and body size organized early. Thus far the evidence points to sonic muscle size as a strictly activational phenomenon, because it can be type-reversed in adults by hormone treatment. Presumably it doesn't spontaneously reverse because the gonads are fixed and maintain their hormonal profiles. The fact that the hormonal profiles of the two male types are different does not fit Moore's model, however. How the behavior of the animals fits into the scheme is not yet clear. If behavior is like sonic muscles (is type reversed by 11-ketotestosterone manipulation), then the stability of adult behavioral type would arise not because the behavior was organized but because the gonads are organized. If such manipulation does not work, then an organizational hypothesis for the behavior is more likely.

Although genetic determination of male types is known to occur in teleosts and occurs in the ruffs, nothing is known yet about any genetic determination of the two male types in *Porichthys*. Why does it matter whether male types are genetic? In some theories of the evolution of male types, a relationship is predicted between the spatial and temporal structure of the physical and social environment, the presence or absence of predictive environmental cues, and the likelihood of evolving genetically fixed vs. developmentally plastic alternative male mating strategies (Sinervo 2001; Shuster and Wade 2003). An alternative to genetic polymorphism is a developmental switch (which in Moore's model would involve hormonal organization) that is thrown in a condition or environment dependent manner. The analog for sexes is environmental sex determination by the temperature or pH of the water. Tactic switches by individual adults are widely recognized to have this condition- or environment-dependent property, but strategies that arise through a permanent developmental process such as organization could as well. The HPG and HPA axes of vertebrates are functional in early life, providing one route from the outside world to the gonads and adrenals, and from there to an organizational hormone effect on adult phenotype. If the process is truly top-down rather than gonad-up, it would need to be shown that some brain mechanism like aromatase is responsive to the internal or external environment in very young fish.

The other teleost examples of male types that have been studied are predicted to be entirely activational according to Moore's model, because type change can occur in adulthood. As they go through their reproductive lives, male peacock blennies (*Salaria pavo*) change from sneakers that mimic female courtship behavior to larger nest-holders (Oliveira et al. 2001a). Treatment of sneakers with 11-ketotestosterone stopped the female mimicking behavior but didn't stimulate nest-holder behavior, providing partial support for an activational steroid hypothesis. As the researchers speculated, perhaps something else (another hormone?) is needed along with the 11-ketotestosterone, or more time is needed for the nest-holder behavior to appear. On the other hand, in another blenny (*Parablennius sanguinolentus*) treatment of sneakers with methyltestosterone increased their time in nests and the attraction of females to those nests, as if the sneakers were turning into nest holders, supporting an activational hypothesis (Oliveira et al. 2001b).

Let's look at the bluehead wrasses again, with their three male mating types. Initial- and terminal-phase males differ in appearance and (in the case of territorial terminal-phase males) in behavior as well. Territorial and nonterritorial terminal-phase males differ only in behavior. The transition from nonterritorial to territorial behavior occurs very fast. This dissociation between appearance (terminal-phase males always look like terminal-phase males regardless of which behavioral tactic they are pursuing) and behavior (terminal-phase males can do either, whereas the smaller initial-phase males are limited to the nonterritorial tactics) points to different activational mechanisms for appearance and behavior (Semsar et al. 2001). In the sex change experiments described in the previous section, the change in behavior didn't have anything to do with the gonads, whereas a change in appearance required gonads (Godwin et al. 1996).

There is some evidence that peptides might be mechanisms for this and other adult male type switches by some teleosts. Male types sometimes differ in peptide (GnRH or vasotocin) neuron phenotype or gene expression in the preoptic area or elsewhere in the brain (Grober and Bass 2002; Miranda et al. 2003). Territorial and nonterritorial terminal-phase male *Thalassoma* differ in levels of vasotocin mRNA in the brain, and vasotocin manipulations change their behavioral phenotype (Semsar et al. 2001). Thus, the change in appearance when initial-phase males become terminal-phase males is an activational effect of gonadal hormones, whereas the change in behavior as nonterritorial terminal-phase males assume territories is an activational effect of brain peptides.

Organizational hormone actions are best understood in mammals. Are there any cases of multiple male types in this lineage to which Moore's theory might apply? Males do vary, but usually along a continuum, rather than as discrete categories. In some group living primates with pronounced male dominance hierarchies, younger subordinate males may look less "masculine." This is probably socially mediated through mechanisms that inhibit the HPG axis (Dixson 1997). If so, it is an activational effect and a plastic, not fixed, phenotype, in agreement with Moore's categories. Orangutan males are either flanged (possess cheek flaps), sedentary, and territorial, or unflanged, mobile, and female-seeking (Utami et al. 2002). The unflanged males have low androgen and aggression levels, but have normal FSH levels, are fertile, and engage in sexual behavior (Maggioncalda et al. 1999). They are not simply males with an inhibited HPG axis or a very delayed puberty (Galdikas 1985). This is conceivably a genuine case of two male types with different mating strategies in a mammal. Males can change from unflanged to flanged, again suggesting a plastic and activational phenomenon. Thus far, there is no hint of any hormonal organization of male types in mammals. Fixed male types seem to be rare in both mammals and birds, the two groups where hormonal organization of behavior has been most clearly established. Is nature being ironic, or is she telling us something important?

Females are conspicuously absent from much of the literature on within-sex types even though female types occur in a number of arthropods (Shuster and Wade 2003). In addition to female castes in eusocial insects, there are types (phases) of both sexes of migratory locusts that are produced by different population densities, possibly mediated by maternal effects (Pener 1991). A number of teleost fish have more than one kind of female (Henson and Warner 1997). Subordinate naked mole rat females are markedly different from the "queen" due to reversible activational hormone effects (chapter 3). The side-blotched lizard *Uta stansburiana* has multiple morphs of both sexes that differ in throat color and reproductive strategies (Sinervo and Zamudio 2001). Females with orange throats have larger clutch sizes, suggesting a possible common hormonal basis of an organizational sort for throat color and clutch size. Both sexes of white-throated sparrows (*Zonotrichia albicollis*) occur as two morphs, tan or white, a system based on genetic polymorphism combined with disassortative mating (Tuttle 2003). White birds are more aggressive and sing more than tan birds of the same sex. Analyses of parentage suggest that the morphs are different solutions to reproductive trade-offs. The developmental basis of the behavioral differences between morphs is unknown, but white birds have larger song system nuclei than tan birds of the same sex, suggesting hormonal involvement (DeVoogd et al. 1995).

All told, Moore's theory provides an interesting way to think about within-sex types and continues to be a good source of testable hypotheses about their hormonal mechanisms. Given how scientific theories evolve and endure, it is not surprising that over a decade later it needs some modification. The part of the model that predicts that fixed types will have no adult hormonal dimorphism has not held up well, and was a weak prediction to begin with, because it predicts the null hypothesis (no significant difference). There is precious little evidence so far for any hormonal organization of any behavioral within-sex dimorphism in any vertebrate lineage. Is that because it hasn't been discovered yet, or because most hormone related behavior in lizards and teleost fish is an activational phenomenon? Rapid switching of male tactics doesn't find a home in Moore's theory (Oliveira et al. 2001a). There are two solutions. One is to decide that the concept of "types" doesn't include rapid tactic switching— that a theory of types is not intended to explain those. The other is to expand Moore's two category framework by adding a third category of fast-acting activation by peptides or by steroids acting through membrane receptors (Rhen and Crews 2002). Such a scheme would continue Moore's parallel between the developmental and temporal scale of type onset and duration and the nature of the mechanisms producing them.

There are still many mysteries about within-sex types. Type II male teleosts are in some respects feminized males except for their testes, although a good phylogeny is necessary to know which type is derived (Grober and Bass 2002). How do you get a male that is feminized for some characters but not others? Or is the concept of the feminized male a red herring? If we knew more about sexual differentiation in teleosts with two stable sexes and no fixed types, these questions would be easier to answer. Amazingly, sex determination and sexual differentiation are poorly understood even in the zebra fish (*Danio rerio*), the fruit fly of the vertebrate world (Corley-Smith et al. 1996). In addition to all the unanswered questions about mechanisms, there is plenty of unrealized potential for greater integration between function and mechanism. Whether and when animals evolve strategies and exercise tactics depends on the costs and benefits of the various options, and the costs depend in part on the mechanisms that make them happen.

Comparative Overview

Looking back over what has been discovered about hormones and the development of behavioral sexes and types inspires both "glass half full" and "glass half empty" moods. The good news is that there are well-supported organization/activation theories for eutherian mammals and for galliform birds that account for a large proportion of the behavioral differences between the sexes that have been studied, coupled with exciting progress identifying the neural underpinnings of steroid based sexual differentiation. In many vertebrates (including mammals and birds) sex differences in some behaviors are now known to be largely activational. Hormonal mechanisms acting during development to produce sex differences have been identified in *Xenopus* and probably soon will be for a few lizards. The sex of several teleosts can be readily reversed by experimental steroid treatment. There is a theory (albeit one in need of modification) for thinking about mechanisms producing multiple types, along with several important discoveries of steroid and peptide mechanisms likely to be responsible for sex or type change in lizards and teleosts. Further cause for celebration is the progress made in understanding sexual differentiation in crustaceans and in those workhorses of developmental biology, *C. elegans* and *Drosophila*.

The bad news is from two fronts. First, after an initial period of promise, zebra finch sexual differentiation continues to resist explanation in spite of its most striking outcome, a highly dimorphic brain. The entire songbird lineage, a speciose group of exceptional importance in organismal biology, is a puzzle beyond knowing that song dimorphism in some species is mainly activational. Second, there is a lot of diversity in patterns of sexual and type differentiation but without a higher order principle to make sense of it. There isn't even the right sampling of species (other than mammals) to know what the taxonomic distribution of patterns looks like. A more general principle would be beneficial for realizing the promise of a fuller integration of mechanism and function. How should the search for such a principle proceed?

As is often the case, the framing of the question biases the path taken. Because mammals are best understood, the mammalian pattern (eutherian version) has tended to be the stepping off point, producing an agenda devoted to testing the generality of the mammalian pattern among vertebrates. What if, however, mammals are the oddballs? Instead of possessing the most widespread pattern, what if they are the only lineage with gonad-up organization where the testis has an organizing (not just activational) role in producing behavioral sex differences? There is little evidence against this alternative. The mammalian testis-determining gene *Sry* arose recently, probably from an X-chromosome gene expressed mainly in the brain (Graves 2001). That raises the interesting possibility that the state ancestral to mammals was brain-directed sex determination. The reframed question then becomes, why did some behavior of mammals come to be organized by the testes? The diversity of patterns needs to be viewed in a phylogenetic framework to see what's ancestral, what's derived, and what other changes have been associated with origins of new patterns. That's why the fact that so few major vertebrate lineages have been studied at all is so problematic.

Insects, most of which have cell autonomous sexual differentiation, suggest the need to understand why some animals use this rule exclusively while others (both invertebrates and vertebrates) use hormonal mechanisms for some or all of their bodies and brains. The usual suspicion about insects is that they are so tiny and develop so fast that there isn't developmental time for regulatory networks involving slow-acting hormones. Such a hypothesis implies that the insect method is derived. Perhaps cell autonomous differentiation evolved multiple times in invertebrates, whenever miniaturization occurred (as in *C. elegans*). A few vertebrates such as gobies and some salamander lineages are unusually tiny (miniaturized) and conceivably have gone down a similar path.

When the research with birds is placed in a phylogenetic framework, it is immediately apparent that the two large branches where there is experimental evidence for ovarian organization—galliforms, where both morphology and behavior have been studied, and ducks (anseriforms), where only morphology has been studied—are relatively basal lineages, whereas the zebra finch belongs to one of the relatively apical lineages. Furthermore, galliforms and anseriforms are sister groups and could reflect a single origin of a sexual differentiation pattern. At any rate, the quail pattern is more likely to be ancestral than the zebra finch (non-) pattern. Pinpointing the place in the avian clade of the origin of the zebra finch pattern will require study of more avian lineages. With current knowledge there is no way to know whether it is representative of songbirds generally (in which case it might have arisen along with the telencephalic song system), representative only of estrildid finches, representative of nonseasonal opportunistically breeding birds (Wade 1999), or limited to a single species. The best evidence that a locally produced neurosteroid is a differentiation signal comes from the zebra finch (estradiol's effect on the HVC to RA connection) but it is not known whether this mechanism is unique to the song system, to the nidopallium, to the telencephalon, or to the avian brain. Sexual differentiation of behavior and brains needs to be studied in avian lineages such as the ratites (emus and their relatives, which have a phallus and some machinery for using it), hummingbirds (which have dimorphic courtship and song), dimorphic finches of the nonestrildid sort (for example, house finches or sparrows), and dimorphic songbirds that are not finches (such as red-winged blackbirds).

Viewing the teleost research in a phylogenetic framework reveals that gonadal sex reversal by experimental steroid treatment has been found in several lineages, including some that are relatively basal (cyprinids), apical (cichlids), and in between (salmonids), increasing confidence that this is a general teleost feature. Similarly, sex reversal of behavior in adults by hormone treatment has occurred in a broad array of species. Whether these are ancestral vertebrate states is unclear, however, because little is known about more basal vertebrates such as sharks. The phylogeny of the spontaneous sex changers (sequential hermaphrodites) studied so far suggests that this particular variety of sexual plasticity has arisen numerous times independently but mainly in relatively apical lineages (Helfman et al. 1997).

Such phylogenetic frameworks don't explain why different lineages have different patterns of sexual development, of course, but they do reveal how many times different patterns evolved and when. What are the features of the biology of animals that could be relevant to the search for a more general principle about why and when? There are several possibilities but few if any have been subjected to a comparative test, mainly because the necessary information about sexual differentiation isn't there to do one. As is usually the case with complex phenomena, more than one hypothesis might be right, or might be the same root hypothesis but expressed in the language of a different level of biological organization.

One hypothesis connects the pattern of organization to oviparity and viviparity. Mammals have both internal gestation and testicular organization. It has long been speculated that this association reflects the need to avoid feminizing all the male fetuses by the mother's hormones. In the years since then it has been recognized that other viviparous vertebrate embryos are also in intimate physiological contact with the mother and also that embryos developing in yolked eggs are not altogether sheltered from the mother's hormones (Moore and Lindzey 1992). Viviparity has arisen multiple times in vertebrate evolution, but nothing is known about sexual differentiation in most viviparous groups to see if testicular organization is more common in viviparous lineages. The relevant groups would be those whose dimorphism is laid down prior to birth. The aptly named lizard *Lacerta vivipara* has male embryos that are very easily feminized by experimental exposure to estrogens; this and other evidence from this genus points to ovarian organization (Dufaure 1966; Adkins-Regan 1981). That contradicts the hypothesis as applied to vertebrates generally, although it could still be relevant to the evolution of the mammalian lineage.

The ovaries of birds grow faster earlier in development than the testes and produce higher steroid levels, whereas the testes of mammals grow faster earlier and produce higher steroid levels (Mittwoch 1998). This suggests the hypothesis that vertebrates with adult ovaries that are larger than adult testes have ovarian organization (ovarian primacy), whereas those with adult testes that are larger than ovaries (such as mammals) have testicular organization (testicular primacy). If true, this would provide an interesting link to mating systems, because testis size reflects the degree of sperm competition faced by males. Good candidates for ovarian primacy would be oviparous species that have been selected for high fecundity.

Overall bodily growth could be part of the picture, for example, whether growth is indeterminate (continues throughout life, as in plenty of fishes, snakes, and crocodiles) or determinate (slows down or stops altogether at the onset of reproductive maturity). Indeterminate growth is an essential precondition for the evolution of adult sex change in most models. Moore's (1991) model for within-sex types might work better for animals with relatively determinate growth. Animals face trade-offs between allocating energy to growth versus reproduction (chapter 6). The solutions to this trade-off seem to have a lot to do with the evolution of male types in teleosts. Territorial male *Thalassoma* defend a mating site and thereby achieve more matings but have smaller gonads and don't grow. Initial-phase (nonterritorial) males engage in fewer matings but have larger gonads and grow fast. In species with temperature sex determination, the temperature during early development has a lasting impact on growth rates and adult body size. Putting growth and size in the foreground also raises questions about when (in what kinds of animals) adult size is subject to hormonal organization. In garter snakes, there is some evidence that the early hormonal environment might influence later growth rates and contribute to the adult size dimorphism in which females are larger (as in most animals other than birds and mammals) (Lerner and Mason 2001).

Different parts of the animal that are not serially homologous (gonads, posterior anatomy, head ornaments, brains) undergo sexual differentiation. Another hypothesis would be that homologs will be likely to follow a similar developmental rule, whereas nonhomologs are "free" to use different rules (Wade 1999). The difference between body size dimorphism and anatomical dimorphism in birds and mammals is an example. In the case of the wallabies, much of the male and female reproductive anatomy does follow the Jost model. It's the pouch and scrotum that don't. The marsupial scrotum is the homolog of the pouch, not the homolog of the eutherian scrotum (Renfree et al. 1995). The gonads of zebra finches are masculinized by early treatment with fadrozole just as those of chickens are. It's their behavior that seems to follow different rules. Zebra finch singing has telencephalic components that have no homologs in galliforms. The brain-directed adult sex reversal of teleosts is coming from the preoptic area, also a part of the telencephalon. The insight of Phoenix et al. (1959) was that behavior (brains) might differentiate like bodies, but when brains (or parts of them) don't, we could seek an explanation in terms of the different developmental mechanisms for different brain parts.

Other speculations and hypotheses are specific to one particular group of vertebrates, such as songbirds, although they too might eventually lead to a general principle. The zebra finch puzzle has been downright inspirational. Robert Goy once commented, tongue in cheek, that it must be occurring through divine intervention. Kendrick and Schlinger (1996) distinguished between sexual and other social behavior, proposing that social behavior like singing, which is sexually selected, follows different rules than sexual traits like copulation, which are naturally selected. Since then, however, there have been two experiments in which early treatment with fadrozole did not result in mounting by females (Adkins-Regan et al. 1996; Wade et al. 1999), casting doubt on whether we know the rule for zebra finch copulation.

In the research with mammals it doesn't seem to matter if the behavior is sexual or social, so why would it matter for nonmammals? Copulation is based in basic brainstem mechanisms shared with other vertebrates, which might be why it responds to early estradiol the way it does in quail (elimination in males). If brain division is a predictor, then brainstem components of vocalization should follow the copulation rule (whatever it is) rather than the telencephalic song system rule. The syrinx, syringeal muscles, and hindbrain motor nucleus for the syringeal muscles (nXIIts) are sexually dimorphic in size in zebra finches, but effects of early hormone manipulations are just as confusing as they are for the telencephalic part of the song system, because steroids produce alterations but steroid antagonists or synthesis inhibitors are without effect (Wade et al. 2002).

Nottebohm (1999) proposed that open-ended song learners (those that continue to learn new songs as adults, such as canaries) have song system dimorphism that is produced by adult activational effects of hormones, whereas closed learners such as zebra finches require hormonal organization in addition. Although adult testosterone treatment does largely eliminate the sex difference in singing in budgerigars, starlings, and canaries (all open-ended), it also does so in white-crowned sparrows and several other closed learners (see Table 1 in Balthazart and Adkins-Regan 2002).

Teleosts inspire hypotheses about why extreme forms of sexual plasticity seem to occur more frequently in this than other vertebrate groups. It's not just a matter of numbers; birds are the next most speciose group of vertebrates yet sex change as a normal reproductive strategy is absent. There are several features of teleosts that are suspected to make it easier or more advantageous (better cost/benefit ratio) to head down a path toward adult sex change or within-sex types. First, many teleosts don't have the kind of elaborate sexual "plumbing" that other vertebrates tend to have. Species that spawn masses of eggs and sperm are not very dimorphic in their reproductive anatomy (or behavior). Some of the best evidence for hormonal organization of anatomy and behavior comes from the viviparous internally fertilizing highly dimorphic *Gambusia affinis*. Second, all vertebrate gonads are bipotential, but teleost gonads seem to remain bipotential until much later in development (Yamamoto 1969). Their gonads can be viewed as sequentially hermaphroditic to some extent, starting as predominantly ovarian tissue. Protogyny is then similar to development of testes in males of "regular" species. It simply happens later. Shapiro (1992) expressed a similar idea and provided some evidence that in protandry, the same process occurs but still later ovarian tissue predominates once again. The question then becomes, why do teleosts have more hermaphroditic gonads than other fishes? A problem with this hypothesis is that frogs also retain bipotential gonads until late in development yet most have stable sex. Third, teleosts retain bipotential brains in adulthood along with their bipotential gonads, and behavioral dimorphism is created largely by activational hormone actions (Francis 1992; Shapiro 1992). But so do some other vertebrates that are not sexually plastic. Fourth, the control of the pituitary by the brain occurs through a different mechanism in teleosts than in other vertebrates, including other fishes (chapter 1). The anterior pituitary is directly innervated by neurons from the preoptic area. It is not obvious why this mechanism would make brain-directed gonadal sex change more likely to evolve compared with peptide regulation of the anterior pituitary, however. Fifth, teleosts make the nonaromatizable androgen 11-ketotestosterone, an activational mechanism underlying male type differences. But couldn't male types be created in other vertebrates by taking advantage of the dual pathways from testosterone, one through dihydrotestosterone and one through estradiol? Again, the association is tantalizing, but the mystery remains. Sixth, many teleosts have indeterminate growth and a great deal of adult neurogenesis, much more than birds and mammals (Zikopoulos et al. 2000). This might make it easier to evolve replacement of one sexual phenotype with another. This hypothesis assumes that behavioral sex change involves replacing something in the neural sense, which seems to contradict the bipotential brain concept, however. Perhaps hormonal organization that is limited to an early critical period is found only when neurogenesis and formation of connections between new neurons are mainly early developmental phenomena. It will be interesting to see how much adult neurogenesis there is in sexually stable vertebrates with indeterminate growth, such as snakes.

To the extent that any of these six teleost features are part of the story, they are, of course, not sufficient by themselves to account for the evolution of sexual plasticity in teleosts. Rather, they might have facilitated or accompanied such evolution in response to features of life such as high reproductive skew (Grober and Bass 2002). They might also help explain why high reproductive skew hasn't led to this kind of plasticity in other lineages. Framed this way, one can ask if hormonal organization has reduced the likelihood of within-sex types evolving in birds and mammals. One problem with all these teleost hypotheses is that they fail to explain why most of the plasticity in response to social conditions occurs in the Perciformes when other kinds of teleosts have these same putatively predisposing characters.

What can we conclude from this overview about patterns of sexual differentiation? The points made can be summarized in the following multipart hypothesis. Hormonal organization of adult behavior is more likely to occur when behavior and morphology (including gonad size) are notably dimorphic, when there is determinate growth, when there is developmental time for a regulatory hormonal mechanism, and when there is selection for developmental primacy of one gonad (ovary or testis) over the other. Arnold (2002) proposed that birds and mammals use some kind of hormonal organization because they need tight correlations between all the male or female characters (including a fixed relationship between gonadal sex and the sexual phenotype), because with hormonal organization it is easy to evolve changes in sexual dimorphism by simply changing steroid receptor expression, and because it is too expensive to create all the dimorphism through direct genetic means. Although it is unclear why this wouldn't apply to many other vertebrates and invertebrates, it is an interesting way to think about the evolution of developmental mechanisms for sexual dimorphism, one that is explored further in the next chapter.

5 Evolutionary Change and Species Differences

In previous chapters the relationships between steroids and social behavior have been treated in a largely static manner, as if species are discrete categories eternally frozen in time with no history prior to their own ontogenies. This chapter brings into the picture the evolution of phenotypes over relatively short timescales, as in speciation or population divergence. Large-scale evolution (comparisons of entire lineages to understand divergence and convergence) will be taken up in chapter 7.

Several questions are addressed: How does evolutionary change occur in hormones and in hormone-based behavior? How do different kinds and degrees of sexual dimorphism in behavior evolve, and how do hormones alter this process? How can a consideration of hormonal mechanisms help bridge the gap between genes and social behavior in understanding behavioral evolution? What conceptual frameworks can be used to think about evolutionary change in hormones and behavior and to generate questions and hypotheses? Contemporary interpretations of the basic Darwinian "equation" are essential for any evolutionary thinking and this chapter will begin there.

Heritable Phenotypic Variation: Individual Differences and Their Basis

Evolutionary change can occur when selection acts on heritable phenotypic variation. Some variants are more fit than others—they survive and reproduce more and sooner, leaving more descendants than others do. Some degree of phenotypic variation is a universal feature of organisms. Wherever anyone has looked, there is plenty of behavioral, neural, and physiological diversity among individuals and populations of the same species (Lott 1991; Crews 1998; Ishikawa et al. 1999; Spicer and Gaston 1999). Let's look at the nature and magnitude of this variation, its genetic basis, and its heritability. Then let's ask whether steroids are a mechanistic basis of individual differences in social behavior.

First some clarification about this word "heritability." The essence of heritability is the degree to which the trait has a genetic basis so that it will be passed on to the offspring and respond to selection. The statistic "heritability" in the broad sense, or h^2, is the proportion of the total trait variation in the population in that particular environment that is due to additive genetic variation, or V_G/V_P. Because it is a proportion, it ranges from 0 to 1. In practice heritability is estimated in one of three ways. The most frequently encountered method in population genetics is pedigree analysis, especially comparisons of offspring with their parents. Heritability is the slope of the least-squares regression of offspring against parental values. The second method, which is common in research with domestic or laboratory animals, is determination of realized heritability, the response (R) to an artificial selection differential (S). The equation $R = h^2 S$ yields h^2. A third method is used in species such as humans and cattle that produce both monozygotic (MZ, "identical") and dizygotic (DZ, "fraternal") twins. MZ twins should be more similar than DZ twins if the trait has a genetic basis. The difference between the MZ twins correlation and the DZ twins correlation, when doubled, is an estimate of h^2. The first two methods yield similar h^2 values after a few generations (Roff 2001), but obviously pedigree analysis is easier to do in wild populations. Where there is internal gestation or parental care, it may be necessary to cross-foster or do embryo transfers to rule out parental effects. Also, for some traits the fact that h^2 reflects only additive genetic variation is a serious limitation.

Back to behavior. Social behavior is an unusually interesting collection of traits from a Darwinian perspective. The distribution of individual variation in social behavior can be either discrete and multimodal, as with bimodally distributed sex differences or within-sex types, or continuous and more or less normally distributed, as with the variation within one sex or one type (Brockmann 2001). Social acts such as mating or feeding young have immediate and obvious fitness consequences. But more than other behaviors with immediate fitness consequences, such as running into a burrow to avoid predation, the consequences of social behavior depend on what other conspecifics (the other parties) are doing and how they are reacting. Social behavior is subject to frequency-dependent selection, including frequency-dependent sexual selection. It's usually dynamic rather than static. Sexual selection leads to traits in one sex but not the other, even though they are the same species. With sexually selected traits that fit a handicap model, part of the point of the traits is how wasteful they are, not how efficiently they are tuned to the physical environment.

These same interesting properties of social behaviors make them less than ideal traits for traditional genetic analysis. In addition, the evanescence of behavior makes it hard to capture in a single number compared to, say, body mass or limb length. As a result, many studies of the genetics and heritability of social behavior focus on well-defined motor acts such as calling or singing that can be quantified by counting the number of vocalizations produced in a given time period, or on scores (for example, latency to attack an intruder introduced into the home cage) obtained in a controlled test given at a standard age. Even the style or tone of an animal's engagement with the physical and social environment—its temperament or personality—can be captured in numbers provided there is good interobserver agreement about scoring.

No matter what the social behavior measure and how it is obtained, there are always individual differences. Where does this variation come from, and is it heritable? As with other traits, there are alternatives to a genetic basis such as maternal effects or diet. Sexually selected displays, like other sexually selected traits, are thought to be particularly variable, but to the extent that some of this is because of their sensitivity to condition (chapter 2), not all the variation would be heritable. In addition, behavior is famous for plasticity due to experiential influences, various forms of social and other learning, and tactic switching. Plasticity means sizable norms of reaction. Wherever environmentally induced phenotypic plasticity is this great, individual and population differences cannot be safely assumed to have a solely genetic basis, and responses to selection will be slow. Even genetically identical animals are quite variable in their behavior (Archer et al. 2003), a testimony to the probabilistic and epigenetic basis of development, which in mammals likely includes lasting consequences of interactions between littermates and with the mother prior to weaning.

Nonetheless, when the heritability of social behavior is examined carefully, the most common result is evidence for some genetic component or predisposition but with nongenetic factors also making an important contribution, as would be expected for any phenotypically plastic trait. In both laboratory animals and wild populations, a wide array of social behaviors, including mating preferences and behaviors known or suspected to be steroid based, have been found to have a significant genetic basis, with heritabilities typically in the 0.3 to 0.7 range for laboratory vertebrates (Greenwood et al. 1979; Endler 1986; Bakker and Pomiankowski 1995; Kölliker and Richner 2001; Plomin et al. 2001). Few of the responsible genes have been identified, and some (most?) of the relevant genetic variation probably concerns whole networks of genes, not single genes. *Mus musculus*, the house mouse, is the current focus of cutting-edge vertebrate behavior genetics and has also been the subject of much research on steroids and social behavior. These mice are territorial and aggressive toward intruders/strangers and therefore well suited to a goal of revealing the genetic basis of territorial aggression and its regulation by steroids. The aggressive behavior of both laboratory and wild populations is variable and heritable (van Oortmerssen and Bakker 1981; Benus et al. 1991).

Measures of dominance have also been found to be heritable in several species, including Japanese quail (Boag 1982; Nol et al. 1996). This might seem odd, because dominance is based on group dynamics and dyadic interactions, rather than being a fixed property of an individual (Chase et al. 2002). That's true of most social behavior, however. To the extent that heritable physical or behavioral traits like size, strength, or persistence contribute at all to outcomes, then dominance scores will be statistically heritable.

Hormones and brains also differ between individuals and populations. Some of this variation is heritable (Shire 1976; Kroodsma and Canady 1985; Airey et al. 2000; Finch and Kirkwood 2000; Williams 2000). Counts of neurons in well-defined brain nuclei are quite variable. Circulating steroid levels of adult individuals of the same sex and breeding state can vary by an order of magnitude even when assays are used that introduce little variability. Spratt et al. (1988) found a huge range of testosterone levels in medically normal men living in Boston. Standard errors of group means of steroid levels of birds are notoriously variable even when the birds are captive and singly housed. There is more than enough variation to provide the raw material for selection. But is any of this hormonal variation genetic? Steroids themselves, like behavior, have properties that are important for the animal but poor for genetic analysis (Garland and Carter 1994). Both glucocorticoid and sex steroid levels fluctuate markedly, depending on a whole host of internal and external factors, including diet and condition, time (of day, year, of ovulatory cycle), and the behavior the animal has engaged in recently. The concept of "baseline" or "basal" levels is an attempt to capture steroids as traits independent of recent events. But even baseline levels sometimes have fairly poor repeatability, especially in wild animals where there is nearly always something important going on and in small-bodied animals with fast steroid clearance rates, such as birds. To be heritable and respond to selection, the between-individual variation would have to be greater than the within-individual variation. It is hardly surprising then that there are very few studies of the heritability of circulating steroid levels in animals other than primates and large mammals of economic importance.

Studies comparing human MZ and DZ twins suggest that baseline glucocorticoids and production rates of testosterone and dihydrotestosterone are heritable (Inglis et al. 1999; Meikle et al. 1988; Bartels et al. 2003). There are also marked differences between human populations in putatively baseline testosterone and glucocorticoid levels, but because of the well-established influence of diet on steroid levels this cannot be assumed to be genetic (Bribiescas 2001). In tropical breeds of cattle in northern Australia heritabilities of the testosterone response to GnRH stimulation were 0.52 (Davis 1993). Reproductive hormone levels of pigs had small to moderate heritabilities (Bates et al. 1986). Testis size in domestic sheep, cattle, and mice has a relatively high heritability but not all of testis size is relevant to hormone production (Land 1984).

In studies of adrenal responsiveness, both baseline glucocorticoids and the magnitude of the glucocorticoid elevation in response to a standardized threatening/challenging stimulus vary quite a bit between individuals (Orchinik 1998; Cockrem and Silverin 2002). As with sex steroids, however, some of this variation is secondary to nonheritable individual differences in condition or recent events (Smith et al. 1994; Schoech et al. 1997). In primates, adrenal responsiveness is a relatively stable characteristic that is correlated with behavioral measures of an important temperamental component of personality. In groups of free-ranging rhesus monkeys, some individuals are very shy and respond to environmental or social change and novelty with fear and withdrawal, whereas others are more relaxed and even curious. Timid monkeys have higher peak cortisol levels and a longer-lasting elevation to the change than relaxed monkeys; they are more reactive hormonally. Here it is plausible to hypothesize that there is a causal relationship between the two kinds of individual differences—that hormones (the HPA axis) might produce the individually distinctive temperaments. Both the behavioral and the hormonal responses are highly heritable, raising the interesting possibility that these bimodally distributed styles are strategies with different costs and benefits resulting from balanced polymorphism (Suomi 1991). Other animals, including birds, also have noticeable individual differences in a similar temperamental component of personality, and as in the monkeys both the behavior and the steroidal stress reactivity are heritable (Koolhaas et al. 1999; Mormède et al. 2002; Satterlee et al. 2002; Drent et al. 2003).

A number of single-gene mutations have been discovered in laboratory rodents and humans that affect steroids and their pathways of action, usually because the mutated gene fails to produce a critical steroidogenic enzyme or steroid receptor protein (Shire 1976; McGinnis et al. 2002). For example, mutation of the gene for the androgen receptor results in a reduced number or absence of androgen receptors and, in genetic males, a failure of normal sexual differentiation even though testes are present, a condition called testicular feminization (Tfm). These mutations are very important for discovering the steroid mechanisms necessary for normal function. Spontaneous mutations at these sites would be strongly selected against, and, indeed, these conditions are fairly rare. These gene loci are probably not relevant, however, to understanding the source of individual differences within the normal range, where everyone is producing all the steroids and receptors, but in differing amounts (Shrenker and Maxson 1983). Similarly, the development of "knockout" methods has been useful for testing hypotheses about the behavioral functions of specific gene products such as steroid receptor subtypes (Nelson 1997; Ogawa and Pfaff 2000; Scordalakes and Rissman 2003), but their "all-or-nothing" nature makes extrapolation to naturally occurring individual differences difficult. There are polymorphisms for steroid receptors in humans but their functional significance is not clear (Finch and Rose 1995).

The sexual behavior of laboratory rodents is individually variable and heritable, and the relation between steroids and sexual behavior is well understood for several species. Is there any evidence that individual differences in sexual behavior are due to individual differences in hormones? In a classic study, Grunt and Young (1952) found sizable and stable individual differences in the copulatory behavior of male guinea pigs in spite of standardized housing and testing conditions. Some were "duds" and others were "studs." Following castration and testosterone replacement (the same dosage for all the subjects), the "duds" were still "duds" and the "studs" were still "studs." In other words, the individual differences seemed to have a source other than deficiency or excess of circulating androgens. While it is conceivable that individual differences in hormone clearance rates meant that the "studs" actually had more circulating testosterone than the "duds," within the limitations of the techniques available at the time the hypothesis that individual differences in behavior were caused by individual differences in adult hormone levels was resoundingly rejected. As already discussed in chapter 2, once assays were developed that made it possible to measure circulating steroids accurately, it became clear that more often than not there is no significant correlation between circulating sex steroid level and measured sexual behavior once steroids are above a threshold level, a conclusion that holds not just for mammals but for other vertebrates (e. g., lizards [Crews 1998]).

In laboratory rodents there are strain, as well as individual, differences in endocrine characteristics and hormone–behavior relationships even though these were not a deliberate part of the artificial selection criteria that produced the strains (Shire 1976). McGill (1978) found that F1 male mice produced by crossing two strains continued to copulate for a year following castration. This retention of behavior in the F1s is not due to residual circulating steroids following castration (Clemens et al. 1988). In a comparison of multiple mouse strains along with F2 generation crosses, correlations between copulatory behavior and testosterone level were negative, not positive, both between strains and within the F2s (Batty 1978).

Does this research on individual differences in laboratory animals mean that sex steroids are not a basis of individual differences in social behavior in the "real world"? Not necessarily, because there are several hormone-related alternatives to consider. Under conditions in the wild, where life is more difficult, there might be individuals whose steroid levels have fallen below threshold temporarily, individuals who because of energetic costs of maintaining higher levels are physiologically "opting out" of social life until they achieve a better nutritional or health status. Also, different social behaviors might have different steroid thresholds (Wingfield et al. 1990). Individual differences in behavior and hormones might be more likely to be correlated where the threshold is high, where levels don't have to fall very far to drop below threshold.

Another possibility was recognized by Grunt and Young (1952). Perhaps what differs between individuals is not hormone level, but sensitivity of the body's tissues to hormones. W. C. Young was the head of the same research group that developed the concept of organizational hormone actions on behavior, and Grunt and Young saw such actions as a potential mechanism not just for determining differences between the sexes, but also for determining differences between individuals of the same sex. The subsequent discoveries of intracellular steroid receptors and steroid metabolizing enzymes in the nervous system provided a concrete instantiation of the concept of sensitivity. There is abundant evidence for hormonal organization of sex differences in adult steroid receptor and enzyme levels. Thus, the set of possible hypotheses about the sources of individual differences can be expanded to include the reception as well as production end of hormone action acting both early in development and in adulthood (Shrenker and Maxson 1983).

Do individual differences in steroid receptors or enzymes in the adult brain account for individual differences in behavior? Such mechanisms are a powerful explanation for some differences between sexes and seasons, but after a period of initial promise it is still not clear to what extent this accounts for individual differences within the same sex and breeding stage. It's not that the hypothesis has been clearly rejected, but rather that it is difficult to test so as to establish causation. For example, receptors and enzymes would have to be measured with a high degree of precision and only in the relevant neurons or brain regions. They would have to be manipulated somehow (but remain within the normal range) so as to assess the effect of the manipulation on variation in behavior. If the hypothesis does turn out to be wrong, we are left with two possibilities: (1) the explanation lies further downstream in the steroid-initiated cascade, in processes transpiring after steroids are metabolized and bind to receptors, possibly including regulation of peptides and their receptors), or (2) the explanation lies in cascades that don't include steroids.

The idea that steroid receptors might underly some kinds of strain differences in mouse aggression has received some correlational support. In a study by Simon and Whalen (1986), one strain (CD-1) responded to androgens but little to estrogens, one (CFW) responded mainly to estrogens, and the third (CF-1) responded to either. The strain differences in responsiveness to estrogens were correlated with estrogen receptor binding in the hypothalamic–preoptic–septal region. The mouse strain differences in retention of copulatory behavior following castration are not related to numbers of brain estrogen receptors, however (Clemens et al. 1988).

Chapter 4 asked whether hormonal organization accounted for multiple male types. Is there any evidence that it generates within-sex continuous variation? Again, this is a difficult hypothesis to test. One would have to measure individual differences in steroid levels during the critical period for organization of the behavior, then measure behavior in adulthood, then (if there was a correlation) equalize and exaggerate individuals' hormone levels during the critical period, and then see if individual differences are reduced and enlarged, respectively, in adulthood. But when animals are very small, as they often are early in development, measuring steroids may require sacrificing the animal so that later behavior cannot be assessed. Accurately manipulating steroid levels in tiny animals so that they stay within the normal range would be a challenge as well.

A good hint that the early steroid environment might be a source of individual differences in adult social behavior is the discovery, mentioned in previous chapters, of intrauterine position effects in litter-bearing house mice, rats, and gerbils. Fetuses positioned between two males in a uterine horn are slightly masculinized behaviorally as adults compared to fetuses of the same sex positioned between two females (vom Saal 1981; Drickamer 1996; Clark and Galef 1998; Ryan and Vandenbergh 2002). The relevance for evolution of this otherwise interesting phenomenon is not clear, however, because the variation thus produced would not be heritable. Female wild house mice that were between two males had larger home ranges than females between two females, but survival and reproductive success were not different (Zielinski et al. 1992).

Other hints come from the extensive literature on effects of early experience on later HPA reactivity and sexual characteristics in rats and monkeys (Suomi 1991; Ward et al. 1994; Meaney 2001; Levine and Mody 2003). A variety of manipulations of either the infants or the mother have a lasting influence on HPA reactivity of the offspring that reflects organizational effects of adrenal hormones, and early stress can alter sexual differentiation of behavior as well. These effects are likely to be one source of individual differences and to have consequences for the survival and reproductive success of the offspring. Important effects of maternal stress on their offspring have also been detected in fish (McCormick 1998). As with other kinds of maternal effects, the consequences for evolution are still unclear.

In humans there is substantial variation in fetal circulating sex steroid levels and also in preoptic area neuron number after birth. These are conceivably causally related (Swaab et al. 1992; Finch and Kirkwood 2000). Organizational effects of steroids are one of several current hypotheses to explain individual differences in sexual orientation in humans (Gooren 1990; Ellis and Ebertz 1997). The most common version is that homosexual men have been exposed to lower androgen levels than heterosexual men prenatally (the time of the critical period for sexual differentiation of human anatomy) and homosexual women have been exposed to higher androgen levels than heterosexual women. It has also been proposed that the source of these variations in prenatal steroid levels is maternal stress during pregnancy. So far such hypotheses have received only very limited support in humans. Male domestic sheep sometimes prefer other males sexually. Such rams have lower aromatase levels in the medial preoptic nucleus and fewer estrogen receptors in the amygdaloid complex (Perkins et al. 1995; Roselli et al. 2002). It is not yet known, however, whether these two traits (sexual preference and brain characters) are causally related, nor whether organizational hormone variations have produced either of them.

Leopard geckos (*Eublepharis macularius*) of same gonadal sex that hatch from eggs incubated at different temperatures differ in adult social behavior and in the size of steroid target brain areas such as the preoptic region and ventromedial hypothalamus (Crews 2000). For example, females produced at male-biased temperatures are less attractive to males than females produced at female-biased temperatures. But females from male-biased and female-biased temperatures are equally attractive to males if gonadectomized and given estradiol replacement, suggesting that these individual differences have an activational rather than organizational basis (Rhen and Crews 2000).

All told, the hypothesis that steroids or their mechanisms of action are a source of evolutionarily significant heritable variation in behavior remains just that, a hypothesis, and not an established conclusion. There is more than enough heritable variation in both the hormones and the behavior. It is still an open question, however, whether the same mechanisms that are clearly so important for creating differences between sexes and types account for the continuous variation seen within sexes and types (Crews 1998). There are plenty of other interesting questions to ask as well. Are organizational effects of steroids responsible for the higher coefficients of variation of sexually selected traits compared with other traits? Do hormones magnify individual differences, producing a bigger norm of reaction than they would otherwise have? But answers are in short supply.

Reproductive Success and Differential Fitness

In the end, it all comes down to how successfully and when, compared to other individuals, the animal has produced genetic offspring that are likely to reproduce themselves. Surviving to adulthood (which many don't) is not enough, for reproduction is fraught with perils, as when bluebirds in nestboxes routinely lose their first clutches to cold and rain. Fitness itself is heritable. Fitness is zero for many individuals, and so if the population is not declining, then a few individuals of both sexes are very successful indeed.

Success rates of gametes are also low. The overwhelming majority of sperm never encounter an egg, most ovarian follicles of birds and mammals never result in a fertilized egg, and in some species a significant percentage of fertilized eggs fail to produce a viable fetus, chick, or larva. An impressive array of complicated hormonal and other mechanisms underly successful reproduction, and so many things have to go right that it's amazing it ever works.

To what extent is fitness a function of steroid-based social behavior or steroids themselves? Clearly the answer depends enormously on the particular component of fitness and hormone. Differences in juvenile survival in a solitary species with high predator-induced mortality would be minimally related to sex steroids, but adrenal steroids and their behavioral effects could be quite important. Differences in annual reproductive success of adult territorial birds that are due mainly to whether the nest is discovered by a predator are also unlikely to be closely related to steroids. Instead it seems more likely that hormones would be related to such fitness components as mating success, fertility, and (in species with hormonally influenced parental behavior) offspring survival due to effective care.

Sex steroids are obviously essential for vertebrate fertility (production of mature gametes) and mating success because they are among the critical mechanisms for gametogenesis and mating behavior. But this has been the case for many millions of years. In thinking about their contribution to fitness in relation to evolutionary change on a more "micro" scale, what we want to know is (1) whether the natural range of variation in sex steroid levels in wild animals is correlated with variation in reproductive success, (2) what the function that relates them looks like, and (3) why this is the function. These questions really need to be answered in wild free-living populations, which is difficult, however. In some species, mating behavior is hard to observe and often goes undetected. Luckily it now seems that mating success is too poor a proxy for reproductive success to be worth the trouble, because of uncertainty of parentage due to extrapair matings, sperm storage, and sperm competition. Determining reproductive success requires the use of molecular methods to establish parentage, which can be achieved in some free-living populations. It remains difficult to keep track of reproductive success in mobile or cryptic animals. Offspring may disappear but without any way to know if they died, left the area, or are hiding nearby.

On theoretical grounds some predictions can be made about what the function relating a steroid like testosterone to individual differences in fitness should look like (fig. 5.1). As always we need to bear in mind that within individuals the relationship between steroids and behavior can be a step function (chapter 2). A simplistic view that testosterone equals maleness equals a good thing predicts a positive correlation over the entire normal testosterone range, resulting in strong directional selection. The theory that testosterone is an indicator of a male's quality because of its handicapping costs (chapter 2) predicts a function in which fitness is lowest at the lowest testosterone levels, then gradually rises (directional sexual selection over this part of the range) to a sharp peak followed by a drop off as the costs become too great, with a higher drop-off threshold for higher quality males. Stabilizing selection (selection against extremes but with a broad range of fit values), on the other hand, predicts an inverse U-function with a broad and level plateau (Feder 1987). Notice that whenever there are individual differences in thresholds (either at the low end, for hormonal activation, or at the high end, when a drop off occurs) averaging results from multiple individuals will mask the true shape of the functions, producing something more nearly resembling a graded function (Hews and Moore 1997) (see fig. 2.2C). According to the challenge hypothesis/

A **B** **C**

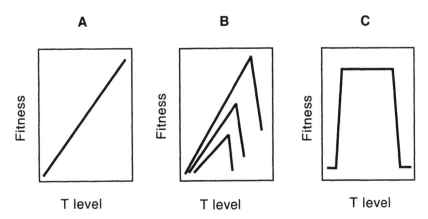

FIGURE 5.1. Three hypothetical functions relating testosterone (T) level and male fitness. In (A) more testosterone is always a good thing, and there is a dose-dependent relationship between hormone level and fitness. In (B) an individual male's fitness increases up to a point beyond which the costs of the testosterone cannot be borne. This point occurs at a higher T level for higher quality males. In (C) a broad range of hormone levels is fit and the extremes are selected against (Feder 1987).

trade-off framework discussed in chapter 3 the function should look different in monogamous vs. polygynous birds to the extent that necessity for male parental care also differs. Reproductive success will drop off at a lower androgen level in monogamous than polygynous species. Real data are messy and might not be sufficiently precise to discriminate between alternative functions.

In principle, one way to find out if variation in steroid levels is related to reproductive success would be to measure steroid levels of a number of individuals and then see how many offspring those individuals produce. It seems highly likely that we would find a strong positive correlation between steroid level and reproductive success if we looked at young animals just reaching sexual maturity or animals just coming into the breeding season, times when steroid levels are ramping up from a nonbreeding to a breeding state but are not yet at asymptote. Some individuals will be ahead of others. If the breeding season is short, late maturing young might miss out altogether on a chance to breed. Among adult seasonal breeders, it is often those that get an early start that are more successful. There is no doubt that there are important fitness consequences of this kind of hormonal variation that have been a powerful shaper of age at puberty and timing of breeding over both short- and long-term evolutionary scales.

Now let's ask if steroid levels predict reproductive success among adult animals in a breeding state who have reached their individual baseline steroid asymptotes. Very few researchers have had the fortitude and patience to do such difficult research. There are a few reports of positive correlations between male testosterone levels and display characteristics or mating success (for example, Borgia and Wingfield 1991; Alatalo et al. 1996) but mating success is not reproductive success. In studies with males of two species of territorial and polygynous blackbirds, levels of testosterone and corticosterone were positively correlated with eventual fledging success in red-winged blackbirds (*Agelaius phoeniceus*) but not in yellow-headed blackbirds (*Xanthocephalus xanthocephalus*) (Beletsky et al. 1989; Beletsky et al. 1990). In socially monogamous greylag geese (*Anser anser*), which have long-term pairbonds, fecal testosterone levels of pairbonded free-living males and females were positively correlated with each other; the greater the correlation, the greater the long-term reproductive success of the pair (Hirschenhauser et al. 1999).

Another approach is to experimentally elevate steroid level through implants and then determine the consequences for reproductive success (offspring produced) compared to controls with empty implants, an example of the use of phenotypic manipulations to test hypotheses about current fitness consequences (Sinervo and Basolo 1996; Wingfield and Silverin 2002). Such experiments have mainly been done in birds. In skilled hands the steroid implants produce levels within the high end of the normal range or extend the period of time over which males have their peak hormone levels without increasing the peak level. (In principle one could also decrease steroid action through implants containing steroid synthesis inhibitors or receptor blockers, but in practice it is hard to create animals operating at the low end of the normal range instead of well below it.) The first such experiment was carried out with pied flycatchers (*Ficedula hypoleuca*), a species in which some males are polygynous, defending a secondary territory where there is another female and nest (fig. 5.2) (Silverin 1980). Males with experimentally prolonged testosterone were not more likely to acquire a second female, and produced fewer fledged young on their territory than controls because they fed the young less. On the other hand, male sparrows (*Zonotrichia leucophrys* and *Melospiza melodia*) with elevated testosterone levels were more likely to acquire a second female mate and nest (Wingfield 1984). Male house sparrows (*Passer domesticus*) with elevated testosterone resembled the pied flycatchers, however, losing more of the chicks in their nests to starvation because they were poor parents (Hegner and Wingfield 1987). In contrast, testosterone seemed to benefit red grouse. Territories of testosterone-implanted males produced more chicks in a poor year (Moss et al. 1994). Such species differences in the effects of this manipulation are to be expected. Any depressant effect of testosterone on paternal care will reduce chick survival more in species where such care is obligatory (where the female cannot or will not compensate for its loss). Red grouse have precocial young with female-only parental care.

These studies assumed that offspring produced on a male's territory were his own, a reasonable assumption at the time but now known to be very risky. Raouf et al. (1997) implanted male dark-eyed juncos with testosterone but

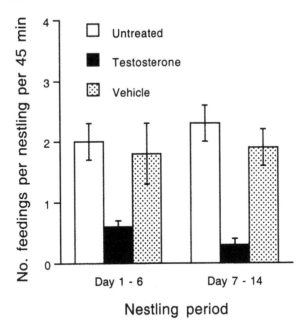

FIGURE 5.2. Male pied flycatchers given a long-acting injection of testosterone fed nestlings less than untreated or vehicle control males, both early (days 1–6) and later (days 7–14) in the nestling period. Testosterone-treated males also had fewer fledglings per hatched brood, and were the only males that sang at a high rate during the nestling period. This was the first experiment to show that experimentally elevating testosterone in free-living male birds decreases behavior related to parental effort while increasing behavior related to mating effort, a phenomenon that seems to be widespread (although not universal) in birds with biparental care. Redrawn from Silverin, B. 1980. Effects of long-acting testosterone treatment on free-living pied flycatchers, *Ficedula hypoleuca*, during the breeding period. *Anim. Behav.* 28: 906–912. © 1980 Ballière Tindall (Elsevier).

in addition applied DNA fingerprinting and paternity analysis to the entire population. Total (net) reproductive success of testosterone-implanted males was not different from control-implanted males. What differed was the proportion of their total arising from chicks in their own nest versus chicks in other nests. Testosterone-implanted males had fewer fledglings in their own nests and correspondingly more elsewhere. The authors interpret these results as indicating a shift in reproductive strategy from "at-home" chick care to flying afield to engage in extrapair copulations, and proposed that testosterone is a mechanistic basis for individual differences in the solution to a mixed reproductive strategy. Testosterone-implanted male starlings underwent a similar shift but without any reduction in fledging success "at home," possibly because their paternal care is not essential (De Ridder et al. 2000).

In these studies annual reproductive success was the outcome measured rather than lifetime reproductive success. If testosterone increases annual success when first administered but reduces subsequent survival or reproductive success, that greatly changes the picture. There have been a few attempts to determine survival through return rates, with mixed results (Wingfield 1984; Saino and Møller 1995; Ketterson et al. 1996; Casto et al. 2001). Reduced survival in hormone-implanted animals is difficult to interpret if the implants are not removed at the end of breeding and/or if the implants interfere with molt (Nolan et al. 1992). In the red grouse, the breeding density drops as the treated males expand their territories, producing a local population decline (Mougeot et al. 2003). Because what matters is differential (relative) fitness, the treated males may be doing even better than it would seem just from their offspring production. On the other hand, if a critical population size is needed for individuals to continue function normally, even the high-testosterone males' genes might plummet to extinction.

What can we conclude then about the function relating testosterone levels of male birds to reproductive success? The few studies suggest that in biparental species experimentally elevated testosterone either leaves reproductive success unchanged or reduces it. In species with female-only care, testosterone might enhance reproductive success, but there are no studies showing this in which genetic paternity was determined. Male house sparrows with flutamide implants did as well as controls, and so there is no evidence yet that males at the low end of the testosterone action range are at a disadvantage (Hegner and Wingfield 1987). It will be hard to tell what the function is relating testosterone and reproductive success without providing more than one dosage. The results with the paired geese raise the exciting possibility that what matters in biparental birds with long-term pairbonds is not so much either party's absolute levels but the match (synchrony) between them.

Steroid levels of females in relation to reproductive success have been examined in a few species (mainly domestic or captive animals) in addition to the pairbonded geese above. Sheep breeds that tend to ovulate two or three eggs have higher circulating estradiol levels than breeds that ovulate one egg (Cahill et al. 1981). In captive female black-footed ferrets (*Mustela nigripes*) fecal estradiol was positively correlated with litter size (Young et al. 2001). Such correlations between mammalian litter size and estrogen make sense on mechanistic grounds, because the developing follicles are a major source of estradiol and estradiol exerts positive feedback on the LH surge that triggers ovulation. Presumably the difficulty that "real-world" females would have in gestating and lactating for a greater number of young sets a limit to how much of a good thing more follicles producing estrogen can be. Female laboratory mice with experimentally elevated estrogen lose more fetuses during gestation (Mahendroo et al. 1997). In pigs treatment of pregnant females with progesterone combined with a dose of estradiol lower than the effective estrous dosage (1.25 µg) increased litter size but the same combination with a higher dose of estradiol (18.75 µg) decreased it (De Sa et al. 1981). (Notice how small these amounts are for such a large animal.) In female starlings supplemental estrogen increased neither clutch size nor egg mass, contrary to expectation (Christians and Williams 1999), and in female canaries variation in fecal estradiol and progesterone during laying was not related to clutch size (Sockman and Schwabl 1999).

Thus, the very limited information suggests that in female mammals intermediate levels of estradiol might be best for one component of reproductive success, litter size, a function suggesting stabilizing selection. Little is known about hormones other than estrogens in females or, for that matter, hormones other than testosterone in males. In two of the rare exceptions, experimental elevation of testosterone in free-living female dark-eyed juncos increased the time between completion of the nest and appearance of the first egg, a likely negative fitness effect (Clotfelter et al. 2004), and experimental administration of prolactin to free-living female willow ptarmigan increased the number of surviving chicks one week after hatching (Pedersen 1989), a positive effect. It would be interesting to know how circulating estrogen is related to reproductive success in male horses and pigs, which have high circulating estrogen, and how circulating androgens are related to reproductive success in females of species that have high levels or that compete for males or nest sites. There might be stabilizing selection for androgens in female birds if higher testosterone inhibits egg laying, as it does in birds (Brahmakshatriya et al. 1969; Searcy 1988; Clotfelter et al. 2004). The functions might be very different for female vertebrates with different reproductive modes.

Responses to Selection

For many organisms, including vertebrates, only a small percentage of individuals ever survive and successfully reproduce. Intense selection is occurring on every generation. It is sobering to realize that in a wide array of animals (best documented in birds and humans) most individuals in a population have no remaining descendants at all five generations later (Partridge 1989). Even though selection is powerful stuff, changes in traits because of selection would not always be expected. If both high and low trait values result in low fitness (stabilizing selection), the trait will not change. Sexually selected traits, on the other hand, are thought to be subject to directional selection and to change relatively rapidly, at least up to the point where the costs become so great as to put a stop to the process. Lists of male traits under sexual selection include behaviors known or suspected to be hormone based (Andersson 1994).

Studies of selection for behavioral traits in wild animals generally find evidence for weak directional selection for most traits but strong selection for a few (Endler 1986; Kingsolver et al. 2001). By comparing closely related congeners, Civetta and Singh (1998) found evidence for fairly strong directional selection on genes involved in mating behavior in *Drosophila* and *Caenorhabditis*. More relevant to steroids is the evidence of positive directional selection at the locus for the alpha subunit of the androgen-binding protein in mice of the *Mus* genus (Karn and Nachman 1999).

Much of what is known about selection in relation to hormones and behavior has come from artificial selection studies, some of which start with genetically wild animals. While artificial selection in laboratory environments, where many of the normal trade-offs are absent, can't be extrapolated to real-world evolutionary change, nonetheless responses to artificial selection produce valuable insights (beyond just determining heritability) and excellent opportunities to discover what else "comes along for the ride" that might limit responses to selection in the wild.

Populations of wild white-footed deermice (*Peromyscus leucopus*) in the United States show interesting variation in seasonality and photoperiodic responses that are very likely adaptations to climate and food supply. Connecticut populations are photoperiodic and Georgia populations are not photoperiodic. Laboratory bred mice from the two populations retain these traits (Gram et al. 1982). A Michigan population was variable and responded after only one generation of artificial selection for photoperiodic responsiveness to yield a high line with 80% responders and a low line with 80% nonresponders (Desjardins et al. 1986). Evidently photoperiodic responsiveness can change rather quickly. The same conclusion is also emerging from studies of birds that have large geographical ranges, marked population differences in photoperiodic thresholds for breeding, and, in some cases, shifts in range accompanied by changes in breeding date onset that have occurred in historical times (Berthold 1996).

Both wild and domestic *Mus musculus* have been artificially selected for high and low aggressiveness. Mouse aggression responds rapidly to selection, producing high and low lines in only a few generations (van Oortmerssen and Bakker 1981; Cairns et al. 1990). High and low lines differ in several features of the HPA axis as well as in behavior (Veenema et al. 2003). The aggressive behavior of male mice is steroid dependent, but thus far attempts to determine whether hormonal mechanisms differ in high and low lines of wild mice, and specifically whether organizational actions of early testosterone exposure determine the line differences, have not led to a clear outcome (Compaan et al. 1992; Miczek et al. 2001).

In rare and valuable selection studies in a genetically wild bird, great tits (*Parus major*) have been selected for their response to novel objects (slow versus fast exploration) in tests conducted when the birds are about 40 days old (fig. 5.3) (Drent et al. 2003). This is a measure of the coping style component of personality (reactive and timid versus proactive and relaxed) referred to earlier. When males from the slow and fast lines experienced staged confrontations with aggressive resident males, the fecal corticosteroid metabolites of the slow

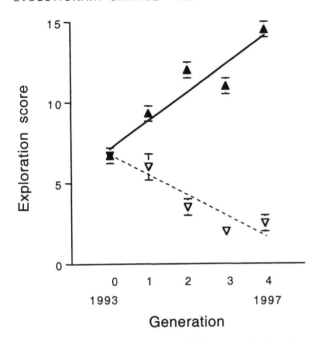

FIGURE 5.3. Response of great tits to artificial selection for juvenile exploratory behavior across four generations of selection from 1993 to 1997. Up-selection (filled triangles) and down-selection (open inverted triangles) quickly led to divergent responses, showing that this behavior has significant heritability. The lines are linear regression lines. Redrawn from Drent et al. (2003). © 2003 The Royal Society.

males showed a greater rise after the confrontation than those of the fast males, who showed little response at all (Carere et al. 2003). Selection for personality style brought along with it increased HPA reactivity. As in the monkey research, it will be interesting to find out what the direction of causation is here. Does increased HPA reactivity cause behavioral reactivity and timidity or vice versa? Or do both result from selection on a third factor such as CRF?

The most extensive and systematic program of artificial selection for sexual behavior was carried out by Siegel and colleagues with Japanese quail. This is an ideal domestic species for such research because egg production is high, time from hatching to sexual maturity (6 weeks) is unusually short for a bird, and sexual behavior is quite dependent on gonadal sex steroids. Male quail were selected for forty generations (Yang et al. 1998). The selection criterion was the cumulative number of completed matings (CNCM) made by the males in eight 8-minute observation periods during a four-week period in young adulthood. After forty generations mean CNCMs were 59.4 for the high line and 2.8 for the low line, a huge difference. The realized heritability was low (0.06 to 0.09), however, as if nonadditive genetic variance were accounting for much of the marked divergence between the lines. A subsequent study in which the two lines were crossed confirmed the sizable nonadditive genetic contribution (Yang et al. 1999). When selection was relaxed, CNCM in the high line rapidly decreased to its original level, showing that high CNCM is negatively correlated with fitness in the domestication environment of ordinary artificial selection. In fact, one high line had gone "extinct" before forty generations. With relaxed selection, the low line stayed low, suggesting that genes predisposing to low CNCM had gone to fixation.

Male Japanese quail have a proctodeal gland, a large androgen-sensitive external structure that produces foam that increases a male's fertilizing success (Cheng et al. 1989; Adkins-Regan 1999a). The size of this gland also responded to selection for CNCM (Yang et al. 1998). In a similar artificial selection study with chickens, crowing frequency was a correlated response to selection for mating frequency; it was higher in the high mating frequency line (Benoff and Siegel 1977). Fertility was not a correlated response, however; in the chickens the lines did not differ in fertility either in natural or artificial insemination matings (Bernon and Siegel 1981). Nor did selection for early versus late sexual maturity in chickens result in any correlated change in mating frequency (Craig et al. 1977).

The high and low chicken and quail lines did not differ in circulating levels of the three major androgens or in aromatase activity in the brain (Cohen-Parsons et al. 1983; De Santo et al. 1983). Birds from high and low mating frequency lines retained the phenotype even after castration and testosterone replacement (McCollom et al. 1971; Cunningham et al. 1977). The same results occurred when chickens were selected for high and low aggressiveness and then tested following castration and testosterone replacement (Ortman and Craig 1968). These comparisons between artifically selected lines of birds lead to the same conclusion as Grunt and Young's (1952) study of individual differences in male guinea pigs. High and low mating frequency males differ not in hormone level, but in some mechanism related to responsiveness to hormones. What the mechanism is, and where it is in the cascade of events in the brain, is still unknown, but progress should be facilitated when the chicken genome is sequenced. The correlated increases in proctodeal gland size and crowing suggest some mechanism closely related to androgen, not estrogen, action.

Japanese quail have also been selected for high and low levels of fear defined by tonic immobility ("playing dead") to a standard stimulus when tested as young chicks. Correlated responses included not only larger corticosterone responses to a novel object in the high line (which is not unexpected) but also better copulatory performance as adults in the males of the high line (Burns et al. 1998; Faure and Mills 1998). Several studies have successfully used HPA reactivity itself as the selection criterion. In a different research program quail were selected for low corticosterone responsiveness to brief restraint stress. Correlated responses include earlier puberty, larger testes and proctodeal glands, and faster and more efficient copulation (Satterlee et al. 2002; Marin and Satterlee 2003). This negatively correlated response (higher sex steroid related measures with lower adrenal steroid measures) to selection across generations reminds one of the widespread tendency for HPA activation to depress the HPG axis on a within-individual basis in vertebrates (chapter 1). Notice that reproductive performance was correlated in opposite directions, depending on the stress stimulus and on whether the selection criterion was fear behavior or corticosterone level. In chickens, artificial selection for a high or low corticosterone response to social stress works, in that the lines diverge (Gross and Siegel 1985). Similarly, the magnitude of the cortisol response to stress is heritable in rainbow trout and artificial selection yields high and low responding lines (Pottinger and Carrick 2001).

What happens if sex steroids are the basis for the selection? Selection for continuous egg laying (which would also select for the relatively high estrogen and relatively low prolactin levels that are associated with laying) has eliminated incubation behavior from some chicken breeds. Experiments in which sex steroids have been measured and then used as the selection criterion are rather few, however. In pigs, selection for high and low levels of the volatile androgenic steroid (5α)-androst-16-en-3-one (the boar pheromone) yielded a heritability of 0.56; correlated responses included testosterone and conjugated estrogens and age of puberty in females, which occurred later in the low line (Willeke et al. 1987). In pigs selected for high or low cholesterol (the precursor of steroids), the low cholesterol line, paradoxically, ovulated more ova and had a larger litter size. The lines also differed in progesterone level, with the high line having higher progesterone (Wise et al. 1993).

As Darwin (1859) recognized, domestication is artificial selection carried out with varying degrees of formality over many generations to produce animals that will have characteristics valued by humans and high fitness in human created environments. Basic features of social organization and mating systems have been altered in some species (for example, dogs) but not others (for example, pigeons). Comparisons between domesticated animals and their wild counterparts have proven to be quite interesting from a behavioral endocrinology standpoint. As Richter and Leopold first systematically described in rats and turkeys, domestic mammals and birds have larger gonads but smaller adrenals and brains than wild conspecifics, are tamer (less wild in temperament), less aggressive, more sociable (better able to live in dense groups), have an earlier puberty, higher fertility (larger and more frequent litters or clutches, less seasonality), and are more sexually active (Leopold 1944; Richter 1949; Bronson and Rissman 1986). Leopold's (1944) studies of turkeys showed that the wild phenotype has higher fitness in the wild but lower fitness in captivity. In more recent studies, domestic dogs were found to have larger testes than wolves even taking allometry into account, and domestic ducks had higher testosterone levels than wild ducks even though LH levels were similar (Haase

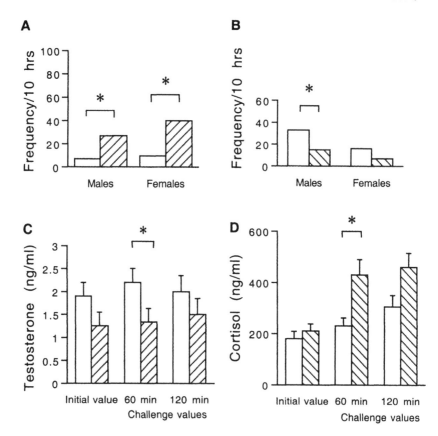

FIGURE 5.4. In a comparative study, domestic (clear bars) and wild (diagonally striped bars) guinea pigs differed both behaviorally and hormonally. Wild guinea pigs of both sexes were more aggressive (A), and wild males showed less positive social behavior such as social grooming (B). Behavioral values shown in (A) and (B) are medians. When placed singly into a unfamiliar cage, wild males had lower testosterone levels than domestic males (C), and wild but not domestic males showed an acute elevation in cortisol 60 minutes later (D). Similar differences between wild and domestic forms are seen in most other species that have been studied. Redrawn from Künzl, C. and Sachser, N. 1999. The behavioral endocrinology of domestication: a comparison between the domestic guinea pig (*Cavia aperea* f. *porcellus*) and its wild ancestor, the cavy (*Cavia aperea*). *Horm. Behav.* 35: 28–37. © 1999 Academic Press (Elsevier).

and Donham 1980; Haase 2000). Domestic guinea pigs (*Cavia aperea*, formerly *porcellus*) have slightly higher testosterone levels than genetically wild cavies, and have similar basal cortisol levels but lower cortisol reactivity to a novel environment (fig. 5.4) (Künzl and Sachser 1999).

In most cases, domestication has been haphazard, uncontrolled, and undocumented selection, not scientific study. One exception is a long-term domestication study of foxes (*Vulpa vulpa*) (Trut 1999). Over a period of decades, silver foxes were systemically bred for tamability, defined as the response to humans (friendly approach rather than fearful withdrawal) in controlled tests conducted at a particular age. The experiment succeeded, in that the strain is now tamable and rather dog-like, as if the original domestication of the wolf has been repeated in another canid species. This behavioral change has been accompanied by a delay in the onset of the juvenile increase in corticosteroid levels.

These parallel changes in behavior and hormones under domestication suggest a causal relationship but this hypothesis has seldom been tested, nor is it known how much of the difference in behavior is due to organizational or activational effects of adrenal and gonadal hormones. If wild animals are given the hormonal profile of domestics (or vice versa), does the behavioral phenotype change as well? It has been suggested that domestication not only changes hormone levels, but also results in behavior that is less hormone dependent than in wild animals (chapter 2). It is certainly the case that domestication has reduced or eliminated strict seasonality of breeding in some domestic breeds, so that the HPG axis is "on" all the time. But it does not follow that reproductive behavior is therefore less hormone dependent. It would be interesting to know more about how brain steroid mechanisms have changed with domestication. Chicks of a feral population of the domestic chicken were more behaviorally sensitive to testosterone (showed more attacks and precocial copulatory behavior) but comb growth was less sensitive, as if a few generations back in the wild might have led to more brain androgen receptors but fewer comb androgen receptors (Astiningsih and Rogers 1996).

What can we conclude from these selection studies? There is little experimental evidence so far that selecting for sex steroid-dependent behavior results in changes in circulating sex steroid levels. In contrast, selecting for adrenal steroid-related behavior, including aggression, does change steroid (glucocorticoid) levels, HPA responsivity, and sometimes reproduction-related traits as well. Selection for hormone levels themselves produces the expected change (they are heritable enough), but with the important exception of selection for different photoperiodic thresholds it is not known if this would be the case in wild populations. Both selection studies and comparisons of wild and domestic conspecifics point to a trade-off between gonadal and adrenal activity. A higher gonad to adrenal ratio is more fit in human managed environments but a higher adrenal to gonad ratio is more fit in natural environments, where there are more energetically demanding challenges to meet. These studies also suggest a possible causal relationship between HPA activity (particularly reactivity) and a fundamental dimension of behavior related to approach and avoidance contributing to tameness and wildness and to the personality or style of both domestic and wild animals. The fact that domesticated animals have smaller brains yet higher reproductive rates than their wild ancestors shows that large brains don't just result from good condition (plentiful food supply) but are being selected for in the wild by something other than reproduction in the narrow sense.

Correlated Traits, Hormones, Costs, and Evolutionary Change

Noticing how traits change together can yield insights into evolution. Traits show positively or negatively correlated responses to selection ("genetic correlation") for two reasons (Arnold 1987; Wilkins 2002). One is when the traits are multiple effects of the same genes. In the language of developmental biology, this means the gene product is part of a cascade or network producing multiple features or the gene product is employed in several different kinds of tissues. The second is linkage disequilibrium, which occurs when the traits are based on genes located close together that are transmitted together. Because many gene products are indeed deployed in multiple ways and at different periods and body parts in development, the first type of pleiotropy is quite common. If building the traits requires simultaneous use of a gene product in limited supply, the traits will likely be negatively instead of positively correlated, reflecting a trade-off. These genetic correlations are regarded as an important determinant of the rate and type of evolutionary change. If traits are genetically correlated, then they cannot change independently, at least over short-term evolutionary scales (Fisher 1958; Lande 1980). If both traits are fitness enhancing, the rate of evolutionary change could be enhanced by a positive correlation (reinforcing selection). But if the correlated trait is maladaptive, or it is beneficial but the correlation is negative, then the genetic correlation might slow down a change in the first trait (antagonistic selection). Expressed in the language of adaptive landscapes, genetic correlations can speed up or delay getting to a local adaptive peak (Lande 1980). With sufficient evolutionary time, the correlations can be broken, because they too are subject to selection (Reeve and Sherman 1993). Thus, genetic correlations are more relevant to understanding evolution within a species or between closely related species (recent changes and divergences or failures to change) than the evolution of clade differences originating far back in time. This is one reason why within-species trait correlations won't necessarily generalize to comparisons between species or clades.

Hormones and their mechanisms of action enter the picture because they are a major class of physiological and developmental mechanisms that link arrays of behavioral and other traits. When they are gene products themselves (for example, oxytocin family peptides or steroid receptors) they are a source of pleiotropy pure and simple. When they are steroids, they are not gene products but they are made by and act through gene products, serving in effect as the physiological analog of pleiotropic genes (Ketterson et al. 2000). Hormones link behavior and morphology in both naturally selected behavior and sexually selected signaling and could influence the rate of evolution of both kinds of traits (Hayes 1997; Hews and Quinn 2003). If selection on a steroid-based behavior results in a change in sex steroid level, all other traits that also depend on that hormone would be affected. This applies to any physiological mechanism, but sex hormones are also a mechanism of gametes and fertility, which are dangerously close to fitness (Ketterson and Nolan 1999). As discussed in previous chapters, potential costs of the hormones (correlated traits with negative fitness consequences) are thought to put a limit both to how high the hormone levels can evolve and to how far an otherwise desirable hormone dependent trait can go (to place limits on perfection [Ketterson et al. 2000]). The word "contraints" is sometimes used to refer to these and other processes thought to slow down or prevent change. Upon close analysis, however, "constraints" is a difficult concept, semantically loaded with unfortunate connotations (Maynard Smith et al. 1985; Wilkins 2002). In the language of adaptive landscapes, hormones are hypothesized to be one of many reasons that animals fail to reach a nearby adaptive optimum even after otherwise sufficient evolutionary time (Foster and Endler 1999).

How good is the evidence that hormones might influence the rate of evolutionary change, increasing or decreasing the odds that the animal will make it to an adaptive peak? In chapter 3 and the section above ("Reproductive Success and Differential Fitness"), research was summarized in which experimental increases in testosterone in free-living male birds and lizards increased some components of fitness but decreased others. These reveal traits that are correlated (positively or negatively) because of a common hormonal mechanism (Sinervo and Basolo 1996; Ketterson et al. 2000). Such "phenotypic engineering" (Ketterson and Nolan 1992) has the virtue of being an experimental approach to correlated traits that gives much insight into what is possible in individuals of a single or a few generations. If only there were a comparable method to change hormone levels over many generations.

Let's look again at the correlations that emerged (or didn't emerge) in the artificial selection experiments. There were two statistically positively correlated responses that very likely resulted from a common androgen-related mechanism: high mating frequency and larger proctodeal gland size in Japanese quail, and high mating frequency and high crowing frequency in roosters. The selection criterion in both studies was mating frequency only. It would be interesting to know whether these correlations are symmetrical—whether mating frequency would be a correlated response if the selection had been on crowing frequency or proctodeal gland size, because it is crowing and proctodeal gland function that are hypothesized to be under sexual selection in the wild ancestors. The fact that selection on mating frequency, a completely hormone-dependent behavior, didn't change hormone levels means that costs of the hormones themselves will not necessarily get in the way of a response to selection.

It is interesting that steroid-related behavior and fertility are poorly correlated if at all, at least in domestic animals. Among farm animals male fertility (sperm quantity and quality) is uncorrelated with dominance status, producing a big problem if the dominant male in the group, who gets most of the matings, happens to be infertile (e. g., Pizzari et al. 2002). No behavioral or morphological trait of young roosters has been found so far that will predict later fertility (will predict which males should be saved to be breeders) reliably and across strains (McGary et al. 2002). That means that selection on steroid-based behavior will not necessarily alter fertility, either positively or negatively.

Alas, little is known about any of this from wild populations. Correlations between traits among the individuals in a population can be determined, but without selection studies to complement them this kind of static snapshot is hard to interpret. For example, correlations might result mainly from environmental causes and reveal little about underlying genetic correlations. How do hormonal mechanisms (not just those operating in adulthood but also those acting during development) alter the probabilities of reaching local adaptive optima? How are these probabilities different for hormone-based traits compared with other traits? These are important questions that need answers.

If hormones really do constrain evolutionary change, then behavior that is less hormone based should be more evolutionarily labile, all other things being equal (which they never are!). Parental care isn't as closely linked to hormones as some other reproductive behaviors (chapter 2). It has evolved and disappeared multiple times in vertebrates, and in birds and mammals both paternal and alloparental care are quite labile. In some species there seems to be a sexual selection component to paternal care (Freeman-Gallant 1996; Tallamy 2000). Perhaps the looser relation to hormones has helped contribute to the rate at which this behavior comes and goes.

Hormones, Sexes, and Sexual Selection

Males, females, and sexual selection raise a particularly fascinating version of the correlated traits and constraints issue along with many of the same conceptual and empirical difficulties. Some sexually selected traits, both male typical and female typical, are steroid based. Yet most of the genome is the same in males and females, and the sexes have the same hormones, just (at best) different levels of them. This has inspired two related claims: (1) there might be hormone-related constraints on the evolution of sexually selected traits (including behaviors) in one sex due to the likelihood that the trait will pop up a bit in the other sex but have negative fitness consequences there; and (2) "masculine" traits in females or "feminine" traits in males might be functionless epiphenomena resulting from shared genes and hormones.

The first claim is a hormonal version of the more general idea that genetic correlations between male and female traits ought to slow down the evolution of sexual dimorphism, just as genetic correlations between two traits can slow down the evolution of one of them (Lande 1987). How then has sexual selection resulted in sexual dimorphism as frequently as it has? A way through this impasse is to lower the correlation between the traits by evolving sex-limited expression of traits based on autosomal genes (Arnold 1987). Sex-limited trait expression is, of course, a familiar phenomenon in hormones and behavior and much is known about how sex steroids alter gene expression early in development and in adulthood to create sex differences in behavior and other traits out of a largely monomorphic genome (chapter 4). Genetic correlations between the sexes can be eliminated by organizational and activational hormone actions. A trait would pop up in the other sex only if (1) it was not completely suppressed during organization, (2) adult hormonal dimorphism is not pronounced, and (3) the hormonal threshold for the trait is relatively low. Reducing unwanted genetic correlations between the sexes is a more difficult problem (for human minds, at least) whenever there is direct genetic sex determination, as in *Drosophila*, wallaby pouches and scrotums, zebra finch plumage, and possibly telencephalic song systems of birds. Once highly heteromorphic sex chromosomes have evolved, however, there is chromosomal space for genuine sex-linked dimorphisms (see section on genetic architecture below).

Is there any evidence for behavioral traits that are correlated between the sexes because of hormones? What would constitute evidence? It's the "because of hormones" part that's tricky, because a correlated response in a selection experiment can result from shared mechanisms unrelated to hormones. In the artificial selection experiments with quail and chickens, selection for male mating behavior did not affect female mating behavior. That's actually more surprising than it might seem, because although the two behaviors are different in form and the hormone action takes place in different brain regions (the preoptic area for male mating and the ventromedial hypothalamus for female mating) both behaviors depend on aromatase (in the gonads or brain) and on brain estrogen receptors. Evidently that's not what was selected for in the males. Instead, the within-male correlated responses suggest that androgen-related mechanisms were altered. Selection for low pheromone production in male pigs produced later female puberty, presumably for some hormone-related reason, but it's not obvious what since females don't produce the pheromone steroid at all. Selection for male aggressiveness in genetically wild mice had no effect on female aggressiveness (van Oortmerssen and Bakker 1981).

Little else is known about correlations between the sexes related to behavior and hormones in selection experiments, including how selection on the level of a steroid in one sex affects the level of the same steroid in the other. Natural selection, of course, isn't on hormone levels themselves, but some sexual selection comes mighty close, as when fish and mammals are attracted to mates by their steroid excretion products (chapter 2). Ketterson et al. (2000) emphasized that once something is known about how selection might be acting on testosterone-dependent male traits, then determining the hormonal basis of female traits and how that selection is acting on testosterone-based female traits is the next important step. Is the selection on males and females antagonistic? Reinforcing? Independent? The answer helps you understand what has happened and will happen in both sexes and how fast (Shuster and Wade 2003). What makes this step interesting in the juncos is that females and males have similar peak testosterone levels. When females were given testosterone implants that elevated their testosterone levels slightly but not hugely, they showed some negative fitness effects on body mass, molt, and time to lay the first egg, but did not show the negative effects on nest and incubation behavior that males show when their testosterone levels are elevated (Clotfelter et al. 2004). Thus, selection on males for higher testosterone, if accompanied by increased levels in females (which is unknown), could lead to some fitness effects on females, but not necessarily of a behavioral sort.

The second claim is that shared hormones are the mechanistic basis of some correlated traits that have been sexually selected in one sex (usually males) but are neutral in the other sex (Lande and Arnold 1985; Lande 1987; Bakker 1994). The observation that females can share some of the male's fancy ornamentation generalizes to behavior, as when female birds sing but not as much or as elaborately as males. The "neutral" part of this claim raises the difficult issue of how to test the alternative hypotheses "neutral" vs. "adaptive." It certainly isn't safe to assume without evidence that anything is neutral, especially anything that might affect reproduction. Perhaps females need a little maleness to be fit and vice versa. Also, initially neutral traits can later become fitness enhancing (Reeve and Sherman 1993). Mounting by female mammals serves as a good behavioral example of the difficulty. Both wild and domestic female ungulates and primates have been observed to mount other individuals. The behavior looks a lot like the copulatory mounting of males. In ungulates the behavior is more common when the female is in the hormonal and behavioral state of estrus. A neutral hypothesis would be that female mounting is an epiphenomenon of hormonal elevations combined with overlap in the steroid mechanisms underlying male-typical and female-typical mating behavior. Several different adaptive hypotheses have been proposed, for example, that female mounting might signal to distant males that the female is in estrus, attracting such males (Dagg 1984; Vasey 1995; Billings and Katz 1999). Whether the behavior increases female fitness is not known, however, nor have other predictions of such hypotheses been tested, and so there is no empirical basis yet for favoring either a neutral or adaptive alternative.

The hormonal part of the claim is not much easier. If we assume an activational hormonal basis (as evolutionary biologists tend to do), what is required is (1) finding out the hormonal basis of the trait in males, and (2) deleting (or blocking) and then restoring that same hormone(s) in females and showing that the "rudiment" disappears and then reappears. The are few such experiments for male-typical behaviors that occur at a low level in females. Low levels of female songbird singing can be increased by testosterone administration (for example, in house finches, *Carpodacus mexicanus* [Bitterbaum and Baptista 1979]), but that is not the same thing as showing that naturally occurring female singing is produced by endogenous testosterone (or estradiol, which stimulates male singing along with androgens).

Putting Hormonal Mechanisms in the Foreground

A deeper integration of evolutionary biology and endocrinology to achieve an understanding of the evolution of hormone-related behavior would have the two as equal partners. Instead of starting with evolutionary processes and concepts, let's start with some facts and concepts of hormone action from previous chapters and explore their implications for evolutionary change, especially change in sexually selected behavior and sexual dimorphism. Sometimes this will lead to the same place and sometimes it won't.

ACTIVATION AND ORGANIZATION

Chapter 4 had plenty of examples of sex differences in behavior due to activational hormone actions alone. One can imagine an evolutionary change in the degree of dimorphism (either an increase or a decrease) occurring simply through an adjustment in adult hormone levels, at least up to the point where the levels are too high to be afforded or too low to maintain gametogenesis. There are rather few comparisons of closely related species differing in dimorphism to tell how often this might have happened. Among an assortment of socially monogamous birds from different families and genera, the ratio of male to female testosterone levels tends to be higher in dimorphic species than in monomorphic species (Wingfield 1994a).

Emerson et al. (1997) tested an activational hypothesis for the evolution of fanged frogs. The southeast Asian fanged frogs are a derived lineage in the family Ranidae. They differ from North temperate ranids in that males have lower testosterone and dihydrotestosterone levels, males have no male-typical characters such as thumb pads, vocal sacs, or advertisement calls, and males engage in parental care. The hypothesis is that the mechanistic basis of these evolutionary changes is lower circulating androgens and lower androgen receptor sensitivity in the targets (thumb skin, vocal sac, brain centers for calling and for inhibition of parental care) (fig. 5.5). Injection of dihydrotestosterone into adults of the fanged frog species *Rana blythii* induced North temperate style thumb pads, as if this character has been lost solely because of the evolution of lower androgen levels. This is a nice example of how a phenotypic manipulation experiment can give insight into the origin and loss of traits and not just their current maintenance. It also illustrates how a whole correlated

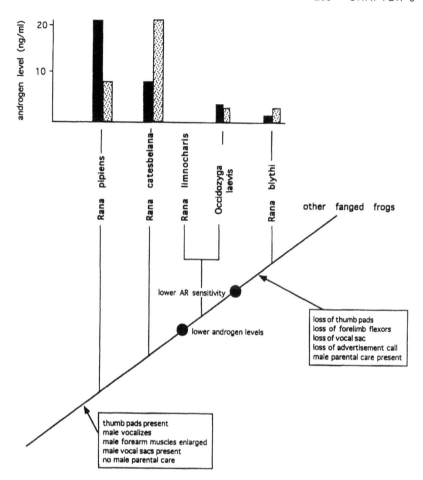

FIGURE 5.5. Interpretation of species differences in androgen levels among frogs in a phylogenetic context helps to formulate hypotheses about the direction and origin of evolutionary changes in hormone–behavior relationships. Behavioral, morphological, and hormonal differences between the fanged frog *Rana blythii* and other frogs suggest that a change in androgen levels and androgen sensitivity may have been the mechanistic basis for the loss or acquisition of several morphological and behavioral traits in the fanged frog clade. Reprinted from Emerson, S. B., Carroll, L. and Hess, D. L. 1997. Hormonal induction of thumb pads and the evolution of secondary sexual characteristics of the southeast Asian fanged frog, *Rana blythii*. *J. Exp. Zool.* 279: 587–596. © 1997 John Wiley & Sons. Reprinted by permission of Wiley-Liss Inc., a subsidiary of John Wiley & Sons, Inc.

suite of characters might have changed together because of physiological plei-otropy in the form of a common hormonal mechanism. The selection pressure for this change may have been high predation pressure in their tropical habitat requiring male parental care (Emerson et al. 1997). What is surprising is that the change in hormone level (an activational hormone mechanism) seems to have been more important than a change at the target, the same conclusion reached by Rose (1996) in studying evolutionary transitions between metamor-phosis and nonmetamorphosis in salamanders and newts. This is not the usual conclusion from vertebrates (see next section). Emerson (2000) interprets the ability of amphibians to evolve such striking new morphologies through acti-vational mechanisms with reference to cell fate in target tissues. It has long been appreciated by experimental embryologists that critical periods are the times when the tissues are being formed, when cells are multiplying and taking on the form and arrangement particular to that organ. Cells that are born later or remain pluripotent longer (as in amphibians, some of whom can even regen-erate organs) can be reshaped later in life than those whose fate is sealed earlier.

An androgen hypothesis has also been proposed for the repeated indepen-dent evolution of female mating vocalizations in frogs (Emerson and Boyd 1999). Female frogs have high androgen levels as ovulation nears, androgen treatment induces mate calling even in females that don't normally call, and females of several species in different frog lineages do call. The lack of sexual dimorphism in androgen levels near the time of mating is proposed to have facilitated the repeated evolution of female calling when environmental condi-tions made calling advantageous.

Hews and Quinn (2003) examined evolutionary transitions between mono-morphic and dimorphic aggressive behavior and signaling in a phylogenetic and developmental framework in *Sceloporus* lizards. Territorial aggressiveness and blue abdominal patches tend to co-occur, dimorphism in both characters is the ancestral state in the genus, and multiple independent transitions to mon-omorphism have occurred, some due to loss of the characters in males and some due to gains in females. Results of hormone studies thus far point to hormonal organization as more likely than hormonal activation to account for dimorphism in *S. undulatus*. This suggests two related organizational hormone hypotheses for the species differences, one that the species differ in their andro-gen levels earlier in development, and the other that they differ in the number of androgen receptors in the brain and the abdominal skin.

Changes in singing dimorphism in birds are potentially good material, as illustrated by the four *Thryothorus* wrens that have been studied. In one species only males sing, in one species females also sing but have fewer song types, and in two species both sexes sing together (duet) using equally large song repertoires (Farabaugh 1982; Levin 1996). The species with greater song di-morphism have greater dimorphism in the number of neurons in song nucleus RA, and those that duet have much less dimorphic song systems than other birds (Brenowitz and Arnold 1986; Nealen and Perkel 2000). Enough is known about how song system dimorphism develops (especially the connection from HVC to RA needed for singing) to be able to imagine (as these researchers have) how a small developmental change, for example, in the timing of estrogen action or in the estrogen target cells in HVC that project to RA, might result in a shift from monomorphism to dimorphism or vice versa. There are two alternative phylogenies for this genus (Sibley and Ahlquist 1990; Nealen and Perkel 2000). In one duetting and its associated song system monomorphism have evolved once from a more dimorphic system; in the other duetting and a monomorphic song system evolved and then a reversal occurred back to greater dimorphism in singing and the song system.

This scenario assumes that it is an organizational mechanism that has been tweaked in this genus. If one were to "evolve" a new kind of canary where females as well as males routinely sing, tweaking either an activational mechanism (an increase in female testosterone level) or an organizational one (producing an increase in adult female androgen receptor numbers in the song system) would work. Male typical crowing in quail is a largely activational phenomenon. Could a species in which females also crow evolve simply by increasing adult androgen levels in females or would egg laying invariably be inhibited at any level adequate to produce crowing? Females, like males, do have a loud long distance mate attraction call, but it sounds different and it is strictly estrogen dependent. Males given estrogen also give this call, and so as with crowing, the dimorphism is activational. The females have achieved their communication need through a different path, not by evolving crowing.

An organizational mechanism has been proposed for the evolution of sexual dimorphism in hyenas (Frank et al. 1991). Hyenas have conventional mammalian external genitalia with one intriguing exception. From birth through adulthood spotted hyena (*Crocuta crocuta*) females have quite masculine appearing genitalia. Females are also aggressively dominant to males, a phenomenon that is apparent right from birth as the females in the litters try to dispatch their male sibs. The hypothesis is that both of these traits (the anatomy and the behavior) are organized by exposure to elevated androgens during the critical period for sexual differentiation—that an increase in fetal female exposure to those androgens has been responsible for the evolution of these spotted hyena traits from more conventional hyena traits. This is an attractive hypothesis with a solid foundation in mammalian sexual differentiation that promises to link evolution, hormones, and development. Unfortunately, after an initial period of promising results, it is now looking less likely to be true (Forger 2001). The fetuses are indeed exposed to quite a bit of androgen, especially maternal ovarian androstenedione (Licht et al. 1992). However, there is no sex difference in fetal androgen exposure to account for the greater aggressiveness of the females. Prenatal treatment with antiandrogens (finasteride and flutamide) does slightly demasculinize the genitalia of both sexes of spotted hyenas, but it does not prevent masculinization altogether. Instead, the males end up with phalluses like those of the females, and neither sex ends up looking like females of other hyena species. Either the genitalia are extremely sensitive to androgens, or the masculinized genitalia arose through a mechanism other than an elevation in fetal androgen exposure (Drea et al. 1998). Everything that is odd about spotted hyenas is still a mystery (Muller and Wrangham 2002). They aren't the only mammals with genitally masculinized females, and in no case is this phenomenon understood either functionally or physiologically (Hawkins et al. 2002).

It is hard to compare the costs of the activational and organizational methods for producing sex differences but presumably they are not the same. In the canary example, an organizational mechanism limited to the song system might have fewer costs than the activational route, which would flood the whole system with higher androgens. We can speculate that organization occurs when adult hormone levels can't be used to change dimorphism because the costs of that are too great, or when a little cross-sex behavior popping up has such negative fitness consequences that it must be suppressed completely and permanently. These need to be added to the list of factors discussed in chapter 4 ("Comparative Overview") that might predict the evolution of hormonal organization. These also reinforce the need to know whether male-typical traits in females and female-typical traits in males are strictly activational phenomena. It is well established in the study of mammalian organization that male typical and female typical behavior are two independent dimensions, not ends of a single continuum, and that early hormone manipulations can change one without changing the other (Whalen 1982).

STEROID-BINDING PROTEINS, RECEPTORS, AND ENZYMES

It is not going to be possible to understand how steroid-based traits evolve without taking seriously how and where steroids act. The mechanisms of action that need to be considered begin with the binding proteins in the blood. These are often ignored but this might be a risky assumption (Breuner and Orchinik 2002). If the only active steroids are the free (unbound) molecules, then an increase in steroid production means something very different, depending on whether binding protein production has also increased. Experimental administration of testosterone increases corticosterone levels in juncos but it also increases corticosteroid-binding globulin (CBG). Furthermore, in birds plasma testosterone also binds to CBG (Deviche et al. 2001). Differences between target tissues in their ability to "pull" the steroids from the binding protein will also be important.

Steroid receptors and their tissue-specific expression have obviously undergone some evolutionary change, because the exact expression pattern is taxonomically variable, not only when entire lineages are compared (chapter 7) but even when species in the same family are compared (Gahr et al. 1993; Shaw and Kennedy 2002). As the artificial selection studies and the hyena research show, what changes in evolution is more likely to be on the receiving end of the hormones than in their structure or level. Changes in the expression patterns of genes for steroidogenic enzymes and steroid receptors in different parts of the brain and body are a likely way that behavior and sex differences can evolve (Young and Crews 1995). The change does not have to occur in all the components of the neural circuitry. During development groups of neurons affect those that project to them and those to which they project through trophic relationships. A change in one nucleus will alter several others.

Expression patterns themselves are subject to both organizational and activational steroid influences. Are steroids and their receptors correlated traits? Steroidogenic enzymes and steroid receptor proteins are products of different genes, not pleiotropic effects of the same genes. Independent changes in enzymes and receptors would allow hormone–behavior links to be created or disconnected (Crews and Moore 1986) and would allow hormone-based trait correlations to be broken both within and between sexes. This is why behavioral endocrinologists are skeptical about the idea of hormones as constraints just because they can link traits (Hews and Moore 1997).

Conversion of steroids to other forms in the gonads and at the targets also needs to be taken into greater account in thinking about how evolutionary change can occur. The expression of these enzymes is also tissue specific and in some tissues it is regulated by circulating steroids. The use of different metabolizing pathways for different traits or sexes is one way (in principle) that antagonistic selection can be overcome. We need to know not only how selection acts on testosterone-dependent behavior of females, but also how it acts on estrogen-dependent behavior of males. The crowing and strutting of male quail relies more on conversion of testosterone to dihydrotestosterone and on androgen receptors than on aromatization and estrogen receptors, whereas mating behavior relies more on the latter pathway (chapter 2). It should be possible for a change in behavior using one pathway to occur without altering behavior using the other pathway.

The most likely cases of evolutionary change due to these mechanisms thus far comes from studies of nonbehavioral traits. Deer illustrate how monomorphism can evolve from dimorphism by having females become more like males. Reindeer (caribou) are the only deer species in which females as well as males have antlers. In "regular" deer with male-limited antlers, testosterone is the antler steroid; in reindeer estradiol is the antler steroid (Lincoln and Tyler 1999). What has changed is probably the expression of genes for the two kinds of receptors in the antler tissue and perhaps expression of the aromatase gene as well. Avian plumage and head ornaments provide additional examples. Plumage can be estrogen dependent, testosterone dependent, luteinizing hormone dependent, or independent of any hormonal basis (Goodale 1918;

Witschi 1961; Kimball and Ligon 1999). The mechanisms tend to be conserved within avian lineages. Seabright bantam chickens (an experiment of nature) show how monomorphism could arise from dimorphism by having males become like females (Owens and Short 1995). A change in the genome of this breed has caused males to have female-typical aromatase levels in the skin (Wilson et al. 1987). High skin estrogen produces female-typical plumage so that these males look like females except for their male-sized combs. The combs are androgen, not estrogen, dependent, and this is the mechanistic reason that plumage and combs are not correlated traits. Phalaropes are a case study in the evolution of reversed dimorphism. Females have the brighter more colorful plumage. In Wilson's phalarope (*Phalaropus tricolor*) female skin has higher activity of the testosterone metabolizing enzymes 5α- and 5β-reductase (Schlinger et al. 1989).

Taking enzymes and binding and receptor proteins into account is probably also essential for understanding the costs of hormones. Even whopping hormone levels can be tolerated if the receiving and metabolizing machinery is right (Haig 1993). Another advantage of going straight to these mechanisms in thinking about evolution is that these are gene products, unlike the steroids themselves, making it possible to model how many and what kinds of genetic changes might suffice for a particular kind of change in a hormone related trait.

The Significance of Organization by Testis or Ovary

Most sex differences in mammals that are hormonally organized are produced by actions of testicular hormones on males. Because of the possibility that variation in early steroid exposure does contribute to adult behavioral variability, it has long been thought that hormonal organization should produce greater variability in males than in females. In those birds that use ovarian organization, on the other hand, variability due to organizational hormones might be greater in females. Adult male signals would be more likely to reflect conditions experienced during early development for hormonal reasons where testicular hormones organize them.

Ketterson et al. (2000) recognized that whenever masculine traits are completely demasculinized in females by early ovarian organization (as for quail mounting), those traits will be insulated from selection on males. What also might matter in addition to the organizing gonad is whether the trait is organized by enhancing versus suppressing gene expression. In pursuing these lines of thought it is important to remember not to equate testicular organization with masculinization and ovarian organization with feminization. In mammals, testicular organization can create differences between males and females either by masculinizing a trait in males (for example, mouse aggression) or by defeminizing a trait in males (for example, female typical receptivity). The same could be true for birds as well. That is, the ovary could either demasculinize females or feminize them.

There have been several attempts to link hormones to evolution by proposing that the default sex in organization is the more evolutionarily primitive and the organized sex more derived (Crews 1993; Owens and Short 1995; Ligon 1999). Building such bridges is admirable but there are several problems with this particular idea. First, some of the traits referred to in such discussions are not hormonally organized. Second, such proposals assume that the testis is the organizing gonad, even in birds where there is no evidence for any organization by testes. Once the correct organizing gonad is substituted, we have male galliforms and anseriforms as more primitive than their females. Finally, both sexes are the same species with the same genome and so it doesn't actually make sense to think that one sex is more primitive than the other. A particular trait might resemble the same trait in an ancestral species more in one sex than the other but the whole sex doesn't. What is most valuable here is the idea that the nature of organization might have something to do with what is more or less likely to happen in evolution.

The Implications of Direct Genetic and Hormonally Based Sexual Differentiation

Drosophila and probably many other insects have cell autonomous (direct genetic) sexual differentiation. With pure environmental sex determination (for example, embryonic sex determination by temperature), there are no genetic differences at all between males and females. These are the sex-determining methods that produce the kinds of genetic correlations between the sexes that are supposed to slow down evolutionary change. Mammalian and some avian sex differences, however, arise through hormonal means, including organization. What difference might this make? Insect castes can be either genetically or hormonally determined, and this may have consequences for the evolution of social behavior in social insect colonies (Volny and Gordon 2002). Is it actually harder to evolve sex-limited trait expression (for example, changes in sexual dimorphism/monomorphism) with genetic sexual differentiation? Perhaps insects can pull it off anyway because the generations turn over so rapidly. Whether longer-lived animals would so frequently reach highly dimorphic adaptive peaks without hormonal organization or activation is another matter. From this perspective it's no accident that so many sexually selected traits have a hormonal basis. Hormonal organization seems like such a handy way to prevent unwanted correlations between the sexes that one wonders whether this is a reason that it evolved. Because its role seems to be greatest in mammals and some birds, we can ask if there's something about these kinds of animals that makes avoiding negative correlations imperative. On the other hand (to dump cold water on this line of thought), songbird plumage seems to be pretty labile with respect to acquiring or losing dimorphism even though plumage dimorphism is not always a particularly hormonal phenomenon.

If some traits are hormone based and others aren't, this should reduce the likelihood that they are positively correlated. This may be part of the mechanistic reason why behavior and body size do not always respond the same way to sexual selection. Sexual dimorphism in body (skeletal) size in chickens and humans occurs through direct genetic differentiation while dimorphism in the rest of the anatomy (and in chickens, in reproductive behavior) is hormonally organized and activated (chapter 4). Male song complexity and plumage ornamentation are negatively, not positively, correlated in finches (e. g., Badyaev et al. 2002).

Genetic Architecture and Hormonally Based Sexual Dimorphism

When animals have sex chromosomes, it is important to take into account whether the genes for traits are on sex chromosomes or autosomes in thinking about the evolution of sexual dimorphism in relation to hormones. The location of genes for traits influences their evolutionary potential, the likelihood that they can evolve (Wilkins 2002; Reeve and Pfennig 2003).

A simple way that a dimorphic trait could evolve is through a mutation in a gene on a sex chromosome—through sex linkage. The bright pigmentation of some male fishes is Y-linked (Lindholm and Breden 2002). Colorful male guppies are preferred by females and achieve greater mating success (Brooks and Endler 2001). With Y-linkage, it will be more difficult for females to evolve the bright colors (should there arise some selective advantage to doing so), because only sons inherit a Y chromosome. With hormonal regulation, however, sexually dimorphic traits do not have to be sex-linked and most aren't in vertebrates.

Protected invasion theory provides an explanation for why location on sex chromosomes versus autosomes might matter for the evolution of sexual dimorphism that is more compatible with hormone-based sex-limited gene expression (Reeve and Shellman-Reeve 1997; Reeve and Pfennig 2003). The theory addresses the origin of hard-to-evolve traits, the early difficult stages when chance is important and extinction is a likely outcome. Early extinction must be avoided (invasion must be protected) for sexual selection on males to result in a move toward a distant adaptive peak of elaborate behavior or ornaments. The odds that this initial high hurdle will be successfully cleared by males is greater when they have ZZ chromosomes (are homogametic) than when they have XY chromosomes (are heterogametic), because such alleles are expressed more frequently, hence exposed to positive selection, than are alleles on autosomes or X chromosomes. For elaborate male traits to evolve through intersexual selection, a female preference for the trait also has to evolve.

By simultaneously considering the origin of the male trait and the female preference for it, the theory explains how Fisherian sexual selection can get started and why that has happened more often in species with homogametic males. The basic idea is that a female passes Z-linked preference alleles to all of her sons (rather than just to half, in the autosomal case, or none, in the X-linked case). Thus, the allele is especially likely to be increased in frequency when the elaborate male trait allele spreads due to female choice. It explains why there are more fancy male birds and butterflies than fancy male mammals, and why the fanciest male vertebrate brains belong to birds, instead of a clade with XY males. The theory applies not only to extreme sexually selected traits but also to other uncommon sex biased behavior such as alloparental care (Reeve and Shellman-Reeve 1997).

What does all this have to do with hormones? Traits that are differentiated or activated by gonadal sex steroids are ultimately regulated by genes on sex chromosomes and therefore are also protected in the homogametic sex during the invasion phase. The protection of new mutations against extinction is extended to all the many autosomal locations where gene expression is steroid regulated. Thus, according to protected invasion theory, a steroid basis for behavior should make a difference for evolution, but of a different sort than is conveyed by theories of hormones as costs and constraints.

The Perspective from Evolutionary Developmental Biology

Evolutionary developmental biology, or "evo-devo" for short, has as its goal explaining what changed during development that produced the diverse morphologies of organisms by integrating molecular developmental biology (the descendant of experimental embryology) with the modern approaches to phylogenies (relationships between animals) that are necessary to identify evolutionary origins. When living species differ in hormone-related behavior or dimorphism, this approach offers a way to understand at a deep mechanistic level what might have been tweaked or co-opted in the nervous system and elsewhere as they diverged to produce the selected changes in phenotype.

This vision of development at the level of genes and molecules rests importantly on concepts such as regulatory and structural genes, feedback linkages, developmental (regulatory) gene networks, hierarchies of levels, induction (conditional specification), modularity (compartmentalization), and the distinction between mosaic and regulated development (Gerhart and Kirschner 1997; Hall 1999; Graur and Li 2000; Carroll 2001; Wilkins 2002). These concepts are necessary because there is no simple one-to-one relationship between genes and traits or between molecular and morphological change. Even a word like "cascades" sounds too linear. Strange and wonderful nonlinearities happen when you go from one level to another. What look like big genetic changes, such as mutations or polyploidy, don't necessarily produce a big phenotypic change, whereas simple changes in developmental mechanisms can have enormous consequences. Phenotypic complexity seems to be more related to regulation of gene expression than to gene number (genome size). The recognition that gene expression can be regulated by the external as well as internal environment (a recurring theme of this book) provides a link to ecology as well (Hall 1999; Gilbert 2001).

The most widely publicized successes thus far have concerned major body plans and limb morphologies of entire lineages (orders, phyla), and how these can be conserved within lineages while maintaining the latent flexibility to change under the right circumstances. This kind of large-scale evolution is most relevant to chapter 7. One can already see, however, the tremendous potential of this approach for understanding differences between more closely related species.

Some of these successes have flown in the face of earlier ideas about development and evolution (Wilkins 2002). Modules give stability but also allow a mutation to produce sizable changes that are sensible, not deleterious. It was long thought that mutations with effects early in development would be unlikely to spread because their effects would surely be too disruptive to the whole organism, precluding any possibility of rapid changes in morphology and constituting one kind of developmental constraint on evolution. It is now apparent, however, that the upstream (early) triggers for cascades can change as well. Another big surprise is how conserved all the molecular developmental mechanisms are, beginning with the genome itself. This paradox of diverse phenotypes created by conserved mechanisms highlights the importance of understanding how the mechanisms work together (how the networks operate).

Evolutionary developmental biology gives substance to some of the concepts of evolutionary theory and provides less semantically loaded language for others. The presence and absence of genetic correlations between traits makes more sense when developmental mechanisms are known. Who would have expected sex and smell to be genetically correlated? But they are, and the same mutation produces both Kallmann's anosmia and hypogonadism in humans (Bouloux et al. 1992). Antagonistic pleiotropy trade-offs might reflect a conflict over gene expression across the whole genome (Stearns and Magwene 2003). Understanding the evolution of complex structures starts to seem possible, providing a way out of an excessive focus on single traits that may not even be traits with respect to how nature carves the joints. Embryonic neural crest cells not only become parts of cranial nerves and their sensory ganglia, but also mediate development of other head structures, including bone, cartilage, muscles, and pituitary. Tweaking genes for neural crest cells will change all these traits in ways that preserve a sensible match between them (Noden 1992; Olsson et al. 2001). Here is a conserved yet flexible module for evolutionary change that is relevant to the evolution of behavior, sexually selected head morphology, and more. Complexity is not constraint. Rather than talking about developmental constraints, evo-devo asks what the important construction rules are, how the components they produce have actually changed during evolution, and how the rules were used to change the components (Oster et al. 1988). It asks, what is the space of all possible organisms, and what initial conditions (historical starting points) and developmental rules account for the occupied and empty regions now and in the past (Goodwin 2001)?

Brains and Their Development

As the most complex organ of vertebrates (anatomically and functionally) the brain and its evolution is a ripe target for an evo-devo approach. A significant percentage of the genome is devoted to products expressed only in the brain. The brain is also famously plastic, but that's not unique to brains or even to animals (plants are rather plastic too). Brain targets have to be part of any effort to link genes to behavioral evolution. The prospect of tackling behavioral evolution through developmental molecular neuroscience with real phylogenies as underpinnings is exciting. Imagine, for example, finding out exactly what has changed in the nervous system each time that manakins (Pipridae) have evolved mechanical sound production through wing snapping (Prum 1998; Schlinger et al. 2001). The hard part is that the relevant dimension of the nervous system for small-scale evolution (species divergence, for example) is not usually its gross morphology but rather its functional networks, most of which have not yet been discovered. The proper development of those networks depends on what the animal does, is exposed to, and experiences (Zhang and Poo 2001).

Neurons are born, live or die, migrate, differentiate into subtypes, and form functional connections with other neurons. Their fate depends not only on their intrinsic properties but also on their trophic relationships with other neurons. These developmental processes create sexual dimorphism and species differences in behavior (Kelley 1988). Much neurogenesis goes on early in development, but neurogenesis continues in adulthood to some extent in all vertebrates and to a major extent in some (Zupanc 2001). Although steroids are not known to affect neurogenesis, they do affect whether the new cells live and what they become and so could be mechanisms at any point in ontogeny for evolutionary change in behavior. The molecular mechanisms that can be tweaked to change rate, timing, and place in the nervous system are being discovered (Ross and Walsh 2001). Some of the developmental rules for size, positional patterning, and segmentation are becoming clear (Rubenstein et al. 1994; Finlay and Darlington 1995; Rallu et al. 2002). The ways in which the cortex programs itself through activity-dependent processes are beginning to be understood (Krubitzer and Kahn 2003). Just as for other parts of animals, in going from the genome to the nervous system to behavior, we can't trust our instincts about

the consequences of a change at one level for other levels. Polyploid vertebrates act like their relatives, not like some other form of beast, and salamanders with unusually large genomes have the simplest, not the most complex,
of all vertebrate nervous systems yet remarkably competent behavior (Roth et
al. 1997). It is possible to get a big change in the brain (for example, greater
overall size) with a small developmental change (Finlay and Darlington 1995;
Chenn and Walsh 2002). That means that big changes in behavior might be
possible with relatively few genetic changes.

SEXUAL DIFFERENTIATION AND THE EVOLUTION OF SEXUALLY DIMORPHIC BEHAVIOR

Let's see how this looks from the perspective of evolutionary developmental
biology. The distinction between direct genetic differentiation and hormonally
regulated sexual differentiation is the distinction between mosaic and regulated
development. The first developmental genetic pathways to be determined were
the sex determining pathways of *C. elegans* and *Drosophila* (Wilkins 2002).
It's precisely because their development is mosaic that these pathways were
easier to analyze. Hormonally regulated cascades are more difficult. Expressed
in the language of evo-devo, they are modules. They can change without altering other modules. Female spotted hyenas may have masculinized genitalia,
but their other organs are those of any other hyenas. Teleosts can evolve sequential hermaphoditism without messing up other aspects of their biology
(such as excretion) or ceasing to be proper looking fish. In vertebrates the sex-
determination cascade can be captured by different initial triggers such as
genes on a particular chromosome or temperature, with the rest of the cascade
conserved (Graves 1995; Wilkins 2002). *Sry*, the initial trigger for male-typical
development in mammals, is evolving rapidly, as if it is under strong selection.
Once again, earlier developmental mechanisms are not harder to change.

One part of the sexual differentiation module that could be important for
evolutionary change is expression of the P450arom gene, across developmental time or in different body and brain parts. The molecular mechanisms
that could make that happen, such as control of the expression of the gene by
tissue-specific promoters and alternative splicing, are beginning to be understood (Lephart 1997). What is still elusive is the molecular cascade for brain
sexual differentiation. In male mammals there is good evidence that expression
of genes for aromatase and ERα in the hypothalamus is critical for its organization by testicular steroids but it is not known what genes the estrogen receptors
are activating (McCarthy 1994). Other biochemical mechanisms that are involved that may lead to discovery of the genes include increased phosphoryla-
tion of CREB, increased GABA (a neurotransmitter) production, and activation of L-type calcium channels (Auger et al. 2000, 2001; Perrot-Sinal et al.
2003). The songbird song system might be a separate module from other com-

ponents of songbird sexual differentiation, allowing the repeated evolution of female birds that sing and have large telencephalic song nuclei without being masculinized or defeminized in other ways (Arnold 1992).

There is no dearth of examples where changes in sexual dimorphism of steroid-related behavior have occurred in evolution in response to selection. But there are few cases where the developmental mechanisms responsible for the change have been investigated or even hypothesized, fewer yet with hypotheses at the molecular or even cellular level, and fewer still with an explicit phylogenetic framework. A phylogeny is required to know the direction of the change (from lesser to greater dimorphism? from greater to lesser?) and who has changed (the female? the male? both?) as we saw with the *Thryothorus* wren example earlier.

The songbird song system is ripe for an evo-devo approach and likely to yield to its power eventually. Quite a bit is known about the developmental neurobiology of the song system at the cellular level (Arnold 1992; Nordeen and Nordeen 1994; Balthazart and Adkins-Regan 2002). Sexual selection on male singing is associated with an increase in the size of the telencephalic song nuclei but female nuclei are small if females don't sing. Greater dimorphism can be produced either by lower neuron death rates in male nuclei or by greater death rates in female nuclei or both. The size of the projection from HVC to RA is sensitive to locally produced estrogen, and so tweaking gene expression for estrogen production there and for androgen receptor production in song system nuclei would change singing dimorphism.

What also makes song system evolution attractive for study is that something is known about the heritability of the system in one species (Airey et al. 2000). In zebra finches, nuclei involved in song production (the posterior pathway HVC, RA and nXIIts) had substantial heritabilities (about 0.4, 0.7 and 0.5, respectively). Also, song nuclei sizes were correlated with overall telencephalon size, which is sexually dimorphic, and with song repertoire size (DeVoogd et al. 1993; Leitner et al. 2002). These observations raise the intriguing possibility that a simple developmental mechanism responsible for telencephalon size is varying that carries along with it larger nuclei within the telencephalon. The same developmental mechanism could account for both species and sex differences in song nuclei sizes. If the only developmental mechanism available that will generate larger forebrain structures is an increase in overall forebrain brain size, then the costs of a larger song repertoire could be serious indeed (Airey et al. 2000).

Species Comparisons in Hormones and Behavior

Comparisons between two or more related species to get a glimpse of the evolution of hormone behavior relationships are a time-honored approach in hor-

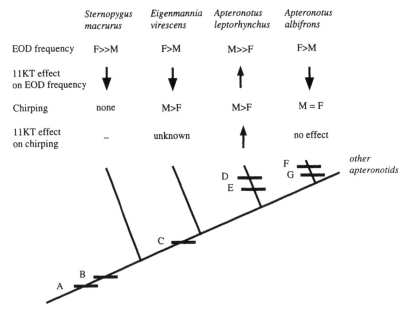

	Sternopygus macrurus	*Eigenmannia virescens*	*Apteronotus leptorhynchus*	*Apteronotus albifrons*
EOD frequency	F>>M	F>M	M>>F	F>M
11KT effect on EOD frequency	↓	↓	↑	↓
Chirping	none	M>F	M>F	M = F
11KT effect on chirping	–	unknown	↑	no effect

FIGURE 5.6. A phylogeny based hypothesis for the evolution of sexual dimorphism in and andro-gen (specifically 11-ketotestosterone) effects on electric organ discharge (EOD) frequency and chirping in four species of wave-type gymnotiform weakly electric fish. Letters A to G on the cladogram indicate the hypothesized series of evolutionary changes. Redrawn from Dunlap, K. D., Thomas, P. and Zakon, H. H. 1998. Diversity of sexual dimorphism in electrocommunica-tion signals and its androgen regulation in a genus of electric fish, *Apteronotus*. *J. Comp. Physiol. A* 183: 77–86. © 1998 Springer Verlag.

mones and behavior. Such comparisons have cropped up repeatedly in this and other chapters, yet their interpretation is not always straightforward. Proposals that species X is different from species Y because it has W hormone level or brain steroid mechanism instead of Z are statements of a correlation with an *N* of 2. However plausible such proposals, they are hypotheses that need to be tested either experimentally or by looking at additional pairs, enough pairs for the correlation to be have some statistical power. Now that some conceptual frameworks have been examined for approaching the evolution of species dif-ferences, let's look at four of the better known two-species comparisons that have not already been discussed to see how they have been interpreted.

WEAKLY ELECTRIC FISH

Two congeneric species of South American gymnotiform weakly electric fish, *Apteronotus leptorhynchus* and *A. albifrons*, both generate electric organ dis-charges (EODs) for communicative purposes but sexual dimorphism in signals differs in the two species (fig. 5.6) (Dunlap et al. 1998). In *A. leptorhynchus* males "chirp" more than females and have a higher electric organ discharge frequency, whereas in *A. albifrons* there is no sex difference in "chirping" and males have a lower electric organ discharge frequency. The two species respond to steroid treatment differently as well. In *A. leptorhynchus* testosterone increases chirping in females (Dulka and Maler 1994), whereas in *A. albifrons* implants of androgens in gonadectomized males and females did not affect chirping but dihydrotestosterone and 11-ketotestosterone decreased EOD frequency. The mechanistic basis of the species divergence seems to be a change in the steroid sensitivity in the components of the discharge-producing system responsible for the two EOD features. The phylogenetic relationship between them suggests that *A. leptorhynchus* changed the direction of EOD frequency dimorphism since divergence and *A. albifrons* lost dimorphism in chirping due to loss of androgen sensitivity since divergence. The hormone treatment experiments further increase confidence in the interpretation.

Whiptail Lizards

Cnemidophorus inornatus is a diploid sexually reproducing species, whereas *C. uniparens* is a triploid unisexual (all female) relative that reproduces parthenogenetically. The biparental species is the ancestor of the unisexual species, a rare case where the ancestor is a known extant species. Females of both species become sexually receptive when preovulatory estradiol is high, and *C. uniparens* females mount other females when postovulatory progesterone is high (Moore and Crews 1986; Young and Crews 1995; Crews 2000). Being mounted increases the fecundity of the mountee, as if need for sperm has been lost in the parthenogen but not the need for the stimulation of being mounted. The changing sex steroid profiles accompanying ovulation are similar in the two species, as if the evolutionary change in mounting lies at the receiving end, not in the hormonal signal. Progesterone stimulates mounting by some castrated male *C. inornatus*, inspiring the hypothesis that female *C. uniparens* inherited this hormone behavior link from their male ancestors (Lindzey and Crews 1988). This scenario assumes, however, that female *C. inornatus* would not mount if given progesterone, which does not appear to have been tested. Furthermore, testosterone is more effective than progesterone for stimulating mounting, and when given to females testosterone stimulates mounting equally well in the two species (Wade et al. 1993). Thus, an alternative scenario is that female *C. uniparens* retained a mounting system already present in female *C. inornatus*.

Another difference between the two species is that the threshold for estrogen-induced receptivity is lower in *C. uniparens*. *C. uniparens* has much lower estradiol levels but higher estrogen receptor gene expression in the preoptic area and greater estrogen-stimulated upregulation of estrogen receptor mRNA in the ventromedial hypothalamus. Here the evolution of a quantitatively different hormone–behavior relationship involving a decrease in hormone level is compensated for by an increase in the sensitivity of the central nervous system mechanisms (or vice versa). If the change in estradiol levels drove the change in the brain, the question then becomes, what selected for lower estradiol levels in *C. uniparens*? Perhaps with no need to impress males, there was no selection for traits signaling high estrogen.

DWARF HAMSTERS

Phodopus sungorus and *P. campbelli* didn't diverge very long ago, but nonetheless females differ in both behavior and hormones (Wynne-Edwards 2003). The endocrinology of the estrous cycle and pregnancy in *P. sungorus* is similar to that of other rodents, whereas *P. campbelli* has a whole suite of novel features in its prolactin and progesterone profiles. *P. sungorus* requires progesterone for receptivity to occur but *P. campbelli* doesn't. While the reasons for these species differences are not entirely understood, the animals' ecology and parental care arrangements are quite different. *P. campbelli*, the Djungarian hamster, lives where it is so cold and dry that females cannot leave the pups unprotected, even in a nest, whereas *P. sungorus*, the Siberian hamster, lives in a less extreme climate. *P. campbelli* has obligatory biparental care, whereas *P. sungorus* has maternal care. Again, placing the endocrinology and behavior of these two species in a phylogenetic context tells us which are the derived characters (those of *P. campbelli*). That doesn't yet make it easy to understand whether, how, or why the unique features of its endocrinology are related either to the species difference in paternal behavior or to the contrasting environments the animals live in. It does suggest hypotheses to test, however.

VOLES

Previous chapters described how voles of the genus *Microtus* differ with respect to parental behavior (some are biparental, some are uniparental), mating systems and pairbonding tendencies (some are affiliative and form pairbonds, others don't), and aggressiveness toward strangers. These behaviors are regulated by steroids and peptides, and the species differences in the behaviors are correlated with species differences in the distribution of peptide receptors in the brain (in the expression patterns of the genes for the receptors). What is most exciting about these animals for understanding mechanisms of evolutionary change is the promise of being able to apply a new and exciting experimental approach to the two species comparison problem. Prairie voles (*M. ochrogaster*) but not montane voles (*M. montanus*) are monogamous and respond to vasopressin with affiliative behavior. Prairie voles have many more V1a vasopressin receptors in the diagonal band than montane voles do. The molecular basis of this includes a gene duplication in the prairie vole. When mice were created that were transgenic for the prairie vole receptor gene, they had a V1a receptor expression pattern like prairie voles. When vasopressin was administered into the brain they showed an increased tendency to spend time near (affiliate with) ovariectomized (i. e., not sexually stimulating) anesthetized female mice (Young et al. 1999). Control mice did not respond to vasopressin this way.

What is wonderful about this experiment is that it tests a two species comparison hypothesis relatively directly by attempting to literally turn one species into another by altering the mechanism that is hypothesized to underly the evolutionary change. Even better would be to create a transgenic montane vole that is monogamous and affiliative. Now Lim et al. (2004) have come remarkably close to that. They used viral vector gene transfer to increase V1a receptor expression in the ventral forebrain of male montane voles. The treated males spent much more time huddling with a familiar female with whom they had mated than control males did.

Notice that the implicit phylogeny here is that voles such as the montane vole are ancestral and the prairie vole is derived. This is consistent with the gene duplication, but so far the molecular phylogeny of the *Microtus* genus points to a more basal position for the prairie vole than the montane vole (Conroy and Cook 2000). The social behavior of most of the species in the phylogeny is poorly known. Perhaps monogamous voles turned into polygynous voles and vice versa at multiple points in vole evolution. The discovery of intraspecific variation in vole neurochemistry and mating systems suggests that these are relatively labile traits (Cushing et al. 2001).

Conclusions

This chapter has examined closely what has been done and thought about how evolutionary change might occur in hormone-related behavior, drawing on all of Tinbergen's aims. It has come at the problem from several angles, always asking whether a hormonal basis might make any difference, either slowing down change or speeding it up, and why. It has looked on the bright side (the potential, the ideas), glossing over the dim side, the relative lack of knowledge of selection and fitness in the wild in relation to hormones. It has emphasized the importance of (1) greater integration between evolutionary biology and behavioral endocrinology, (2) an experimental approach whenever possible, including manipulations to test hypotheses about maintenance (current adaptive value) and origins (by tweaking mechanisms in related species in a phylogenetic context), and (3) a phylogenetic backbone such as a cladogram. It has argued that a close look at developmental mechanisms and at the nervous system (the "black box" that is often missing) is essential to make future progress and to connect genetic change to behavioral change.

Understanding what can change in behavior and how requires some serious developmental biology and neuroscience. An evo-devo approach to sexually dimorphic behavior is quite promising and would have important implications for the phenomenon of speciation as well (the processes responsible for splitting of one species into two or more). Both sexually selected behavior and evolutionary change in sex-dependent gene expression patterns are increasingly acknowledged as generating species radiations and richness (Ranz et al. 2003; Streelman and Danley 2003). Steroids are a key mechanism producing sex-dependent gene expression patterns in vertebrates. Models will need to incorporate knowledge of developmental and adult mechanisms and processes underlying sexually selected behavior and other traits. For example, Rhen (2000) developed models for the evolution of sexual dimorphism that take into account what is known about genetic mechanisms (including steroid regulated genes) of sex-limited expression of traits such as body size or presence of horns and found that the models predicted the results of artificial selection experiments better than models for correlated trait selection that leave the mechanisms out of the picture. That, in a nutshell, is why mechanisms and evolution should be integrated.

6

Life Stages and Life Histories

Mice live fast and die young, producing large litters in rapid succession if they survive at all. Elephants live slow and die old, producing a single offspring at long intervals. A male anglerfish (*Borophryne*) spends his entire adult life attached parasite-style to a female whose size dwarfs his, and neither sex reaches reproductive maturity until it finds the other, a bizarre but reproductively effective arrangement. These are just a few of the many ways in which species differ in their life histories, passing through different life stages over different timescales. Figuring out how and why diverse life histories are adaptive is a major goal of an evolutionary approach to life history theory. Lack (1968) was instrumental in the development of the field and showed through his studies of birds that it could be an experimental science. Birds are still the main subjects of much empirical work with vertebrates.

At the level of individual animals, life history is lifespan development. As with other aspects of development (chapter 4), steroids are potential mechanisms. This chapter focuses on life history stages and transitions that have known links to steroids and social behavior. Many phenomena that appeared in previous chapters, such as adult sex change and multiple male types in teleosts and alloparenting in birds, can be viewed from a life history perspective. Some of the social behaviors in chapters 2 and 3 have a different relationship to hormones depending on the life history stage in which they occur, for example, the territorial behavior of songbirds during and outside the breeding season. Some of the concepts that were important in previous chapters are central to life history theory, for example, correlated traits and their interpretation (chapter 5).

Life Histories, Fitness, and Hormones

Life history refers to such highly fitness relevant traits as, how long does the animal live? How long does it take to reach reproductive maturity? How many times does it reproduce and at what intervals? How many offspring does it have when it reproduces and how large are they? All these traits are tremendously diverse across animal species. They are somewhat species typical and predictable from phylogeny. They are also variable within species, and best captured by shapes of distributions and not just averages. There is heritable variation, enough to provide the raw material for continued evolution (Newton 1989; Roff 2001). They influence rates of evolution but also themselves evolve. Yet they are rather plastic traits with somewhat low heritabilities. There is substantial nongenetic variation, reflecting each individual's facultative decisions given its energy state and the local environmental (including social) conditions.

Measures of life history traits are not independent of each other, but instead tend to be correlated either positively or negatively. For example, long-lived birds and mammals tend to take longer to reach reproductive maturity and to have fewer young at a time. These traits are also correlated with body size. For example, larger animals tend to live longer. The correlations are not necessarily the same at different phylogenetic scales, however (Daan and Tinbergen 1997; Bennett and Owens 2002). For example, large oviparous fishes tend to produce more eggs than small fishes, not fewer as among birds. Correlations within species can be different than those between species. For example, the usual negative correlation among avian species between survival and fecundity is likely to be positive within species because better quality females both have larger clutches and live longer.

Life history theory posits species level as well as individual level trade-offs between somatic functions (growth and survival) and reproduction that have shaped the evolution of life histories (Stearns 1992; Daan and Tinbergen 1997; Partridge 2001; Roff 2001). These trade-offs are reflected in negative correlations between life history measures. Because resources are finite, those allocated to growth and survival are not available for reproduction and vice versa. The two components cannot be simultaneously maximized but instead are traded off in fitness currency using solutions that vary depending on the individual and the circumstances. Within the overall reproductive allocation path further trade-offs are known or hypothesized, for example, between offspring size and offspring number, between mating effort and parental effort, and between current and future reproductive effort (the "cost of reproduction" [Williams 1966]). Life history (the allocation of effort across the lifespan) represents the solutions to the trade-offs that have worked in the past for that species or lineage.

Hypotheses derived from life history theory are tested in several ways. One is to simply see if the required negative correlation exists (bearing in mind, of course, that it might be negative between but not within species). Experimental methods manipulate one resource allocation path and observe the effect on the other and on the animal's overall fitness, revealing individual level trade-offs (Sinervo and Basolo 1996). Comparative analyses that take shared phylogeny into account allow tests of ecological hypotheses and also yield estimates of when in the past particular distinctive life histories evolved. A combination of methods provides insights into both life history origins and their current maintenance (Bennett and Owens 2002).

Steroid and other hormones are important mechanisms underlying life histories in part for a fairly obvious reason, namely, because they are central to the physiology of growth and reproduction (Schreibman and Scanes 1989; Stearns 1989; Finch and Rose 1995; Wingfield and Silverin 2002; Tatar et al. 2003). All of the life stages or events that involve reproduction require the hormones of reproduction. All the aspects of life histories related to growth involve the hormonal regulators of food intake and caloric allocation to muscle, fat, activity, etc. Life stage transitions are associated with changes in the endocrine system or its mechanisms of action.

Their relevance goes deeper than this, however (Stearns 1989; Dickhoff et al. 1990; Finch and Rose 1995). Hormones have the potential to organize future life stage transitions. The timing of life history transitions must be responsive to the external (including social) environment to be adaptive; endocrine axes and their top-down control by the brain enable this adaptive plasticity. Hormones coordinate entire suites of characters (produce positive correlations) that occur at the same life stage (Moore 1991); for example, they ensure that gametes, reproductive tracts, and sexual behavior are temporally synchronized at puberty, at the onset of a breeding season, and with ovulation. Hormones can be hypothesized to be the mechanistic basis of sex differences in life history characters. They have been hypothesized to be mechanisms mediating trade-offs (negative correlations) between growth/survival and reproduction and between the mating and parental effort components of reproduction (Wingfield et al. 1990; Ketterson and Nolan 1992; Sinervo and Basolo 1996). Trade-offs between hormonal, immune, and metabolic mechanisms are hypothesized to have shaped some life histories (Wikelski and Ricklefs 2001). As always, such hypotheses need to acknowledge that evolutionary changes are just as likely to be due to changes on the receiving end—steroid receptors and the cascades they initiate—as to changes in hormones or their levels (Stearns 1989; Finch and Rose 1995).

Life Stages Prior to Reproductive Maturity

HATCHING AND BIRTH

Hatching and birth are sudden, risky, physiologically momentous life stage transitions. Several species of both birds and mammals have been found to have elevated testosterone and glucocorticoid levels during this period. In chickens and ducks corticosterone starts rising before hatching and the rise is clearly not being stimulated by the mother (as it might be in mammals), because it occurs under artificial incubation (Tanabe et al. 1983, 1986; Carsia et al. 1987). It seems likely that the elevated glucocorticoids could serve a similar function as in other physically and socially challenging situations, mobilizing the energy needed to deal with the challenge, in this case of getting out of the shell fast and being sufficiently alert and interactive to solicit maternal care immediately. In rats and other altricial rodents, where the testosterone rise is sexually dimorphic and the neonatal period is a critical period for steroid organization, a social behavior function for the testosterone rise is well established. But precocial mammals and birds with critical periods for sexual differentiation located well before birth also show sex steroid elevations around the time of birth or hatching (Ford 1983; Corbier et al. 1992a). In mammals it again occurs mainly in males, and hypotheses linking this second elevation to sexual differentiation have also been proposed (chapter 4). In chickens and quail it seems to occur in both sexes in some studies and more in males in others, and the source includes adrenals as well as gonads (Tanabe et al. 1986; Schumacher et al. 1988; Corbier et al. 1992b; Ottinger et al. 2001). Some of the same hypotheses that have been proposed for maternal yolk androgens could be relevant to these elevations in endogenous steroids occurring at hatching.

METAMORPHOSIS

Metamorphosis is a similarly momentous life stage change with profound consequences for the entire behavioral repertoire, including social behavior. The social lives of frog tadpoles, for example, are very different from those of adult frogs. The hormonal basis of metamorphosis in amphibians is a classic success story in endocrinology. It is still yielding exciting new insights into how hormones ensure that the timing of metamorphosis is adaptive and into how evolutionary change in life histories can occur. The change from tadpole to adult has been known for many decades to be induced by the thyroid hormones thyroxin and tri-iodothyronine (Dickhoff et al. 1990; Brown et al. 1995). These are not steroids, but their receptors are produced by the same gene superfamily as the steroid receptors and their mechanisms of action are similar to those of steroids. Thyroid hormone elevation or experimental administration initiates a cascade involving expression of many genes that target the entire body and presumably the nervous system as well, producing noticeable changes in as little as 48 hours. This is a beautiful example of how a simple hormonal mechanism can cause a coordinated change in behavioral and morphological phenotype leading to the next life history stage.

Thyroid hormones also underlYy some of the evolutionary diversity in amphibian life histories with respect to whether and when metamorphosis occurs (Rose 1996). A comparative experimental approach grounded in phylogeny can distinguish between species transitions due to changes in hormone level and those due to changes in tissue sensitivity. Changes in hormonal profiles seem to have been more important in the evolution of this diversity than changes in tissue sensitivity, the same conclusion that was reached for some other amphibian examples (chapter 5). For example, the axolotl (*Ambystoma mexicanum*), a large salamander, does not metamorphose into the typical adult terrestrial stage, but retains its gills and continues to live and sexually mature aquatically. Animals treated with physiologically reasonable doses of thyroxin undergo a remarkably "normal" metamorphosis, indicating that the evolutionary change underlying the axolotl's derived condition is loss of the hormone rather than loss of responsiveness to the hormone (Shaffer 1993). Thyroid hormones are also the mechanism behind the evolution of a so-called "direct development" life history in frogs (Jennings and Hanken 1998). Some frogs have no posthatching tadpole stage but instead emerge from the egg as tiny little froglets. This occurs, for example, in the genus *Eleutherodactylus*, one of the most speciose of all vertebrate genera (so much for "developmental constraints"!). It has not skipped metamorphosis altogether; rather, a thyroid hormone-dependent metamorphosis occurs before hatching, far earlier than the "normal" time (Callery and Elinson 2000). This type of relatively simple activational hormonal mechanism seems to have enabled frequent evolutionary changes in amphibian development.

Experiments by Denver (1997) with Western spadefoot toads (*Spea hammondii*) show that the HPA axis is the crucial link to the environment that makes the timing of metamorphosis plastic and adaptive (fig. 6.1). This species lives in regions where the natal ponds are likely to dry up at unpredictable times. Metamorphosis is accelerated as the ponds dry up, a highly fitness-enhancing response. In the laboratory, experimental lowering of the water level quickly produces elevations not only in thyroid hormones but also in corticosterone. Metamorphic changes are noticeable within 48 hours after the onset of the manipulation. Injection of a CRH-like peptide causes these same changes in hormones and morphology, and administration of a CRH receptor antagonist interferes with induction of metamorphosis produced by lowering of the water level. CRH also accelerates metamorphosis in the salamander *Ambystoma trigrinum* and might be a conserved mechanism in amphibians (Boorse and Denver 2002). It will be important to determine what the critical environmental stimuli are that set this cascade off (what is the specific cue related to water level?), how they are transduced into physiological signals, and how those get to the CRH neurons.

Insect metamorphosis, like amphibian metamorphosis, has big consequences for social behavior. Many insects grow within a rigid exoskeleton of hardened cuticle and thus have a special need for mechanisms to regulate the behaviors required to get out of the cuticle during molting (ecdysis) and out of the pupa at the final metamorphosis (eclosion). This may explain why the ecdysteroids are not found in vertebrates, why insect juvenile hormones do not resemble chemically the hormones of vertebrates, and why eclosion hormone (a polypeptide) does not resemble other peptide families (Truman and Riddiford 2002). Impressive progress all the way to the molecular level has been

A

B

FIGURE 6.1. Morphological changes characteristic of metamorphosis and postmetamorphic development in Western spadefoot toads: (A) larva (tadpole); (B) juvenile (left) and adult. Hormonal responses to changing environmental conditions (drying up of ponds) underlie metamorphic changes in morphology and behavior. Courtesy of Erica Crespi and Robert Denver.

made in determining the mechanisms of these processes and their accompanying behaviors in the moth *Manduca sexta* (Ewer and Reynolds 2002; Truman and Riddiford 2002). A combination of low juvenile hormone levels plus ecdysteroid levels that have first risen and then more recently fallen sets the stage for multiple peptides to produce the stereotyped preprogrammed rhythmic sequence of movements that enables the animal to wriggle out. Once the animal is an adult, juvenile hormones serve reproductive functions, for example, increasing sexual receptivity in females of some species (Nijhout 1994). This is a good example of how the same hormone can have a different function in different life history stages. It is also an example of hormones with a wide array of functions that have unfortunate names based on the first function that was discovered!

JUVENILE DISPERSAL

In cooperative breeding (chapter 3) individuals stay "at home" with the parents and alloparent younger siblings. In other animals juveniles of one or both sexes disperse from the area where they were hatched or born, moving away from familiar places and conspecifics to face dangers and strangers. Dispersal is a dramatic life history transition with respect to social as well as ranging behavior, one with big consequences for the demography of adult groups. Mortality rates are high during dispersal and it has not been easy to figure out what is so beneficial about dispersal that makes the risks worthwhile (Pusey 1987; Clobert et al. 2001). Is its function inbreeding avoidance? Lowering of resource competition? Lowering of intrasexual mate competition? The link to hormones comes from three properties of dispersal (Smale et al. 1997; Belthoff and Dufty 1998). First, the probability and timing of dispersal are often plastic and condition or environment dependent, suggesting that endocrine axis hormones related to social life (sex steroids?) or to food intake, metabolism, or energy stores (glucocorticoids? insulin? growth hormone? leptin?) might tell the animal if and when it's time to go. Second, dispersal is associated with increased locomotor activity, which has been linked to steroids in other contexts such as mammalian estrus. Third, dispersal probability or distance is often strongly sex biased, as if the hormonal processes that create other sex differences are at work. In thinking about the possibility of hormonal organization, it is important to remember that sex differences that occur in juvenile mammals (such as play) can be organized without requiring any activational hormones for expression (chapter 4).

Studies of birds and lizards suggest a possible involvement of glucocorticoids in juvenile dispersal. Willow tits (*Parus montanus*) hatched earlier in the year and implanted with corticosterone in the late summer, when winter flocks are beginning to form, were more likely to disappear from the area (or die?) than controls; controls were more likely to become integrated into a stable winter flock (Silverin 1997). While birds can disappear from an area for reasons other than dispersal, such as death, juveniles treated in late autumn, after they had become members of a flock, did not disappear, suggesting that dispersal had been affected. Both sexes of western screech owls (*Otus kennicottii*) disperse if they are in good condition, and they do so even if the adults are removed and the young are provisioned, ruling out conflict with the parents and inability to forage effectively as proximate causes of departure (Belthoff and Dufty 1998). In captive juveniles both locomotor activity and corticosterone levels peaked at around the time dispersal would have occurred in the wild. The researchers proposed that corticosterone causes increased locomotor activity, which then produces dispersal if body condition is good but increased foraging instead if condition is poor. These hypotheses have the virtue of being experimentally testable. In the common European lizard *Lacerta vivipara*, females treated with corticosterone while pregnant produce offspring that disperse less (as measured in a laboratory simulation), suggesting that the mechanism mediating the effect of maternal condition on offspring dispersal is corticosterone (de Fraipont et al. 2000).

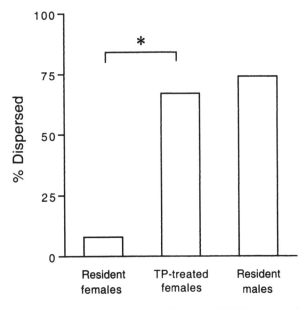

FIGURE 6.2. Sexual dimorphism in juvenile dispersal, a critical life history event, is an organizational phenomenon in Belding's ground squirrels. Females injected with testosterone propionate (TP) within 36 hours of birth were significantly more likely to disperse than untreated (resident) females, and dispersed with the same probability as males. Data from Holekamp, K. E., Smale, L., Simpson, H. B. and Holekamp, N. A. 1984. Hormonal influences on natal dispersal in free-living Belding's ground squirrels (*Spermophilus beldingi*). *Horm. Behav.* 18: 465–483. © 1984 Academic Press (Elsevier).

Belding's ground squirrels (*Spermophilus beldingi*) show a fairly common mammalian sex-biased pattern in which males disperse with much greater probability than females. Holekamp et al. (1984) asked whether this sex difference is due to activational or organizational gonadal steroid effects. The results of the experiments were clear. Animals gonadectomized three to six weeks after birth (well after any critical period for dimorphic behavior in rodents) dispersed at the normal rate, evidence against an activational hypothesis. Females injected with testosterone propionate within 36 hours after birth (within the likely critical period) were dramatically sex reversed, dispersing at close to male rates (fig. 6.2). This is evidence for a classic organizational phenomenon, mammalian style. It would be interesting to know what the process is that was affected. Locomotor activity? Willingness (decreased aversion) to take risks? Increased aversion to the mother?

Birds and mammals have different sex-biased dispersal tendencies (Pusey 1987). Among birds, more species have female dispersal than male dispersal, whereas among mammals, male dispersal is more common. These two lineages differ in their sex chromosomes. As discussed in chapter 4, birds are ZZ/ZW with ZZ males, whereas mammals are XX/XY with XX females. Differences in the sex chromosomal genetic architecture of these two lineages help explain why female dispersal has evolved more frequently in birds (has more often beaten the odds of going to extinction), whereas male dispersal has evolved more frequently in mammals (Reeve and Shellman-Reeve 1997). As Holekamp et al. (1984) pointed out, it would be interesting to know if the sex difference in birds is also hormonally organized and if ovarian hormones are the organizing agents instead of testicular hormones, as would be predicted from the research on sexual differentiation in chicken-like birds and ducks.

Primates are an exception to the general mammalian pattern, in that females are more likely to disperse in a number of species. An activational hormone hypothesis for greater female dispersal was tested and rejected in muriqui monkeys (*Brachyteles arachnoides*), but an organizational alternative has not yet been tested in this or any other primate (Strier and Ziegler 2000).

Could hormone studies help discriminate between alternative hypotheses for the function of juvenile dispersal? Pusey (1987) suggested that an organizational hormone mechanism would be particularly useful when inbreeding avoidance is the function. That is, a pronounced sex difference in dispersal would be particularly likely to evolve if the primary function is inbreeding avoidance, and organizational hormones are a sure-fire (although not the only) way to generate a pronounced sex difference. Moore's (1991) framework for thinking about within-sex types suggests the hypothesis that hormonal mechanisms will be activational when dispersal is facultative and condition dependent and organizational when it is obligatory and sex specific. Functions related to competition for food might be more likely to be based on glucocorticoid mechanisms, whereas functions related to reproductive competition might be more likely to be based on sex steroid mechanisms.

TEMPORAL POLYETHISM IN SOCIAL INSECTS

In advanced eusocial insects, the vast majority of workers never become reproductively mature. The behavior of the workers is not the same throughout their lives, however. Careful study of honeybee workers reveals that the tasks they perform change in a predictable sequence over a period of 8–11 weeks ("temporal polyethism"), beginning with brood care, progressing to hive cleaning and construction, food storage, and removal of the dead, and culminating in foraging outside the hive and returning to communicate nectar locations to other bees. This series of life stages involves significant bodily as well as behavioral change ("cryptic metamorphoses"). The timing of the transition from stage to stage is flexible. The timing varies according to colony need, an indicator of top-down hormonal regulation. Experimental evidence indicates that juvenile hormones are one of several proximate mechanisms for this flexibility (Fahrbach 1997; Sullivan et al. 2000; Schulz et al. 2002). In worker bees juvenile hormones do not serve as gonadotropins and activators of mating behavior the way they do in other insects, but, nonetheless, adult workers do have juvenile hormones. Their levels increase in a socially regulated way as the workers go through their task stages. If juvenile hormones or the juvenile hormone agonist methoprene are administered, workers assume the foraging role at an earlier age (more rapidly than controls). If the corpora allata, the source of juvenile hormones, are removed, workers are delayed in assuming the foraging role. Juvenile hormones have their effects via regulation of the neurotransmitter octopamine. This is an unusually promising system, especially now that the honeybee genome will be sequenced, for figuring out how the bee's assessment of colony need is transduced into hormonal and other physiological changes and how those then alter gene expression so that the different task behaviors occur (Robinson et al. 1997). How the bees assess and communicate colony need is an equally fascinating question that already has yielded some answers (Seeley 1995).

Onset of Reproductive Maturity: Puberty

For many species age at the onset of reproductive maturity (fertility) is a major predictor of lifetime reproductive success (Daan and Tinbergen 1997; Roff 2001). An early start enhances fitness. Age at maturity responds readily to artificial selection, suggesting costs preventing an earlier start in the wild (Drickamer 1981; Bronson 1989). Reproductive onset is regulated by the brain via the HPG axis. In some species it is environmentally regulated and plastic. It deserves close attention here because of its interesting links to steroids and social behavior.

"Puberty" is the term used to refer to the onset of reproductive maturity in mammals, a well-studied group. Age of puberty is usually defined as age at first ovulation or estrus in females and first production of mature spermatozoa in males. Age at puberty differs between individuals of the same sex. This is not surprising, because all traits show individual variation. What is less well known is that mammalian species also differ considerably in the magnitude of this variability relative to total lifespan (Bronson and Rissman 1986). This variation has both genetic and environmental sources.

Primates such as humans are at the end of the spectrum where variability in age at puberty is a small percentage of the lifespan. Individual differences in onset of puberty may be personally momentous to adolescents, but they are quantitatively minor from a comparative perspective. Old World primates and apes show a distinctive developmental pattern in which there is a lengthy juvenile period of low sex steroid levels followed by a dramatic increase in steroid hormones, first adrenal sex steroids (adrenal puberty or adrenarche), then gonadal sex steroids (gonadal puberty). This is the developmental pattern that has inspired hypotheses that puberty is initiated by reaching some critical stage of physical maturation or by some internal clock keeping track of chronological time since birth.

Much attention has been given to the changes in hormones and their regulators that occur in concert with puberty in primates and some other mammals and to the mechanisms that initiate the transition from juvenile to pubertal stages (Plant 2001; Ojeda and Terasawa 2002). Puberty involves changes at all levels of the HPG axis. Its initiation is brain directed, caused by a change not in the GnRH cells themselves, but rather in the neural inputs to them (their control mechanisms). Initiation involves a shift in the balance of hypothalamic excitatory and inhibitory inputs to the GnRH neurons that permits an increase in pulsatile GnRH release. In primates, prepubertal inhibition is pronounced and changes in inhibitory strength are especially important. Perhaps this is because puberty needs to be inhibited for such a long time (months or years), in some cases beyond the attainment of adult body size.

At the other end of the mammalian spectrum are those small rodent species where the age of reproductive maturity is highly variable relative to lifespan and poorly predicted by chronological age (Bronson and Rissman 1986). In some species and individuals, female puberty begins as early as weaning; there is no obligatory juvenile period. Pregnancy in females as young as nine days old (!) has been recorded in precocial rodents (Weir 1970). Under other conditions puberty does not occur until much later. This doesn't sound like any kind of puberty clock at work. Instead, the onset of puberty is regulated by food supply or social environment. It is the brain-regulated nature of the HPG axis that makes this possible.

Female mammals do not go into puberty or ovulate as adults if they are in negative energy balance, that is, if more energy is being used up or lost than is entering the body in food (Bronson 2000). The GnRH pulse generator produces pulse frequencies sufficient for the HPG axis to be "on" only if energy balance is positive. Evidently, the hypothalamic mechanisms regulating the pulse generator are receiving information about energy status through an incompletely understood pathway (Wade and Schneider 1992). Negative energy balance can happen to any mammal, and even female primates do not begin to ovulate if sufficiently starved. A small rodent will reach the starvation point much faster when food levels drop, however.

Experiments have shown that the critical feature of the social environment that influences age of puberty in some female mammals is chemosensory stimuli from adult conspecifics. Adult females or their bedding delay puberty, and adult males or their bedding accelerate it (Drickamer 1977; Vandenbergh 1994). One or both of these phenomena occur in several kinds of rodents but also in cows, sheep, pigs, marmosets, and tamarins. In mice but not pigs the vomeronasal organ and accessory olfactory pathway get the chemical information to the hypothalamus, and in mice the puberty accelerating chemicals (which are in the urine) have been identified (Signoret 1991; Novotny et al. 1999). A striking example of this male induction phenomenon is seen in the musk shrew (*Suncus murinus*) (Schiml et al. 2000). The onset of ovulation is completely dependent on interaction with a male. The interaction begins with the female being very aggressive toward the male but after an hour or less she becomes receptive. An interaction with the male stimulates an increase in GnRH and ovulation anytime after weaning. Odors from the male are part of the set of stimuli responsible for this phenomenon.

There are interesting hypotheses about why social regulation of puberty might be adaptive for rodents (Bronson and Rissman 1986). Perhaps adult females represent competition for space or food, so that it is better for the young female to put her energy into growth or dispersal rather than trying to reproduce, whereas adult males signify a reproductive opportunity that she should take advantage of fast because of her short lifespan. The female's brain is engaged in a calculus designed to turn on the HPG axis when circumstances are propitious (or at least not contraindicated). (See also discussion of these phenomena as signaler–receiver interactions in chapter 2.)

An extreme case of puberty delay because of conspecific stimuli occurs in some cooperative breeders such as marmosets and tamarins where the alloparents are older offspring that fail to become reproductive (chapter 3). In mammals, where alloparents are more often females, this can be viewed as a special case of the more general phenomenon of inhibition of female reproduction when environmental, including social, conditions are not favorable (Wasser and Barash 1983). When mammalian male offspring are the alloparents, they usually do produce spermatozoa (undergo puberty), probably because the costs of reproduction are lower for males so there is less need to keep the system off just because the immediate benefits won't be great (Creel and Waser 1997; Faulkes and Bennett 2001).

Even in species with relatively little variation in age at puberty, there can still be external influences on the variation that does occur. In rhesus males, social rank predicts age at puberty (Mann et al. 1998). In prepubertal rats, males groom their own genital regions more than females do. This self-grooming accelerates pubertal increases in the size of the male accessory sex glands (Moore and Rogers 1984).

The word "puberty" originally referred to the human phenomenon. How well does it capture what happens in other mammals? Not very well, in the case of animals with little or no juvenile period or a highly variable environmentally regulated reproductive onset. In addition, it implies a non-seasonally breeding primate pattern in which there is a sudden transition from "off" all the time to "on" all the time. Sheep are are highly seasonal fall breeders and onset of puberty, like onset of subsequent breeding episodes, is stimulated by decreasing daylength. Lambs are all born at the same time of year, and so this kind of environmental regulation does not produce the degree of variation in age at puberty seen with environmental regulation by food and conspecifics (Bronson and Rissman 1986). From a neuroendocrine standpoint, something like puberty recurs every year. One hopes this is easier on the sheep than it would be on humans! From a life history standpoint, however, the onset of reproductive maturity is an important life history transition regardless of what it is called.

Seasonality and photoperiodism are also important determinants of reproductive maturity in some birds. In the case of spring (long day) breeders, if a young bird is not "ready" later in the summer, it will have to wait (reproduction is put on hold) until the following spring. The age distribution of onsets for seasonally breeding birds is discrete, not continuous. Dawson et al. (1987) reported that thyroidectomy causes earlier puberty in starlings, a highly photoperiodic species, suggesting links between photoperiod, thyroid hormones, and age of puberty in a seasonally breeding bird. Studies of wild birds in which hormones are actually measured during the juvenile period leading up to onset of reproduction are rare. Clearly the profiles would be very different in seasonal versus nonseasonal breeders. In zebra finches, which are more opportunistic than seasonal, sex steroid levels are remarkably constant from hatching until reproductive adulthood in both sexes when they are housed on a constant daylength (Adkins-Regan et al. 1990). One of the best-studied species, the domestic Japanese quail, has an unusually rapid maturation time (six weeks). Androgen levels of males housed on long days rise gradually during the juvenile period (Ottinger and Bakst 1981), but in wild birds changing daylengths probably produce a different profile.

In both birds and mammals the age of reproductive onset relative to the attainment of adult body size varies across species. This phenomenon is particularly striking in birds, which reach adult body size relatively quickly as vertebrates go. Large seabirds and raptors often don't reach reproductive maturity until several years after reaching adult size. This is thought to be due to the additional time required to become a sufficiently adept forager to be independent of the parents. The evolutionary lability in the relationship between the growth and reproductive trajectories may be enabled by the modular nature of development (chapter 5).

Onset of reproductive maturity is quite variable in many fish species. Some of this variation is clearly related to social conditions via the HPG axis (Francis et al. 1993; Hobbs et al. 2004). In studies of sexual maturation in poeciliid species of the genus *Xiphophorus*, age of onset of reproduction (when male growth ceases) varies enormously (from 8 to 104 weeks in some cases), resulting in substantial variation in size of adult males. Some of this is genetic (a sex-linked gene creates alternative maturation timetables) and some is socially produced (Kallman and Schreibman 1973; Borowsky 1978). The maturation cascade is initiated by the nucleus olfactoretinalis, which receives external cues (Schreibman and Magliulo-Cepriano 2002). At maturation major changes occur in the olfactory system and in GnRH cells in the olfactory bulb.

In some salmonid species, male size is bimodal because some males mature precociously and are small at sexual maturation, whereas other males and females mature at a later larger size and age. Genetic males with ovaries (produced by experimental sex reversal) all mature at a female-typical age, indicating regulation of age at maturation by gonadal hormones (Piferrer and Donaldson 1988).

Sex differences in age of reproductive onset are common in vertebrates. Some of this dimorphism (sexual bimaturism) is socially regulated. For example, in cooperative breeders where only one sex alloparents, a sex difference will occur whenever there is socially influenced pubertal delay in the alloparenting sex. In some other cases, however, the sex difference seems to be more "wired in," as in humans, where males reach puberty later than females. Here too social organization is the likely ultimate cause of the sex difference even if it is not the proximate cause of a delay. For example, when there is a size-based male dominance hierarchy and a highly polygynous mating system, young males may be excluded from reproduction so effectively that over evolutionary time male puberty has been postponed to allocate energy to somatic functions instead. Are such sex differences hormonally organized? There is good evidence in sheep that the sex difference in age at puberty is organized before birth (fig. 6.3). Female sheep have a later puberty than males, probably because of energetic limitations of lactation. Wood et al. (1991) treated pregnant females with a form of testosterone. The treated and control male offspring did not differ. Both reached the puberty criterion (a criterion increase in LH) in 5 to 10 weeks. Control females reached the criterion in 27 weeks, but androgenized females reached it in 14 weeks on average, a substantial reduction toward male-typical values. This pattern of results fits the predictions of a mammalian organizational hypothesis. The neonatal elevation in LH and T in male primates has been hypothesized to organize the sex difference in puberty, and there is some evidence that blocking that elevation causes the animals to be hormonally different later on, when they are young adults (Lunn et al. 1994; Mann et al. 1998).

It is well known that behavior as well as physiology changes with the onset of reproductive maturity. There is abundant evidence that the hormonal and behavioral changes that accompany mammalian puberty and quail reproductive maturation are causally related—that the rise in circulating sex steroids is stimulating sexual behavior and other adult steroid-influenced social behaviors. This is not the whole story, however. Chapter 2 already discussed ways that hormones are necessary but not sufficient in mature adults, especially those in social groups. One way to find out if the hormones of puberty are sufficient is to provide juveniles with adult sex steroid levels and see if they start behaving like adults. The result in some species is a marked increase in the occurrence of precocious sexual behavior. The sight of testosterone-treated baby chicks making little high pitched crows and trying to copulate with the

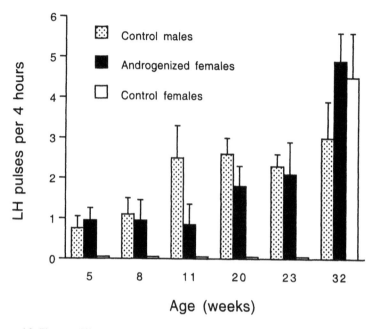

FIGURE 6.3. The sex difference in age at puberty, an important life history transition and a key predictor of lifetime reproductive success, is an organizational phenomenon in sheep. LH pulse frequency at different ages is shown for control males, control females, and androgenized female lambs. All animals were gonadectomized and estrogen treated at the time of the measurements. Androgenized females were born to mothers treated with testosterone cypionate during gestation. Even though the animals were in the same hormonal state at the time of measurement, there was a marked sex difference in the age at which LH pulse frequency increased, indicating the onset of puberty. Puberty occurred much earlier in males. Puberty in androgenized females occurred at a male-typical, not female-typical, age. Redrawn from Wood, R. I., Ebling, F. J., I'Anson, H., Bucholtz, D. C., Yellon, S. M. and Foster, D. L. 1991. Prenatal androgens time neuroendocrine sexual maturation. *Endocrinology* 128: 2457–2468. © 1991 The Endocrine Society.

experimenter's hand is quite humorous (Andrew 1966). The behavior has an unnaturally reflexive quality to it, however, and obviously lacks intelligent mate choice and assessment, as if some additional neural maturation needs to happen. Furthermore, not all species will display precocious sexual behavior when given hormones as juveniles, as if much more has to mature in addition to the HPG axis. In hamsters, the change in male sexual behavior at puberty occurs not just because of rising hormones, but also because of increased sensitivity to those hormones (Sisk 2000). Testosterone activates the behavior much more effectively, and at a lower dosage, in postpubertal male hamsters than in prepubertal animals. What has changed in the animal? Numbers of steroid receptors in sexual circuitry regions of the brain are good candidates, but careful tests of this hypothesis have so far provided no support

(Romeo et al. 2002a,b). The mechanistic cause of this important pubertal change is still a mystery.

When species have a well-defined hormonal puberty, does the increase in sex steroids permanently change the animal's brain and behavior? Is there some kind of hormonal organization occurring at this life history transition? Is this what is missing in the prepubertal hamsters? There are clearly important changes in brains as well as behavior at this time (van Eerdenburg et al.1990; Giedd et al. 1999). Accepting the hypothesis of hormonal organization requires more than just correlated hormone–brain–behavior changes, however. The changes need to be permanent, remaining in place even if hormonal status reverts to the juvenile (nonbreeding) state. The best evidence thus far for an organizational effect of pubertal hormones on behavior comes from experiments with tree shrews (Eichmann and Holst 1999). Male testosterone rises suddenly at puberty and peaks above the adult steady-state level before dropping down to it. Adult males, but not females, increase their scent marking in response to scents from other males. Males were castrated before puberty and given short-term testosterone treatment (to mimic the pubertal rise) either at the normal age of puberty or later. They were then given testosterone again several months later (the hormone replacement needed to activate male typical behavior) and tested for scent marking. The prepubertally castrated males on testosterone replacement only showed the male-typical scent-marking pattern if they also had received the experimental peak of testosterone at the age when normal intact males would be experiencing a peak. Mimicking the peak later did not masculinize the males. These results suggest that the pubertal peak has a long-term effect on the sensitivity of the animal to future testosterone, an effect that is only possible during a critical period corresponding to the age of puberty, constituting evidence for pubertal organization.

All told, the neuroendocrinology of reproductive maturation, especially the key role of the brain's control of the HPG axis, gives deeper insight into how this important life history stage, and the suites of traits involved, can change so readily in response to changing conditions, within the lives of individual animals, across populations over a few generations, and over evolutionary time (Finch 2002).

Aging and Senescence

The developed world is getting grayer fast, demographically speaking. Aging is a front-burner topic in science and the media. Here the concern is with aging as a part of life history and especially with its important connections to steroid hormones and social behavior.

Longevity is a major determinant of reproductive success in some animals, especially those that produce a small number of young at regular intervals (Newton 1989; Finch 1990). Mortality due to predation, disease, or accidents sets the limit to the lifespans of many but not all species and individuals. Some undergo senescence, a suite of changes with negative consequences for reproductive success and overall physical condition. Reproduction may even cease altogether. Isn't this terribly maladaptive? Life history thinking has led to important insights about the evolution of senescence (Medawar 1952; Williams 1957). As an animal ages (chronologically speaking), selection becomes progressively weaker because the percentage of the animal's lifetime reproductive success that remains decreases. Genes that might extend the lifespan a bit more are subject to only weak selection. Genes with deleterious effects later in life will not be selected against so long as their effects on reproductive success earlier in life are positive (antagonistic pleiotropy) (Kirkwood 1977). Hormones are likely mechanisms underlying some of this antagonistic pleiotrophy, as when sex steroids increase the likelihood of future fatal reproductive organ cancers (Finch and Rose 1995).

Viewed in this way, senescence itself is not adaptive. It happens because of other traits that are adaptive in the young. Senescence might be expected to occur in most animals that reproduce multiple times. Nonetheless, lineages are thought to differ in whether and how much they senesce. Congdon et al. (2003) found that females of two freshwater turtles were not only very long lived, but showed no increase in mortality or decline in reproductive ability with age. Birds have slow aging rates compared to mammals and are longer lived at the same body size (Holmes et al. 2001). Although birds senesce, they seem to do so in a less dramatic way than mammals (Bennett and Owens 2002). One difficulty with such a comparison, however, is that it might be harder for humans to notice or measure senescence in some birds than in mammals.

With respect to hormones and social behavior, the questions that need to be addressed are these. When changes in sex steroids and social behavior occur as adults get older, are they causally related? What is the direction of causation? Do effects of social behavior on hormones contribute to individual differences in lifespans? Do they contribute to sex differences in lifespans?

HORMONES AS CAUSES OF AGE CHANGES IN BEHAVIOR

There are remarkably few species for which anything is known about how both sex steroid levels and behavior change with age and whether the one has anything to do with the other. Humans have received a hefty proportion of the attention, reflecting the desire to discover pharmacological ways for the aging to feel and act young again. Sexual behavior, aggressive behavior, and overall sociality decline with age in humans, and in many men and all women who live long enough circulating sex steroids do as well. It is popularly assumed that the hormonal changes are a cause of the behavioral changes.

Before examining the empirical evidence, it is worth remembering a few lessons from earlier chapters. The threshold concept suggests caution in assuming that the hypothesis is correct. Only if sex steroids fall below the threshold would a drop in behavior be expected. This seems much more likely for women (whose ovaries virtually stop producing sex steroids) than men (whose testes continue functioning). Also, when young adult animals are compared with older animals, sexual and other social behavior is often less hormone dependent as the animals get older, possibly because experience comes to play a larger part. And, of course, a change in hormone-influenced behavior as the body ages is just as likely to be due to changes on the receiving end—steroid receptors and the cascades they initiate—as to a change in hormones.

Let's look first at men. How universal is a decline in testosterone levels with age? There are interesting differences between human populations in how much testosterone changes. In some there is little decrease (Ellison and Panter-Brick 1996). Western men, who have higher testosterone to begin with, show a bigger drop, but even here the decrease is not large if the man remains healthy (Bribiescas 2001; Feldman et al. 2002). There are substantial differences between individual men from the same population in whether and how much sexual performance and hormones change with age (Davidson et al. 1983), so that what looks like a correlated gradual decline in grouped data is actually a set of step functions with variation in the age of the step down. Subsequent longitudinal studies have shown that the sexual behavior of men declines before their testosterone levels do, and that testosterone falls below the normal range of younger men only if the man is ill (vom Saal et al. 1994). Lack of sexual function in old healthy men is not due to testosterone deficiency. Androgen replacement would be expected to improve behavior only if sex steroids are abnormally low (for an older man) to begin with, a hypothesis that has received empirical support (Kunelius et al. 2002).

In both men and women levels of dehydroepiandrosterone (DHEA) decline markedly in middle and older age (Laughlin and Barrett-Connor 2000). The same thing happens in both sexes of free-living yellow baboons (Sapolsky et al. 1993). DHEA is a precursor for androgens and estrogens but is only weakly active itself. After decades of ignoring DHEA there has been a flurry of studies administering DHEA to older people. Well-designed experiments with placebo controls concur that there is little or no effect on behavior or self-reported energy. It would appear that the hormonal fountain of youth is not yet in sight.

Male rhesus monkeys in laboratories become sexually sluggish and inactive as they get old but their circulating sex steroid levels do not decline, nor does testosterone supplementation improve behavior (Phoenix and Chambers 1986). What works better is giving them their preferred female sexual partner. Novel females do not have the same invigorating effect. Declines in sexual behavior with age in male rats are due more to changes in the central nervous system than to a decline in peripheral hormones (Gray et al. 1981; Smith et al. 1992). Some male mammals, such as elephants, have increased, not decreased, testosterone with age (Jainudeen et al. 1972).

The only nonmammalian males whose hormones have been studied systematically throughout aging are Japanese quail. Domestic quail undergo a more pronounced reproductive senescence than most birds, even other domestic birds. Males tend to become infertile and cease mating between about 2.5 and 3 years of age, and as they age circulating testosterone, brain metabolism of testosterone, and vasotocinergic innervation to key testosterone targets in the brain all decline (Balthazart et al. 1984b; Ottinger 1992; Panzica et al. 1996a, 1997). Individual males differ markedly; some remain sexually active, while the behavior of others drops to zero. Some of the behavioral decline does seem to be due to the decline in testosterone. Not only does testosterone decline more in the males that become inactive, but if testosterone levels are raised, many of the formerly inactive males become active again (fig. 6.4) (Ottinger and Balthazart 1986). More testosterone is required than would be needed in a younger castrated male, however, suggesting that the behavioral deficit is due not just to lower testosterone levels but also to lower brain sensitivity to testosterone. Furthermore, the behavior of individual males drops before their testosterone levels do (Balthazart et al. 1984). The hormones don't change until the male has stopped reproducing altogether, as if a decline in brain sensitivity (or some other process) occurs first and HPG failure is secondary (Ottinger et al. 1995).

Mention "reproductive aging" and the word "menopause" is likely to be triggered. Human menopause is the cessation of ovarian function, both gametic and hormonal. Provided women live long enough, it is a species-wide phenomenon, not a condition-dependent one. There is currently much interest in whether menopause occurred in the evolutionary past, whether there is (or was) some inclusive fitness benefit to it, and whether it is part of an evolved life history strategy representing a shift from mating and parental effort to grand-parental effort (Hawkes et al. 1998; Shanley and Kirkwood 2001; Lahdenperä et al. 2004).

Gametic menopause is not unique to humans or primates, because the ovaries of all female mammals eventually become depleted of oocytes under captive conditions where lifespans are prolonged (Finch 1990). In the wild, however, there are few reports of cessation of reproduction in otherwise healthy females on a species-wide basis except for baboons and lions (Packer et al. 1998). Hormonal menopause, on the other hand, is not a general mammalian phenomenon even in captive animals, occurring only in a few species (Gilardi et al. 1997; Finch and Kirkwood 2000). Female rats go into a state of persistent estrus rather than lack of estrus.

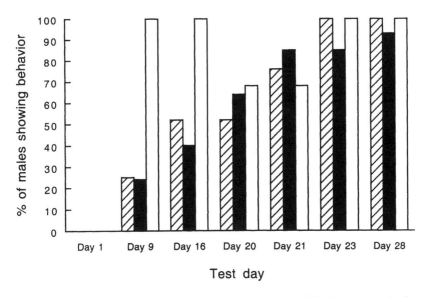

FIGURE 6.4. Percentage of young and old castrated male Japanese quail showing mating behavior after receiving implants of testosterone on days 1 and 14 of the testing period. Young males (clear bars) were 50 weeks old and behaviorally active before castration. Old males were 216 weeks old and were either behaviorally active (diagonally striped bars) or inactive (black bars) before castration. In this experiment castration and testosterone replacement had a restorative effect on the previously inactive old males, in contrast to experiments with mammals where similar manipulations were ineffective. Nonetheless, even these male quail have aging-related deficits in the brain (reduced sensitivity to testosterone), indicating that behavioral senescence is a central as well as peripheral phenomenon. Redrawn from Ottinger, M. A. and Balthazart, J. 1986. Altered endocrine and behavioral responses with reproductive aging in the male Japanese quail. *Horm. Behav.* 20: 83–94. © 1986 Academic Press (Elsevier).

Human menopause tends to be associated with a decline in sexual interest and activity (Myers 1995). The effect size is modest and other factors such as availability of partners and mental and physical health nearly overwhelm it. If menopause is an evolved strategy (which is a big "if"), one could view this decline as a part of the overall adaptive package along with the loss of fertility. Given the costs of sex (for example, sexually transmitted disease), the woman should not continue to be sexually active if a shift to grand-parenting will bring greater genetic returns. Regardless of whether the decline in sexual interest has any function, we can ask if it is related to the drop in ovarian hormones. Well-designed experiments in which middle-aged ovariectomized or naturally menopausal women are given different forms of hormone replacement therapy reveal that estrogen replacement has little effect on any measures of sexual interest or function, whereas adding androgen replacement produces significant increases in several measures (Sherwin et al. 1985; Sherwin 1998).

At menopause androgens decline along with estrogens and progestins and seem to be the "culprit" behind sexual decline.

It is interesting that the few studies of nonhuman females yield a different picture of sexual behavior and sensitivity to hormones than the studies of women and males. Female rats show no decline in sexual behavior with age, and those that go into persistent estrus show persistent receptivity. Older female rats and voles become more, not less, sensitive to estrogen replacement, unlike male mammals (Cooper 1978; Petersen 1986).

Much of the literature on hormones and behavioral aging has been inspired by biomedical concerns. The usual subtext is how everything gets worse with age unless rescued by the miracles of modern pharmacology (ignoring, more often than not, the possible risks and safety concerns of regarding aging as a disease and then drugging it). Life history theory, however, recognizes not only deterioration and pathology at advanced ages (senescence), but also improvements in reproductive success per attempt as animals increase their reproductive effort (Williams 1966; Daunt et al. 1999). They increase their reproductive effort because (evolutionarily speaking) they have less future reproduction left for which to save energy. In mammals, litter size and offspring survival increase with maternal age. The increased sensitivity to estrogen with age seen in rodents could be part of the mechanistic basis for these improvements. With successive litters, maternal behavior becomes less hormone dependent and more readily stimulated by cues from the offspring (Bridges 1996). Parenting and annual reproductive success improve with age in long-lived birds (Newton 1989; Bennett and Owens 2002). It will be important to determine the contribution of hormonal mechanisms (along with experience) to this improvement. Thus far, hormonal aging over a wide age range has been investigated in only one wild bird, the common tern (*Sterna hirundo*) (Nisbet et al. 1999). There was no relationship between age and reproductive hormones in a cross-sectional sample of birds of known ages 2 to 21 years, but only breeding birds (ones that succeeded in nesting and incubating) could be located for sampling, a problem with highly mobile animals. Longitudinal as well as cross-sectional studies are needed.

BEHAVIORALLY INDUCED HORMONE CHANGES AS CAUSES OF INDIVIDUAL AND SEX DIFFERENCES IN SENESCENCE

Longevity has significant heritability, averaging around 0.5, providing the raw material for evolutionary change (Finch and Kirkwood 2000). Nongenetic factors are also important determinants of longevity. Research with laboratory mammals and humans has suggested that one such determinant might be chronic HPA activation (cumulative stress) extending over long periods of time (Sapolsky et al. 1986). As discussed in chapter 3, social instability or status decline can be a powerful stressor. Social environment is also a likely source of chronic stress in wild animals. Whenever chronically elevated steroids take a toll on the animal, there is the possibility that behaviorally induced hormonal elevations could affect longevity. Perhaps frequent engagement in conflict and aggression shortens lifespan. Glucocorticoid levels rise with age and we can ask if individual variation in the rise is behaviorally driven (Laughlin and Barrett-Connor 2000).

This same pathway leading from an animal's behavior through hormonal changes to longevity could also be a factor contributing to sex differences in lifespans. Some of the difference is due to differential mortality caused by hormonally stimulated behavior, as when males have shorter average lifespans due to juvenile dispersal or death from injury during fighting. One sex might be subject to a greater cost of reproduction or more cumulative lifespan-shortening stress than the other. We can also ask if sex steroids organize or activate any of the sex difference through pathways other than energy expenditure during life. It has long been known that castrated men live longer than other men on average (Hamilton 1948) but is this due to different behavior or to some other cost of testicular activity? The best way to test an organizational hypothesis of a sex difference in longevity would be to find a species with substantial between-sex variation relative to within-sex variation and a maximum lifespan short enough to fit into an experimenter's timeframe. There aren't very many vertebrates that fit these requirements.

SEMELPARITY

The semelparous mode of reproduction is undoubtedly the most dramatic kind of senescence and is equally fascinating hormonally and evolutionarily. Many small vertebrates live to breed only once because of high mortality. Semelparity, however, refers to a uncommon mode of reproduction in which the animal breeds once, but in a spectacularly energetically intensive way that is followed by rapid and fatal senescence, suggesting an evolved life history strategy. This strategy is most thoroughly studied in relationship to hormones in several species of salmonid fish and in the marsupial genus *Antechinus*.

Pacific salmon (genus *Oncorhynchus*) populations are often semelparous and are also migratory. After maturing out at sea, males and females swim upstream (avoiding hungry bears) to where they were hatched, engage in a massive spawning event involving much fighting for access to spawning sites and mating partners, and then die. These dramatic events are a familiar staple of televised nature programs. A comparative study of semelparity and iteroparity in salmonids provides evidence that these strategies are different solutions to the trade-off between current and future reproductive effort that can evolve rapidly (Unwin et al. 1999; Crespi and Teo 2002). If adult survivorship is low enough (because of the hazards of migration) and juvenile survivorship is high enough (because of large eggs), all effort should go into the first reproductive event. The somatic and neural signs of senescence begin before arrival at the spawning grounds and coincide with a substantial increase in cortisol, suggesting that high levels of cortisol might be a proximate cause of senescence (Maldonado et al. 2000; Barry et al. 2001). If males are castrated, they don't undergo the big increase in cortisol and they live longer (Robertson 1961).

In semelparous species of *Antechinus* it is mainly but not exclusively the males that are semelparous. Males all die off suddenly before they are a year old after two to three weeks of intense mating and male–male aggression, whereas some females live more than two years (Lee and Cockburn 1985). Male androgen (mainly testosterone) levels rise substantially, adrenals become heavier, and glucocorticoid levels rise markedly (as they do in salmon), but these endocrine changes happen after the mating period begins, not before (Bradley et al. 1980; McDonald et al. 1981). Total cortisol gets very high in females as well, and some of these same hormonal changes are seen in nonsemelparous species, but only in the males of semelparous *Antechinus* does the cortisol-binding capacity of the blood (a measure of CBG level) drop dramatically, resulting in an excessive amount of free cortisol. Are the increases in steroids in males caused by their behavioral interactions? If so, do those increases have immediate benefits such as stimulating yet more sexual and aggressive behavior or reducing fatigue? Which causes the death of the males: the increases in steroids or the aggressive behavior or both? Males brought into captivity before sexual maturity live two years or more, supporting the hypothesis that death is initiated by the activities of the mating season and is not preprogrammed (Rigby 1972). Castrated males returned to the wild don't die young, undergo the changes in adrenals, or undergo the drop in plasma cortisol binding capacity (Bradley et al. 1980). Experimental androgen treatment of castrated males causes a substantial decline in binding capacity (McDonald et al. 1981).

All told, it looks as if death is caused by a combination of rising cortisol and falling CBG, that the fall in CBG is caused by rising androgens, and that the males' behavioral interactions contribute to the elevations in both androgens and cortisol (Bradley et al. 1980; McDonald et al. 1986). In spite of the high glucocorticoids, the males continue to mate and produce sperm. Either the HPG axis is more resistant to inhibition than in other mammals or the time delay between glucocorticoid elevation and HPG inhibition is long enough to permit a reproductively successful mating episode.

The focus in studies of hormones and semelparity has been on fatal senescence, with its heartwarmingly tragic aura. Semelparity can also be seen as likely to produce strong sexual selection (Andersson 1994), but under conditions where the future is irrelevant. To the extent that the hormones supporting sexually selected behavior and signals really do have costs in future fitness currency, those costs are irrelevant. There would be no limit to the evolution of higher hormone levels except the past and current costs of their production.

Thus, we can predict that semelparous species engaging in their one and only reproductive event should have much higher sex steroid levels than related iteroparous species engaging in their first reproductive event. The iteroparous congeners should instead increase effort and sex steroid levels with each successive event as future costs diminish.

Judging by the rarity of semelparity in vertebrates, it's a strategy that is seldom worth it. The complete lack of semelparity in birds might seem to be a mystery (Bennett and Owens 2002) except that the dramatic sort seen in *Oncorhynchus* and male *Antechinus* in which death occurs to all shortly after mating would be unlikely to evolve in any species with obligatory parental care. Most birds have biparental care, and some of the basal lineages have male-only parental care. Adult survival usually far exceeds juvenile survival in birds, the opposite of the demography thought to lie behind semelparity in *Oncorhynchus*. The real mystery is why more small-mammal males have not adopted the *Antechinus* strategy.

Hormones, Social Behavior, and Life History Trade-Offs

TRADE-OFFS BETWEEN REPRODUCTIVE EFFORT AND SOMATIC FUNCTIONS AND BETWEEN CURRENT AND FUTURE REPRODUCTIVE EFFORT

A consensus is emerging that across an array of species there is the predicted trade-off (negative correlation) among individuals between reproductive effort and somatic functions (growth and survival) (for example, Bennett and Owens 2002). Clearly, hormones broadly defined are involved in this trade-off as mechanisms of each of these functions. When does steroid-related behavior seem to be part of the story? The *Antechinus* studies support the hypothesis that it is the male's behavior that shortens his life by causing fatal changes in adrenal hormones. This is an extreme solution to the trade-off but one that has worked adequately for these animals. For iteroparous animals a similar but less extreme phenomenon could be occurring. A less pathology-oriented version of the cumulative social stress hypothesis would posit that glucocorticoids are a mechanism mediating a trade-off between sexually selected dominance/aggressiveness and longevity. What is interesting about having a behavioral mechanism and its hormonal sequelae be the individual level solution to the trade-off rather than some kind of genetic preprogramming is that the animal's nervous system makes a decision about whether and when to engage in the behavior.

The phenotypic engineering experiments in which wild male birds and lizards are given implants to raise their testosterone levels directly address this trade-off (Ketterson et al. 1996; Sinervo and Basolo 1996). In some but not all of these experiments, testosterone-treated males, compared with control males, showed higher levels of behavior contributing to overall reproductive effort (such as territorial defense, home range size) but had poorer survival (Moss et al. 1994; Marler et al. 1995; Sinervo et al. 2000; Casto et al. 2001). When this happens, the treatment has produced the predicted negative correlation between the two life history components. What is not conclusively known, however, is whether such a treatment that affects many traits reduced survival because of an increase in reproductive effort, as opposed to something else affected by the treatment. The "something else" could include not only physiological effects of a nonreproductive sort but also changes in other hormones such as glucocorticoids or pituitary hormones. It is also difficult to pin down whether it is behavioral effort that is reducing survival. This is not critical for life history theory, but it's important for anyone interested in hormones and behavior. One approach would capitalize on the likelihood that some of the effects of testosterone on reproductive behavior are likely to be mediated by aromatase and estrogen receptors, whereas some of the peripheral actions on physiology and metabolism are likely to be mediated by other pathways. If a combination of testosterone and an estrogen receptor antagonist or aromatase inhibitor failed to affect survival under the same conditions that testosterone alone reduces it, that might strengthen the case that it is reproductive effort stemming from the male's behavior that is responsible.

Another manipulation that has been used in birds is increasing the brood size (number of chicks in the nest or brood). This is a particularly good technique for getting at behavioral effort because postlaying investment by birds (with the exception of pigeons and doves who produce crop milk) is mainly behavioral. Parental care by birds is a costly (mortality increasing) activity, and the parenting sex has higher mortality. Increasing the brood size experimentally lowers subsequent survival (Daan and Tinbergen 1997; Bennett and Owens 2002). Because the birds did not themselves produce the larger brood, presumably the effect is occurring via increased behavioral effort (e. g., more parental foraging trips, more chick brooding). It will be important to determine the physiological pathway mediating the effect of behavioral effort on subsequent survival and the extent to which hormonal changes (including changes in glucocorticoids) produced by increasing the brood are involved. In tree swallows (*Tachycineta bicolor*) experimental brood augmentation decreases humoral immune function (production of secondary antibodies) (Ardia et al. 2003).

The hypothesized life history trade-off between current and future reproductive effort is also likely to involve hormones and behavior in interesting ways. This hypothesis goes back to Trivers' (1972) concept of parental investment as occurring at the expense of future reproduction. It does not, however, seem to be as well supported empirically as the trade-off between reproduction and survival. Some of the better evidence comes from the brood size manipulation studies with birds, in which subsequent fecundity is reduced in birds that do survive (Daan and Tinbergen 1997). Is this because subsequent gonadotropin or sex steroid levels are lower? If so, hormone supplements should prevent the fecundity reduction. Hormones related to energy storage would presumably be relevant to the trade-off between current and future reproduction in species that use stored energy for reproduction, such as red deer, more so than for those that rely on daily energy intake, such as small birds and mammals (Stearns 1989). Also, birds and mammals would be likely to have different solutions. Mammals are shorter lived than birds of the same body size. Small rodents have little future reproductive potential and instead of solving trade-offs in a bird-like way, simply go all out, saving no energy for the future, providing they are in positive energy balance to begin with (Bronson 2000).

TRADE-OFFS BETWEEN MATING AND PARENTAL EFFORT

Birds are the parental effort vertebrate lineage par excellence. Eggs may have to be covered nearly constantly, limiting foraging. Chicks rapidly increase in size, requiring frequent feeding and often necessitating biparental care. Anyone who closely watches a songbird nest in the garden soon feels sorry for the poor parents, who seem to be working themselves to death for young that will probably be eaten by the neighbor's cat the minute they fledge. It's hard to imagine that either of the parents could be simultaneously spending a great deal of time and energy on intrasexual competition for additional mates or extrapair mate attraction. What does the research show?

Trade-offs for male birds between mating and parental effort are indeed well substantiated (Wingfield and Farner 1993; Ketterson et al. 2000; Magrath and Komdeur 2003). Both efforts have substantial behavioral components. Negative correlations between behavior functioning to attract and keep mates and fend off rivals (such as singing and attacking territorial intruders) and behavior directed at the offspring (such as bringing food to the chicks) have been obtained in a number of wild populations. Part of the trade-off results from the simple fact that males cannot perform both these sets of behaviors simultaneously—they can't be at the territorial boundaries and feeding chicks in the nest at the middle of the territory at the same moment. Birds are highly mobile, however, and so that's not the whole story. There seems to be a more fundamental trade-off between these two kinds of effort. Most of the behaviors that go into male mating effort are known or suspected to be sex steroid based. This raises the possibility that the two suites of behaviors are hormonally incompatible as well, and that sex steroids mediate the trade-off, shifting the males from the mating effort phase of the breeding effort to the parental phase, a key life history transition for birds. The diverse testosterone profiles of the males of different species then reflect the solution to the trade-off that worked in the past and possibly the present as well (Wingfield et al. 1990; Ketterson and Nolan 1999).

The experiments that tested this hypothesis have been described in chapter 5 (section on reproductive success). Briefly, free-living males of biparental north temperate songbird species given experimental elevations of testosterone increased behavior indicative of mating effort and reduced chick feeding (Silverin 1980; Wingfield 1984; Hegner and Wingfield 1987; Dittami et al. 1991; Raouf et al. 1997; De Ridder et al. 2000). The behaviors that higher testosterone favors are incompatible with good chick care. Small wonder then that testosterone levels of these males are low during incubation and chick-feeding stages of the annual breeding cycle.

These tests raise three important questions. First, do we know that parental effort decreased because mating effort increased? Would the testosterone have interfered with parental behavior even if there had been no increase in mating effort? This is the same question that was posed in the discussion of hormone-manipulation experiments to test trade-offs between reproductive effort and survival, but here it is more difficult to come up with a realistic experimental approach to nail the hypothesis. The experiments do not necessarily show that circulating testosterone itself is the key mechanism mediating the trade-off rather than some other hormone or aspect of physiology that changes following testosterone treatment. It is possible that testosterone treatment lowers prolactin levels, but there is no evidence that the reduction in parental care has anything to do with prolactin (Schoech et al. 1998; Sockman et al. 2004). Also, a hormonal basis for parental care is not required for testosterone (or some other hormone) to mediate the trade-off. As discussed in chapter 2, parental behavior tends to be less hormone dependent than other reproductive behavior. Perhaps for a male exposed to stimuli from chicks in his nest on his territory, all that is required to feed them is the absence of some competing behavioral priority. A drop in sex steroids could remove sexually selected competing priorities.

How else could the hypothesis that testosterone mediates the trade-off be tested? Another approach would be to block testosterone action in males of a facultatively paternal species and see if parental effort increases. Or parental effort could be increased or decreased experimentally by manipulating brood size to determine the effect on testosterone levels and mating effort.

The second important question is how well the hypothesis that testosterone mediates the mating versus parental effort trade-off generalizes, both within and beyond the avian lineage. The hypothesis is intended to account for patterns across species as well as for the solution arrived at by any particular species. On the other hand, with sufficient evolutionary time, negative correlations can be broken (within-species correlations can't be safely generalized to comparisons among species—chapter 5). Does testosterone always drop during paternal care and does elevating it always interfere with paternal care? No, but looking at when it does and doesn't reveals underlying order in the phenomenon (Marler et al. 2003). What is key is whether mating effort and parental effort occur during separate reproductive stages separated in time and whether parental behavior and effort are incompatible with mating behavior and effort. There is a lot of diversity in the timing and magnitude of these two kinds of effort (Clutton-Brock 1991).

The songbirds in which testosterone has been shown to reduce paternal effort tend to be small, with high metabolic rates and limitations on fat storage due to flight (Bennett and Harvey 1987). Their chicks are altricial, with unusually steep growth curves. Both mating effort behavior (singing, aggressively defending the territory) and repeated trips to bring food are energetically demanding activities. Who else among vertebrates shows signs of a similar testosterone-mediated trade-off? *Eleutherodactylus coqui*, one of the few frogs with paternal care, shows a pattern similar to these birds in which androgen drops and mate attraction (calling) stops during the parental phase, although the levels are still high by bird standards (around 20 ng/mL) and no experimental work has been done to show that elevation of androgen disrupts parenting (Townsend and Moger 1987). The fanged frogs also have paternal care. Compared to nonparental male frogs of the same family (Ranidae), they have few morphological characters designed to attract females or fight with males, have lower testosterone levels, and take care of the eggs (Emerson et al. 1997). If Emerson's hypothesis is correct that the mechanism underlying this evolutionary shift is primarily activational (see chapter 5), then males with experimentally elevated testosterone should abandon eggs and instead seek out females and become aggressive toward other males, provided the breeding season is still going on to provide stimuli for those activities. Why can't these paternal frogs parent and attract mates at the same time? Here the risk of attracting predators to the helpless eggs might be more important than the sheer effort involved in parenting. Males of some teleost fish species have steroid profiles suggesting the same hormone-mediated trade-off found in songbirds, but with 11-ketotestosterone substituting for testosterone. In male three-spined sticklebacks levels of 11-ketotestosterone are 34 times higher during spawning than at the end of the parental (egg care) phase, a massive drop indeed! However, giving the males 11-ketoandrostenedione, which can be converted to 11-ketotestosterone, doesn't prevent the shift to egg care, and so the drop in androgen is not the mechanism for the shift nor do androgens interfere with egg care (fig. 6.5) (Páll et al. 2002a,b).

Other fathers manage to combine parenting and mating with seeming ease, so that mating and parenting levels of sex steroids would have to be the same. For example, several basal avian lineages such as ostriches have male-only parental care. The males are large enough that running out of energy quickly is unlikely, and their chicks are precocial and mobile. Males engage in mate attraction and competition with other males on the same days that they are incubating eggs and guarding chicks. Male mammal parents don't fit the song-

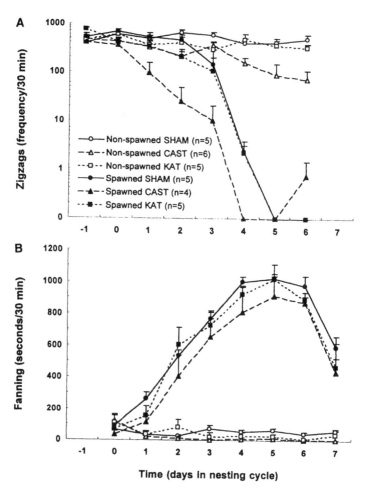

FIGURE 6.5. The effect of sham surgery (SHAM), castration (CAST), or castration plus 11-ketoandrostenedione treatment (KAT) on the courtship (zigzags, A) and parental (fanning, B) behavior of spawned and nonspawned male three-spined sticklebacks. The young hatched between days 6 and 7. In this species spawning is followed by a large drop in courtship behavior, a large increase in parental behavior, and (not shown here) a large drop in 11-ketotestosterone, from over 150 ng/mL to less than 10 ng/mL. 11-Ketoandrostenedione is readily converted to 11-ketotestosterone and both are potent androgens. The postspawning behavioral changes do not, however, appear to be due to changes in these hormones, because castrated males still courted and KAT males still stopped courting and started fanning after "spawning." These results emphasize the importance of testing hypotheses about hormone–behavior relationships with manipulation experiments rather than with hormone–behavior correlations. They also show that the hypothesis that androgens mediate a trade-off between parental and mating effort is supported only in some species (see, for example, fig. 5.2) and not others (see, for example, fig. 6.6). Reprinted from Páll, M. K., Mayer, I. and Borg, B. 2002b. Androgen and behavior in the male three-spined stickleback, *Gasterosteus aculeatus*, II: castration and 11-ketotestosterone effects on courtship and parental care during the nesting cycle. *Horm. Behav.* 42:337–344. © 2002 Academic Press (Elsevier).

bird description either (Creel et al. 1993). Most are much less mobile, lowering costs of fat storage as well as locomotion, they might not have to travel so far to get food (especially because the mother provides the food for the offspring), and some (such as beavers and wolves) are large enough to store a fair amount of energy. All in all, parental effort might be low compared to male birds, generating fewer trade-offs with other energy outlets. Furthermore, some female mammals have a postpartum estrus, in which ovulation and mating occur during the hours following birth. This is another one of those animal reproduction facts that seems quite bizarre to humans. Yet this is the normal time for the next pregnancy to begin for many species. When a postpartum estrus co-occurs with paternal care, mating and parental effort are clearly not incompatible for the male. In experiments with the California mouse (*Peromyscus californicus*), a territorial socially monogamous biparental species with a postpartum estrus, paternal behavior and aggression toward a male intruder are positively, not negatively, correlated, and testosterone restores paternal behavior following castration, rather than decreasing it (fig. 6.6) (Trainor and Marler 2001). In teleosts androgens are more likely to stimulate, rather than inhibit, parental behavior (Liley and Stacey 1983; Borg 1994). Overlap between mating and paternal effort is common, in which males simultaneously guard older eggs and and attract additional females to the nest to add more eggs (a little like the male ostriches). Displays of paternal behavior can attract mates (Reynolds et al. 2002), a phenomenon that probably occurs in some birds and mammals as well (Freeman-Gallant 1996; Tallamy 2000). Even parental behavior can be sexually selected!

There are almost certainly other predictors of negative versus positive correlations between androgens and paternal care as well. In birds it has been proposed that testosterone is more likely to interfere with paternal care when such care is obligatory for egg or chick survival (Van Duyse et al. 2002). Paternal care is not always required for offspring fitness, and only male removal experiments can reveal when it is and isn't. Also, if there are no reproductive opportunities for male birds during the period when they are engaged in parental behavior, androgens should drop to the extent that they are costly. In some colonial seabirds, for example, birds are single brooded and highly synchronized in laying date. By the time most males are incubating, there will no longer be any fertile females left.

The third question is whether hormones are mediating any kind of mating versus parental effort trade-off for females. The sex steroid levels of female songbirds also fall during incubation and chick care (Wingfield and Farner 1993). Males of many species expend more sexually selected mating effort, but females expend some as well, for example, when they defend a jointly held territory against female intruders to keep them away from their mate. Is female parental behavior also reduced by sex steroid elevation? (It wasn't in female reed warblers, *Acrocephalus scirpaceus* [Dittami et al. 1991].) If so, how do

FIGURE 6.6. A pair of California mice with their pups. The hind legs and tail of one of the pups can be seen at the bottom. California mice are monogamous and biparental. Testosterone promotes rather than inhibits paternal behavior, an effect that occurs via aromatization (Trainor and Marler 2002; see chapter 2). Courtesy of Jessica Berndt and Catherine Marler.

male and female thresholds compare? The answers are largely unknown. To read some of the literature on hormones and trade-offs, you'd think all animals are males and testosterone was the only hormone!

Just as for males, whether there is a trade-off and what its solution should be would be different for female mammals than female birds. In species with a postpartum estrus, the hormonal states of estrus and parenting are obviously not in conflict. Female red kangaroos (*Macropus rufus*) can be pregnant, nursing, and mating at the same instant. In some rodents the females are giving birth to multiple offspring over a period of time and are switching rapidly between expulsion and cleaning of a pup and mating with a male. The behaviors cannot be performed exactly simultaneously but they are done on the same day nonetheless. How they are managed behaviorally would vary depending on whether the species is monogamous (males would be in the nest and close at hand) or not (the female would have to leave the nest to attract a male). Mechanisms for switching between mating and parenting would have to be rapidly acting. Another difference between female birds and mammals that is relevant to thinking about hormonal mechanisms for sorting out effort is that high levels of aggression (overt attack, biting, scratching) are a bigger part of maternal (and paternal) behavior in mammals than birds.

Conclusions

Some of the most notable successes in linking steroids to ecologically relevant behavior in an integrative manner have come from studies of the adaptive timing of life stage transitions with major fitness consequences, such as metamorphosis, puberty, and the senescence associated with semelparous reproduction. It will be important to look at hormonal mechanisms of dispersal in animals with different dispersal strategies, ages, and sexes. Possible organizational as well as activational roles for steroids in life histories and their neural basis need to be explored. For example, do hormones at puberty set up the timing of later senescence? Future research will need to incorporate the brain's role in behavior and in regulating the major endocrine axes. Brains are plastic, and multiple features of the brain change in interesting ways during life stage transitions such as birth, puberty, and onset of seasonal breeding.

The hypothesis that steroids are mechanisms mediating some of the trade-offs predicted by life history theory has received some experimental support, although many questions remain about how they do this. Life history theory trade-offs refer both to trade-offs due to time or energy limitations faced by individual animals that are resolved through some kind of decision making (hormonal or neural) and also to solutions that have been arrived at over evolutionary time that vary across species, creating diversity. The testosterone-supplementation experiments with birds reveal trade-offs in the lives of individuals and mechanisms that are currently maintaining observed patterns. The assumption in the literature is that this also operated in the past to give rise to the current patterns, and that evolutionary change occurred through a change in hormone level (see also chapter 5). There is a need for good comparative experimental work on hormonal mechanisms of these trade-offs, because only in a phylogenetic context can evolutionary origin be distinguished from maintenance. Many avian life history characters, and therefore solutions to some trade-offs, are quite old and vary little within lineages (Bennett and Owens 2002). Mating systems and mating effort are more recent and more diverse within lineages. Trade-offs between mating and paternal effort occur for some fathers but not others, and androgens can be either negatively or positively correlated with paternal care. There is a need for more comparative hypotheses and tests with predictions made in advance about whether a hormone-mediated trade-off is occurring.

7

Phylogeny: Conservation and Innovation

Tinbergen's fourth aim and approach for understanding behavior was phylogenetic history. Since his time, enormous advances have been made in phylogenetic methods for analyzing and interpreting both organismal and molecular (genomic) characters. "Trees of life" in the form of cladograms based on shared derived organismal and molecular characters are being constructed that constitute more objective and testable hypotheses about the actual history of life on earth than ever before. Previous chapters have acknowledged the importance of a phylogenetic context for interpreting differences among related species that have arisen over relatively short time scales. Here the camera will zoom out to view larger chunks of the vertebrate tree (fig. 7.1), allowing comparisons between more distantly related clades (monophyletic lineages) with respect to hormonal mechanisms, behavior, and the connections between them. This large-scale view allows us to see what has been conserved over long periods of evolutionary time, to see similarities between clades, to determine whether they are due to common descent (homology) or to some other process, such as independently evolved adaptation to a similar environment (homoplasy), and to see major novelties that have arisen and been maintained (pivotal events in evolution). Why does it matter whether similarity is due to homology or homoplasy? Because in trying to recreate what happened in the past, it is important to try to estimate the number of independent origins and to estimate when and where they occurred, not just tabulate their current frequency and taxonomic distribution (Mindell and Meyer 2001). As evolutionary developmental biology emphasizes (chapter 5), it is also important to figure out the developmental processes that produced stasis (conservation), innovation, and homoplasy.

Large-scale phylogenetic approaches have been important in comparative endocrinology. This chapter begins by revisiting some of the mechanisms in chapter 1, starting with a classic of molecular evolution, the story of the oxytocin family peptides.

Oxytocin Family Peptides and Their Receptors

The oxytocin family of peptides is very old. Structurally related peptides serving regulatory functions are found in invertebrates as well as vertebrates, sug-

Peptides

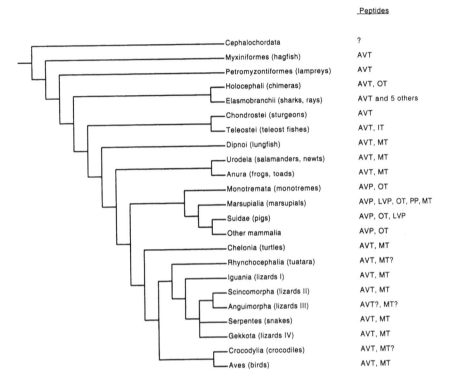

	Peptides
Cephalochordata	?
Myxiniformes (hagfish)	AVT
Petromyzontiformes (lampreys)	AVT
Holocephali (chimeras)	AVT, OT
Elasmobranchii (sharks, rays)	AVT and 5 others
Chondrostei (sturgeons)	AVT
Teleostei (teleost fishes)	AVT, IT
Dipnoi (lungfish)	AVT, MT
Urodela (salamanders, newts)	AVT, MT
Anura (frogs, toads)	AVT, MT
Monotremata (monotremes)	AVP, OT
Marsupialia (marsupials)	AVP, LVP, OT, PP, MT
Suidae (pigs)	AVP, OT, LVP
Other mammalia	AVP, OT
Chelonia (turtles)	AVT, MT
Rhynchocephalia (tuatara)	AVT, MT?
Iguania (lizards I)	AVT, MT
Scincomorpha (lizards II)	AVT, MT
Anguimorpha (lizards III)	AVT?, MT?
Serpentes (snakes)	AVT, MT
Gekkota (lizards IV)	AVT, MT
Crocodylia (crocodiles)	AVT, MT?
Aves (birds)	AVT, MT

FIGURE 7.1. Simplified cladogram of the living vertebrates along with their closest living ancestors, the cephalochordates, showing the oxytocin family nonapeptides of the clades. Peptide acro-
nyms: AVP, arginine vasopressin; AVT, arginine vasotocin; IT, isotocin; MT, mesotocin; OT, oxytocin; LVP, lysine vasopressin; LVT, lysine vasotocin; PP, phenypressin. See table 7.1 for peptide
structures. Information about peptide phylogeny is based on Bentley (1998) and Suzuki et al.
(1995). See Meyer and Zardoya (2003) for a discussion of controversies surrounding
vertebrate phylogeny.

gesting an origin prior to the appearance of the chordates. Connections to behavior and reproduction are old as well. For example, in earthworms and
annelids oxytocin family peptides stimulate egg-laying behavior (Fujino et al.
1999). In vertebrates these peptides are released both from the posterior pituitary (neurohypophysis), where they function as hormones, and in other brain
sites, where they act locally as neuromodulators.

Oxytocin and its structural relatives are peptides with nine amino acids. By
looking at the exact amino acid sequence in different kinds of vertebrates, it
is possible to see what kinds of point mutations have occurred in the DNA that
codes for these sequences and how often these have occurred, and to make
guesses about when they must have occurred—to construct a molecular phylogeny of the oxytocin family (table 7.1 and fig. 7.1). The changes in amino
acids have occurred at only some points in the sequence and not others, and

TABLE 7.1

Amino acid sequences of the principal vertebrate oxytocin family nonapeptides

Peptide	Sequence								
	1	*2*	*3*	*4*	*5*	*6*	*7*	*8*	*9*
Oxytocin	Cys -	Tyr -	Ile -	Gln -	Asn -	Cys -	Pro -	Leu -	Gly -	NH₂
Mesotocin	Cys -	Tyr -	Ile -	Gln -	Asn -	Cys -	Pro -	Ile -	Gly -	NH₂
Isotocin	Cys -	Tyr -	Ile -	Ser -	Asn -	Cys -	Pro -	Ile -	Gly -	NH₂
Arginine vasopressin	Cys -	Tyr -	Phe -	Gln -	Asn -	Cys -	Pro -	Arg -	Gly -	NH₂
Lysine vasopressin	Cys -	Tyr -	Phe -	Gln -	Asn -	Cys -	Pro -	Lys -	Gly -	NH₂
Phenypressin	Cys -	Phe -	Phe -	Gln -	Asn -	Cys -	Pro -	Arg -	Gly -	NH₂
Arginine vasotocin	Cys -	Tyr -	Ile -	Gln -	Asn -	Cys -	Pro -	Arg -	Gly -	NH₂

Note: Cys, cysteine; Tyr, tyrosine; Ile, isoleucine; Gln, glutamine; Asn, asparagine; Pro, proline; Leu, leucine; Gly, glycine; Ser, serine; Phe, phenylalanine; Arg, arginine; Lys, lysine.

Source. After Norris (1996) and Bentley (1998).

the overall sequence similarity is high enough to consider these as homologous molecules coded by homologous stretches of DNA. Here the word "homologous" has its molecular biology structural meaning. The sequence changes provide a window into the underlying genetic changes.

What can we conclude about the vertebrate phylogeny of these peptides? The answer depends on whether we assume that reverse mutations (from a new amino acid back to the old one) have occurred. If they have not, then it appears that mutations have been infrequent—that a lot of the structure has been conserved for many millions of years. The anatomical distribution of gene expression in the brain has been somewhat conserved as well. At the same time there have been some changes not only in structure but also in the number of different members of the family found in different vertebrate clades. As is usually the case in the evolution of signaling systems, there is no simple relationship between the time of origin of the clade and the number of oxytocin family members. For example, sharks and kangaroos have more oxytocin family peptides than primates. Both the lamprey and hagfish clades (the two agnathan groups) have arginine vasotocin (Suzuki et al. 1995; Sower 1998). The most parsimonious conclusion is that the "original" vertebrate form was similar to arginine vasotocin (Urano et al. 1992). Based on fossil evidence for the time of vertebrate appearance on earth, this form must have been present at least 500 million years ago.

The likely mechanism for an increase in the number of oxytocin family peptides in postagnathan vertebrate evolution is gene duplication (Sower 1998). The potential importance of gene duplication for the evolution of vertebrate brain peptides and their functions was first recognized by Ohno (Ohno et al. 1968). When a peptide-coding gene duplicates, mutations in one copy are more likely to be neutral or even favorable than is usually the case for a mutation, because the other copy will continue to make the "old" sequence. The original function will be maintained at the same time that a new or more specialized function can evolve for the new sequence (Ohno 1970). For example, the combination of arginine vasopressin and oxytocin appeared early in mammalian evolution. In living mammals the behaviorally relevant neuromodulator functions of vasopressin and oxytocin are not the same, nor are the genes for them expressed in identical brain regions.

The pattern of sequence novelty in table 7.1 and fig. 7.1 is most often interpreted as indicating that the structural changes in the peptides themselves have been selectively neutral. With a long time since divergence you expect a small number of differences in the amino acid sequence to accumulate—a slowly ticking molecular clock. The peptide may still be critical but the change in one amino acid might be irrelevant (might not be detected by the receptors). The alternative hypothesis is that there is some physiological significance to the exact amino acid sequence, for example, some reason based in natural selection why pigs have lysine vasopressin instead of arginine vasopressin. Such alternatives are difficult to test, however, and so far there doesn't seem to be compelling evidence to abandon the neutral hypothesis (Barrington 1986). Conserved sequences are obviously adaptive in some generic (and possibly not very interesting) sense. There are ways to tell if gene substitutions (DNA sequence changes) and protein changes are neutral or whether selection has gone on (Graur and Li 2000) but this doesn't seem to have been done for the oxytocin family. Some of the social behavior functions are quite conserved, as if the molecular changes have indeed been neutral (Goodson and Bass 2001).

The evolution of a new function for a peptide coded by a duplicated gene will require changes at the receptor end, in the structure, distribution, or cascades regulated by the receptor protein. Here too gene duplication followed by structural divergence due to mutation has been important (van Kesteren et al. 1996). Prairie voles seem to have a duplication in the gene for the vasopressin 1_A receptor that has altered the anatomical distribution of gene expression in the brain in a behaviorally significant way (chapter 5). The existence of multiple receptor subtypes by itself helps account for the multiple functions that have evolved for a single peptide molecule, for example, the difference between the functions of arginine vasopressin as a peripheral hormone (an antidiuretic by virtue of actions on the kidney) and its functions as a brain neuromodulator of social behavior.

Thus, both peptides and their receptors have evolved structurally and functionally. We can speculate that this has allowed the diverse array of combinations of different social behaviors (aggressive, parental, pairbonding) to evolve and to be coordinated with the animal's physiology (for example, its water balance and reproductive mode). Yet peptide ligands and their receptors and

TABLE 7.2

Amino acid sequences of some of the vertebrate GnRH family decapeptides

	Sequence									
Peptide	1	2	3	4	5	6	7	8	9	10
"Mammal GnRH"	pGlu -	His -	Trp -	Ser -	Tyr -	Gly -	Leu -	Arg -	Pro -	Gly - N
"Chicken I GnRH"	pGlu -	His -	Trp -	Ser -	Tyr -	Gly -	Leu -	Gln -	Pro -	Gly - N
"Chicken II GnRH"	pGlu -	His -	Trp -	Ser -	His -	Gly -	Trp -	Tyr -	Pro -	Gly - N
"Salmon GnRH"	pGlu -	His -	Trp -	Ser -	Tyr -	Gly -	Trp -	Leu -	Pro -	Gly - N
"Catfish GnRH"	pGlu -	His -	Trp -	Ser -	His -	Gly -	Leu -	Asn -	Pro -	Gly - N
"Dogfish GnRH"	pGlu -	His -	Trp -	Ser -	His -	Asp -	Trp -	Lys -	Pro -	Gly - N
"Lamprey I GnRH"	pGlu -	His -	Tyr -	Ser -	Leu -	Glu -	Trp -	Lys -	Pro -	Gly - N
"Lamprey III GnRH"	pGlu -	His -	Trp -	Ser -	His -	Asp -	Trp -	Lys -	Pro -	Gly - N

Note. Glu, glutamate; His, histidine; Trp, tryptophan; Ser, serine; Tyr, tyrosine; Gly, glycine; Leu, leucine; Arg, arginine;]
proline; Gln, glutamine; Asn, asparagine; Asp, aspartate; Lys, lysine.
Source: Bentley (1998).

receptor subtypes are not genetically correlated. They have separate origins, are produced by separate gene families, and their co-evolution is not a simple matter of pleiotropy or correlated genes in the sense meant by evolutionary biologists (van Kesteren et al. 1996; Bentley 1998). The tale of how this co-evolution has occurred has not yet been told.

GnRH and Its Receptors

The story of GnRH began in mammals. The hypothalamic peptides regulating the pituitary gonadotropins FSH and LH were named FSH-RH and LHRH and it was hypothesized that other vertebrates would have a similar pair of regulatory peptides. Over time it became apparent that the same molecule can produce both FSH and LH release, thus the name change to GnRH. Direct behavioral functions of GnRH were also discovered (Moss et al. 1975); GnRH is both a hormone and a neuromodulator. What do other vertebrates have, and how has the structure of GnRH changed? As with the oxytocin family, there is both conservation and change in the molecular phylogeny of GnRH. There are also surprises and mysteries.

The GnRH that is the final common signaling molecule for the HPG axis is a sequence of ten amino acids. A similar molecule has this function in other vertebrates but the exact sequence varies, reflecting an occasional mutation in the DNA occurring at long intervals (table 7.2) (Sherwood 1987). These have not disrupted the essential HPG regulation function of the molecule, as if they have been selectively neutral. Having something at the head end that produces something like mammalian GnRH (a homolog resulting from a similar gene) that exerts reproductive control is very old (Gorbman and Sower 2003). There is GnRH-like activity in the CNS of the mollusk *Aplysia*, there is mammalian GnRH in tunicates (urochordates), and there is suggestive evidence for reproductive functions in *Aplysia* (Cameron et al. 1999; Tsai et al. 2003). The origin seems to lie in marine organisms that preceded vertebrates and in a need to convey chemical information in the water detected by the front end of the animal to the reproductive system.

Meanwhile, other molecules that looked like more diverse forms of GnRH were discovered in nonmammals. At first this was dealt with by naming them for the animal, for example, "chicken-II GnRH." The surprise is the enormous proliferation of forms (now 16 or more and counting in the vertebrate brain alone, six of them specific to teleosts) whose taxonomic distribution makes nonsense of the older names. The forms seemed to fall into categories that corresponded to different embryonic origins and brain systems (Muske 1993). Molecular methods have made it possible to go straight to the genome to look for other similar stretches of DNA, where they are expressed in the brain and body, and what amino acid sequences they produce. Most species have at least two forms and all the forms discovered so far arise from three gene families (Sherwood and Parker 1990; Fernald and White 1999). The genes are hypothesized to have arisen by duplication from a single ancestral form. This duplication apparently took place early in vertebrate evolution (500 million years ago) or even before, because lampreys have multiple GnRHs (Sower 1998). Two of the GnRH forms are structurally identical in all vertebrate species. Only the hypothalamic signaling molecule for HPG axis regulation varies. The three gene families are expressed in different (but, in some species, overlapping) brain and body cells and regions that have different embryonic origins. Thus, there is order under the chaos. Fernald and White (1999) proposed a three-category system in which GnRH1 are the hypothalamic forms, GnRH2 is the mesencephalic (midbrain–tegmentum) form, and GnRH3 is the telencephalic form. Even these names are misleading. The gene families produce proteins expressed outside the brain that don't seem likely to be regulating gonadotropins.

The mystery is what brain "GnRH2" and "GnRH3" are doing, and what it means that they have remained completely unchanged structurally for so long, unlike GnRH1. They almost certainly have different functions, but what are they? Answering this requires (among other things) knowing whether they have receptors and where they are (Lethimonier et al. 2004). In those mammals and amphibians examined so far, "GnRH2" has a specific receptor with a similarly wide distribution to the peptide itself that includes brain regions already known to be involved in reproduction (Millar 2003). "GnRH2" quickly activates mating behavior in female musk shrews (*Suncus murinus*) that have been

FIGURE 7.2. A female dark-eyed junco performing a copulation solicitation display. A similar display is seen in females of other passerine species, especially when they are estrogen primed. GnRH-II has been shown to facilitate this display (see fig. 7.3). Image captured from video, courtesy of Ellen Ketterson and Eric Snajdr.

placed on a diet, as if it might be a mechanism involved in coordination of reproductive with nutritional status (Temple et al. 2003). When administered into the brain of estrogen-primed female white-crowned sparrows, it increases copulation solicitation (figs. 7.2 and 7.3) (Maney et al. 1997b). "GnRH3" is found in some auditory regions and song nuclei in two species of sparrows (house and white-crowned), and elicits LH release when administered systemically, providing a new hypothesis for how the song environment might modulate gonadal steroids (Bentley et al. 2004).

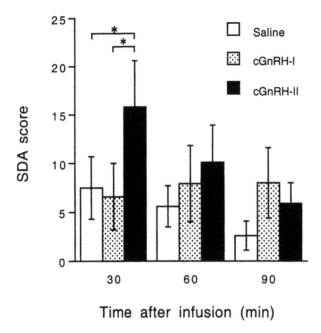

FIGURE 7.3. Solicitation display assay (SDA) scores 30, 60, and 90 minutes after infusion of saline (control), cGnRH-I, or cGnRH-II into the brain of female white-crowned sparrows. Each bird received each treatment. When females were given cGnRH-II, they displayed significantly more 30 minutes after infusion, showing that this GnRH form has a rapid behavioral effect in this species. Redrawn from Maney, D. L., Richardson, R. D. and Wingfield, J. C. 1997b. Central administration of chicken gonadotropin-releasing-hormone-II enhances courtship behavior in a female sparrow. *Horm. Behav.* 32: 11–18. © 1997 Academic Press (Elsevier).

Steroid Receptors

Vertebrate intracellular steroid receptors are proteins, much longer amino acid sequences than the oxytocin and GnRH family peptides. Intracellular steroid receptors are part of a large superfamily that also includes receptors for vitamin D, retinoic acid, and the thyroid hormones. The steroid receptors known or thought to exist in most of the species that have appeared in this book consist of two types of estrogen receptors (ERα and ERβ), a progesterone receptor (PR), an androgen receptor, a mineralocorticoid receptor, and a glucocorticoid receptor, for a total of six receptor types. A second androgen receptor and a third estrogen receptor have been reported recently from two teleosts (Sperry and Thomas 1999a,b; Hawkins et al. 2000). Not all of these names are accurate, because steroid categories don't always map cleanly onto receptor categories, as when glucocorticoids bind to mineralocorticoid receptors as well (chapter 1).

The molecular phylogeny of the genes for these receptors confirms that the vertebrate steroid receptors are indeed a related subgroup of the superfamily (Laudet et al. 1992). The molecular phylogeny of the receptors themselves suggests that they arose early in vertebrate evolution as a result of two successive gene duplications relatively close together in time, diverging in structure and function from a single ancestral form. Because steroid receptors (when bound with a steroid) regulate gene transcription, the genes for them are the type of regulatory genes that could be an innovation with major consequences for all of vertebrate evolution (Baker 1997, 2002). If so, these gene duplications are truly great moments in evolution.

These gene duplications are thought to have been part of a doubling of the entire vertebrate genome. Doubling of chromosome number has occurred in the ancestors of a few living vertebrates, for example, 30 million years ago in the ancestry of the frog *Xenopus laevis*. *X. laevis* has a duplicated ERα gene and the two genes have different tissue expression patterns. The novel gene is expressed in the laryngeal muscles used for mate attraction vocalization (Wu et al. 2003).

Thornton (2001) cloned steroid receptors of lampreys so that the molecular phylogeny for vertebrates would include a basal clade to better elucidate the critical early steps in vertebrate steroid receptor evolution. His phylogeny suggests that ER came first, followed by PR (fig. 7.4). His results also support the hypothesis of two duplications occurring long ago. Based on this phylogeny, Thornton proposed a scenario for how steroid receptors and their ligands co-evolved. If ER appeared first, and if its ligand was an estrogen from the beginning, then the terminal steroid in the biosynthetic pathway was the first to have a receptor. Over time and mutations in the ER gene, the intermediate steroids (androgens and progestagens) came to have their own receptors and became functional as something other than way stations to estrogens. If this scenario is correct, then the intermediate steroids such as progesterone and testosterone did not have physiological roles until later, when their receptors had appeared. In support of this hypothesis, no function for androgens has been found thus far in lampreys, not even in the testis (Sower 1998). An alternative is that the original ligand for ER was not an estrogen, and that the evolutionary innovation was the shift to an estrogen as a ligand, allowing the ER cascade to be coordinated with fertility (Brosens and Parker 2003). A closer look at invertebrates has revealed that ER is extremely ancient (an ortholog is present in *Aplysia*), may have been lost in some invertebrate clades, and may have lost having an estrogen as its ligand in mollusks (Thornton et al. 2003).

Regardless of which particular scenario is correct, it is important to think about steroid–receptor co-evolution and to try to reconstruct the steroid specificity (or lack of it) of the ancestral receptors. As with peptides and their receptors, we know that the gene families and origins are different for the receptors and their ligands (that they are not genetically correlated). The problem is more

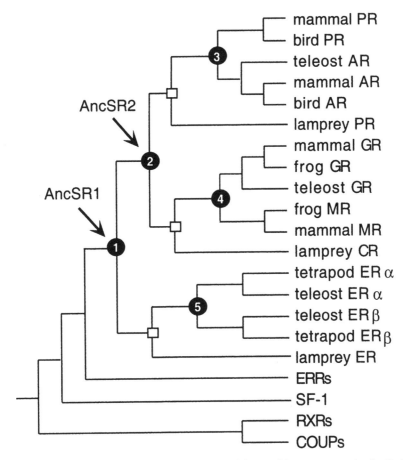

FIGURE 7.4. Thornton's (2001) proposed phylogeny of the steroid receptor gene family. Dark circles indicate gene duplications within the family. White squares mark the lamprey–gnathostome divergence. AncSR1 and AncSR2 are reconstructed ancestral steroid receptors. ERRs, SF-1, RXRs, and COUPs are other receptors that appear to be the most similar to the steroid receptors. Redrawn from Thornton, J. W. 2001. Evolution of vertebrate steroid receptors from an ancestral estrogen receptor by ligand exploitation and serial genome expansions. *Proc. Natl. Acad. Sci. U S A* 98: 5671–5676. © 2001 National Academy of Sciences, U.S.A.

complex in the case of steroids, because they are not gene products, but instead are created by a set of enzymes that are gene products.

The distribution in vertebrate brains of a given steroid receptor type is somewhat but not entirely conserved. Changes in the expression of the gene in different regions have occurred. Innovation and diversity can occur by expressing it in new places or ceasing to express it in old places. As the putatively oldest steroid receptor type, estrogen receptors serve as a good example of what a steroid receptor distribution looks like across different vertebrate clades. Comparative studies using the autoradiographic method established that all vertebrates examined (agnathans, teleosts, amphibians, turtles, lizards, birds, and mammals) had estrogen target cells in periventricular areas that included the septum, preoptic area, and hypothalamus, with additional target cells in the midbrain tegmentum of most and in some portion of the amygdaloid complex of those vertebrates that have an amygdaloid complex (the amniota) (Kim et al. 1978; Morrell and Pfaff 1981). At the same time there were a few places in the brain where one or more vertebrate species had some estrogen concentrating cells that others didn't, such as the trigeminal nucleus of lizards. Immunocytochemical methods give similar but not identical results and lead to a similar conclusion: vertebrates share a common core of brain regions with neurons containing estrogen receptors that is highly conserved. On top of that are some clade differences on both large and small scales. In rainbow trout (*Oncorhynchus mykiss*), unlike the quite unrelated teleost species that had been studied earlier, cells containing estrogen receptors are limited to brain areas known to regulate the pituitary and with direct connections to it (Anglade et al. 1994).

Gahr et al. (1993) used immunocytochemistry to see where the ERα receptors are located in the brains of 26 species of birds in six different orders: Anseriformes, Galliformes, Columbiformes, Psittaciformes, Apodiformes, and Passeriformes. The Passeriformes included three suboscines as well as some oscine songbirds. The distribution in the "limbic forebrain" (hypothalamus, preoptic area, lateral septum, hippocampus, and nucleus taeniae) and in the midbrain and hindbrain was quite conserved, with only a small number of species differences. The rest of the forebrain was another matter. Only oscines had ERα in the caudal nidopallium (including HVC) and the dorsal surround of RA, an innovation that happened only once in avian evolution. There were quantitative differences within oscines in the number of receptors in these regions (for example, zebra finches had fewer receptors in HVC) but the qualitative distribution was quite uniform. Even with such a rich set of information (26 species is a lot for a single comparative neuroanatomy study), it's still not easy to figure out the functional significance of some of the diversity. The oscine innovation is relatively straightforward (see below), but what's the functional significance of the lower number of estrogen receptors in the zebra finch HVC? Is it because they are nonseasonal breeders? Or because they are estrildid finches who are carrying along a neutral mutation? As always, these are hard questions to answer.

Steroids and Steroidogenic Enzymes

Steroids that are similar in structure to vertebrate sex steroids are very old. They are found in invertebrates and even in plants. While their specific functions are usually unknown, there is increasing suspicion that they may be signaling molecules in many multicellular organisms, sometimes with reproductive functions. For example, estrone and estradiol-17β are found in a coral and their production profile suggests that they might be regulating spawning (Tarrant et al. 1999). Gene regulation by steroids probably predates their production by gonads and adrenals. The discovery of 3β-hydroxysteroid dehydrogenase (the enzyme for synthesis of progesterone) in lungfish (*Protopterus annectens*) suggests an early origin for the manufacture of biologically active steroids in the brain (neurosteroids) as well (Mathieu et al. 2001).

Steroids, unlike their receptors, are not amino-acid based. Their structure doesn't change slowly over evolutionary time the way the structure of peptides does. Presumably the evolutionary action lies in the proteinaceous enzymes that synthesize and metabolize them (and in their receptors, of course). The synthesizing pathways for androgens and estrogens in gonads and adrenals are rather conserved in vertebrates. Vertebrate groups differ primarily in the relative amounts of the different steroids produced, not in the kinds of steroids produced. Against this background of conservation, the few innovations really stand out. 11-Ketotestosterone and other 11-oxygenated androgens are unique to teleosts, products of the same enzymes (11β-hydroxysteroid dehydrogenases) that in mammals turn glucocorticoids into weaker forms (Borg 1994). Teleosts also have unique steroids such as 17,20β-P that function in social communication in addition to gamete maturation (chapter 2). Why are such innovations so rare? Given the large total number of different steroids with known functions in vertebrates, why does a new one with a function related to the brain and behavior appear so seldom? Is there enough evolutionary action on the receptor end that the signal seldom needs to change?

The conserved biosynthetic pathways for steroids in gonads and adrenals suggest that the vertebrate steroidogenic enzymes are very old. They are members of enzyme families found throughout the animal kingdom and even in plants and bacteria. For example, aromatase is one of the P450 enzymes, which are 1.5 billion years old! Reconstructing the phylogeny of such complex proteins is challenging. As with other large proteins, the exact amino-acid sequences change even over relatively short evolutionary time. Phylogenetic analysis of sequences for 11β-hydroxysteroid dehydrogenases and 17β-hydroxysteroid dehydrogenases points to a common ancestral origin early in the tree of life and thus homology between the two enzyme types (Baker 1994). Studies in mice suggest that genes in the P450 superfamily are capable of evolving quite rapidly (Aida et al. 1994). Teleosts have by far the highest brain aromatase content of any vertebrates that have been studied; this is thought to be related to the continuous adult growth of their brains (Callard 1984; Menuet et al. 2003). Those teleosts that have been examined (three species from three different clades) have two different aromatase isoforms and two different aromatase genes (Kishida and Callard 2001; Kwon et al.

2001; Tong et al. 2001; Valle et al. 2002). Multiple genes might allow differential expression in brain versus gonads, in different brain regions, or at different times during development.

What about circulating steroid levels themselves? Is there any relationship between phylogeny and steroid levels? Previous chapters have shown the myriad ways that steroid levels vary considerably within individuals, between individuals of the same species, and between related species. Also, the functional significance of a steroid level depends on many other mechanisms, including levels of binding proteins in the blood. All that points to steroid level as a very poor character for any kind of phylogenetic approach. Yet here and there there seems to be some phylogenetic signal showing up amidst all the noise. High or low ratios of cortisol to corticosterone tend to extend across entire clades, with teleosts producing predominantly cortisol and lizards and birds producing predominantly corticosterone (Bentley 1998). One of these two states might be the ancestral, with shifts to a different ratio occurring multiple times in vertebrate evolution for as yet unknown reasons. New World primates have a marked tendency to have higher levels of cortisol than Old World primates or prosimians, even when body size is taken into account (Coe et al. 1992).

Some major vertebrate clades include species producing whopping levels of testosterone and other androgens (over 200 ng/mL in *Bufo*!), whereas others don't, with all species studied staying under 10 ng/mL (Norris 1996). Chapter 3 discussed species differences in birds in levels of male androgens but this variation is tiny in the grand scheme of animals. Even the top-of-the-line males of highly polygynous avian species have very unimpressive androgen levels by toad standards. A zoomed out view of the vertebrate tree reveals that both birds and mammals tend to have low levels of androgens (that is, almost never have high levels) compared to other clades, with the notable exceptions of bats, tree shrews, hyraxes, and elephants (Jainudeen et al. 1972; Gustafson and Shemesh 1976; Rasmussen et al. 1984; Eichmann and Holst 1999). Regardless of whether these high androgens are being compensated for by high levels of binding proteins or low levels of steroid receptors, clearly, androgen production or metabolism (clearance) or both are varying significantly. It is hard to even guess why these clade differences occur, but at least the bat exception rules out the hypothesis that birds have low levels because of something to do with flight! Elephants and hyraxes are part of the same clade of mammals and their shared high levels may reflect homology, especially if the manatees and dugongs, the other members of the clade, are also found to have high levels. As endotherms, birds and mammals have high metabolic rates and, all other things being equal, might be expected to clear steroids from the blood faster. One test of such a hypothesis would be to see if androgen levels of crocodiles, which are ectotherms, are like those of birds (their phylogenetic sister group) or like those of lizards, which are more distantly related but share ectothermy. Sexually mature male alligators (*Alligator mississippiensis*) have maximum testoterone levels that look more like those of other ectotherms (25–40 ng/mL) than those of birds (Lance et al. 1985; Butterstein et al. 2003). What is the allometry of maximum steroid levels? They are reported as amounts per unit volume of plasma or serum, but that's not the same as determining their relation to overall body mass or to the overall mass of the relevant target tissues. When males have very high androgen production, females of those species often do as well (Staub and DeBeer 1997). While there is little evidence that selection on males for higher androgens elevates them in females over shorter periods of time (chapter 5), there does seem to be a correlation over a larger phylogenetic scale.

Behavioral Phylogeny, Brains, and the Conservation Paradox

Darwin (1859) understood that animal species resemble each other because of common descent and are different because of natural selection. The morphology and behavior of species are somewhat but not completely predictable by knowing who their relatives are (their place in a tree of life). Statistically speaking, there are phylogenetic effects on these traits that vary in magnitude depending on the trait. Behavioral traits are famous for being evolutionarily labile (not very predictable from phylogenetic information) but until recently there had been few real quantitative comparisons with other kinds of traits in a phylogenetic context. In an analysis covering a wide array of animal clades, phylogenetic signal (= phylogenetic effects) was lower for behavioral traits but was still there (Blomberg et al. 2003). In a comparative study of 151 species of birds living in the same region, the magnitude of phylogenetic effects depended markedly on taxonomic level, but overall it was greater for morphological and life history traits than for behavioral or ecological niche traits (Böhning-Gaese and Oberrath 1999), the same conclusions reached by Gittleman et al. (1996) in a comparative study of mammals. Bennett and Owens' (2002) review of avian life histories concluded that similarity among distant relatives was associated with shared life history predispositions, whereas diversity within closer relatives was associated with ecological diversity. They identified patterns suggesting that avian lineages with high longevity and low annual reproductive rates are more likely to be socially monogamous with low extrapair fertilization rates and cooperative breeding, whereas lineages with short lives and high annual reproductive rates are more likely to be polygamous with high extrapair fertilization rates. Such phylogenetic signal for an important dimension of social organization does not mean that avian mating systems are not adaptive, or that the animals are "constrained" by the distant past, but rather that if there are several possible adaptive peaks, the starting point may have some bearing on which one is reachable (Reeve and Sherman 1993). At the same time that phylogenetic signals might be detectable, it is not uncommon to find similarity in aspects of social organization such as group size, territoriality, or mating system in extremely distantly related species (even species in different phyla) living under similar ecological conditions (Crook 1970). Such homoplasy was the original evidence that social organization was an evolved adaptation to the environment.

The diurnal primates (those other than prosimians) are a relatively well-studied group and beginning with Crook there have been attempts to understand the social organization and behavior of individual species as adaptations to local ecological conditions. These attempts have seldom been very successful (Kappeler and van Schaik 2002) and in comparative analyses phylogenetic effects seem to swamp ecological variables as predictors (Di Fiore and Rendall 1994). It is not yet clear why social diversity among related species is so much greater (more responsive to ecological circumstances) in some vertebrate lineages than others. In the case of the diurnal primates, group living (the most common lifestyle) combined with relatively large brains are viewed as all-purpose flexible adaptations that will work in any ecological context. The greatest adaptation of primates is to not be too narrowly adapted. A similar all-purpose flexibility idea has been floated for rodents, which are not very specialized either morphologically or behaviorally compared to some animals, yet are spectacularly successful as a lineage both in species richness and in numbers of individuals.

A phylogenetic approach to steroids and social behavior involves several different organismal levels: the genome, the steroids and their mechanisms of action, the nervous system, and behavior. One of the lessons of evolutionary developmental biology is the importance of treating these levels separately in phylogenetic analyses, because there can be changes in one (e. g., DNA) without changes in the others and homologies at one level without homology in another. The discovery that so much more of the genome and of the molecular developmental cascades is homologous than previously assumed does not mean that resemblances in behavior or morphology between distantly related species are more homologous than previously thought. Rather, such resemblances show how a developmental module can be used in flexible ways (Wray 2002). Evolution is opportunistic, and works with what is at hand. Similarity of outcome is interesting developmentally regardless of whether the outcome itself reflects homology or homoplasy. The repeated evolution of a similar outcome tells you something interesting about how these features get built. Either a simple genetic change is sufficient to produce the result or there are several alternative and easily taken developmental paths leading to the same result (Wake 1991; Meyer 1999). Multiple origins point to a mechanism (genetic or developmental) that can be co-opted. Where there is convergence in the underlying neural machinery as well, this tells you something valuable about how the nervous system solves problems, how it produces behavior (Eisthen and Nishikawa 2002). An excellent example of such convergence in both behavior and neural machinery is the use of odors for social communication in insects and vertebrates and the remarkable resemblance between the organization of the insect antennal lobe and the vertebrate olfactory lobe, in which the receptor neurons map onto glomeruli in the same manner (Firestein 2001).

Phylogenetic approaches to these different levels (genomes, steroid mechanisms, brains, and behavior) have led to an apparent paradox. Genomes are surprisingly conserved, steroid mechanisms (including the HPG and HPA axes) have changed a little but not a lot, the basic structure of the brain is remarkably conserved in vertebrates, and yet social behavior, especially social relationships and organization (rather than specific social acts), is rather diverse, with a weaker phylogenetic signal. How can such conversed mechanisms produce such diverse outcomes? Equally paradoxically, there have been a few major brain innovations (Molnár and Butler 2002; Northcutt 2002) but most of them don't correspond to major behavioral change in any obvious way (Striedter 1998; see Wagner and Luksch [1998] for an exception). The telencephalons of ray-finned fishes (Actinopterygii) are "inside out" (develop by eversion) compared to those of other vertebrates (which develop by evagination). Mammals have a six-layered isocortex, whereas the homologous brain region in turtles, lizards, snakes, crocodiles, and birds is unlayered and a chunk of it forms the dorsal ventricular ridge. Evolutionary developmental biology promises to discover the rules for brain assembly that will explain how brains change during evolution, but will this map onto behavioral evolution?

How can this paradox be resolved? Is it an artifact of a tendency for some fields to focus on what is conserved and others to focus on diversity? Although everything about animals is a mix of the conserved and the diverse, different fields of study tend to emphasize one over the other. Because social behavior is diverse over smaller scales (families, genera, and even populations within species), a fine-grained examination of the mechanisms using methods sufficiently sensitive to pick up variation over the same scale is necessary. The brain is probably only highly conserved at the level of gross anatomy, and not at the levels (biochemical, molecular, networks and systems) more relevant to understanding behavior. Another contribution to a solution is the realization that social behavior includes flexible (possibly even learned) tactics. When that happens, species differences aren't the hard-wired kind that would be associated with specialized mechanisms unique to different species or clades. The primate and rodent behavior research community has long thought that these animals rely heavily on more general-purpose problem-solving cognitive and neural mechanisms. Close study of the social life of wild animals has expanded the array of species that seem to use their species typical repertoire as flexible tactics rather than bundles of blind instincts. Species- or clade-specific behavior–mechanism correspondences are more likely to be found at the sensory and motor end of behavior, as in signaling systems.

Steroid-Modulated Vocalization

Vocal communication is not universal among vertebrates but instead has arisen several times independently. Among living species it is found in a few teleost fish, many frogs and toads, some mammals, a few turtles, a few lizards, and most of the crocodile-bird clade. When animals vocalize mainly to locate or attract mates, during courtship, or in intrasexual combat, vocalization tends to be modulated by sex steroids (chapter 2). This is an important example of repeated origins where researchers have asked if similar neuroendocrine mechanisms have been co-opted across these repeated origins.

Vertebrate vocalizations are produced by organs such as the swim bladder in fish, a larynx in frogs and mammals, and a syrinx in birds. The vocal organs themselves are not homologous and are not always anatomically associated with the respiratory system. The vocalizations are controlled by the brain-directed action of muscles associated with the vocal organs that are steroid-sensitive innovations. Vocalizations often have a well-defined rhythmic structure at one or more temporal scales. A single note may be repeated at close regular intervals (as in human laughter or a toad trilling). Individual notes may be produced by rapid contractions of sonic muscles. Bass and Baker (1997) proposed that the nervous system machinery that was later co-opted repeatedly to do this was hindbrain rhythmic circuits with pattern generators for cardiac and respiratory cycles. This machinery first appeared very early in vertebrate evolution and was itself a major vertebrate innovation. The pattern generator neurons have an embryonic origin in two Hox-gene-specified compartments of the hindbrain. The muscles that control the larynx, syrinx, and sonic swim-bladder all have a common embryonic origin that lies in the same Hox-gene-specified compartments of the hindbrain as the circuitry for heart and respiratory rhythms. The neurons that originate there have been the logical (most likely to evolve) mechanism to co-opt whenever rhythmic sound production is needed because of their preexisting electrical properties and their common embryonic origin with the sound organ muscles. What is then needed to complete the story (to produce sexual dimorphism and steroid regulation in vocalization) is expression of steroid receptor genes in the developing muscles and in the brain nuclei for vocalization.

Anatomical and physiological studies of forebrain structures regulating vocalization in two sound-producing teleosts from the same order, *Porichthys notatus* and *Opsanus beta*, suggest remarkable similarities and even homologies with the brain mechanisms for vocalization in other vertebrates, especially mammals, in spite of the fact that vocalization itself is not homologous in these lineages (Goodson and Bass 2002). Except in songbirds and possibly some mammals, the preoptic area and anterior hypothalamus are the most anterior part of the vocal control systems. They are linked to the midbrain and from there to the hindbrain vocal pattern generator. These regions are homologous across vertebrates as is their hormone-regulated contribution to social behavior more generally, independent of whether vocalization is used. An additional neural homology is that vasotocin and vasopressin neurons in the preoptic area and anterior hypothalamus modulate vocalizations in all cases (Emerson and Boyd 1999; Goodson and Bass 2001).

Comparative studies of the neural mechanisms underlying the production of learned vocalizations in birds have also produced important insights into how conserved ancestral features can be used in derived innovative ways. Many animals are capable of learning to recognize vocalizations but the production of vocalizations that have themselves been learned has a much more restricted phylogenetic distribution. Aside from a few mammals, the evidence for vocal learning is limited to three avian lineages: the parrots, hummingbirds, and oscine passerines (Kroodsma 1982). Based on currently available avian phylogenies, these reflect at least two and possibly three independent origins. Here vocalizations of sometimes extraordinary complexity are learned and remembered. Because the oscines are the most speciose group of living birds, the use of learned vocalizations for mate competition and attraction has been hypothesized to have facilitated speciation (Marler and Tamura 1962; Slabbe-koorn and Smith 2002; see Ricklefs [2003] for an alternative view). Although little is known about hormonal involvement in this behavior in parrots and hummingbirds, there is abundant evidence for hormonal regulation of vocal production in some oscines, especially species where males sing more than females and sing on a seasonal basis (chapters 2 and 4).

The oscine neural song system is an interconnected system of telencephalic nuclei with motor output to brainstem structures (fig. 2.11). Some of these brainstem structures produce unlearned vocalizations in other birds and other vertebrates, but the telencephalic nuclei are unique to oscines and a major innovation. Their presence in the oscine brain is associated with two other characters: the presence of sex steroid receptors, especially androgen receptors, in the "nonlimbic" telencephalon (in or near song nuclei) and the presence of a large number of aromatase positive cells in the caudal nidopallium. Suboscine passerines do not have a telencephalic song system but do express aromatase in the caudal nidopallium, suggesting that caudal telencephalic aromatase preceded the song system and could have been itself an innovation that facilitated the evolution of, or was subsequently co-opted by, the telencephalic song nuclei (Saldanha et al. 2000). Nothing is known about why these innovations occurred. What is clear, however, is that an oscine style song system maintained until the present time only happened once in birds, a truly pivotal event (Brenowitz and Kroodsma 1996). Parrots and hummingbirds have telencephalic vocal control nuclei organized in a remarkably similar manner to those of oscines, but they are not homologous to those of oscines (Paton et al. 1981; Gahr et al. 1993; Striedter 1994; Durand et al. 1997; Metzdorf et al. 1999;

Jarvis et al. 2000). There are androgen receptors in two of the hummingbird telencephalic song control nuclei, a convergent mechanism, but not in those of the parrot (Gahr 2000). All three song systems are independently derived neuroanatomically. It has been hypothesized that in each case the evolutionary change from the ancestral state (no telencephalic song system) included the onset of gene expression for estrogen receptors during development in neurons in the telencephalon (Gahr et al. 1993).

How did a telencephalic system for learned songs arise more than once in birds? Ulinski and Margoliash (1990) proposed that song systems are elaborations of older forebrain circuitry shared by all Diapsida, including birds. More recently Farries (2001) proposed that all three avian song systems are specializations of a preexisting avian circuit. The anatomical evidence for this proposal is especially compelling for the motor (caudal, posterior) pathway. Nonoscines have a group of parallel pathways linking the overall region where HVC is in oscines to the area where RA is in oscines, and from there to the brainstem, with input from audition, vision, and the trigeminal sense. This ancestral circuit is not connected to the brainstem vocal/respiratory nuclei. In oscines subparts at each level in the telencephalon are specialized for vocalization, and the brainstem projection includes a projection to the vocal/respiratory nuclei. These are the innovations with respect to function and connectivity. In parrots the posterior system is a specialization of the trigeminal part of the ancestral circuit, that is, a different part of the ancestral system has been co-opted to become specialized for vocalization. The use of the trigeminal is interesting given the nature of the parrot tongue and its use in vocalization. The essence of Farries' proposal is that if evolution is going to engineer a motor song system coming from the telencephalon to the brainstem nucleus controlling the syringeal muscles, this is how it would do it given the preexisting general organization of motor systems in birds. This is the shortest route to the adaptive peak of complex learned vocalization, the one requiring the fewest genetic changes and the least developmental tinkering. It may soon be within the realm of possibility to figure out what those genetic changes and tinkerings were.

That still leaves the selective pressures surrounding these origins and their early maintenance a mystery. There are also unanswered questions about why the system is sexually dimorphic in oscines and not in parrots. Zooming back out to look at all vertebrates, we can ask why an elaborate sexually dimorphic telencephalic system for vocalization happened only in birds. Protected invasion theory (chapter 5) provides an explanation for why fancy sexually selected male-typical behavior is more likely to take hold in a ZZ/ZW genetic architecture (Reeve and Shellman-Reeve 1997). While there are other ZZ/ZW vertebrates, a disproportionate percentage of vocalizing ZZ/ZW vertebrates are birds. On simple probabilistic grounds, a very rare origin event in which a simpler brainstem vocalizing ancestry become elaborated (moved to a distant adaptive peak) was most likely to happen in the avian clade.

Most male birds produce their courtship sounds with the syrinx but a few use other means such as foot drumming or wing snapping. In the suboscine golden-collared manakin (*Manacus vitellinus*), males make very loud wing snaps while displaying on a lek, suggesting that the spinal mechanisms controlling the wings might be steroid sensitive and sexually dimorphic (Schlinger et al. 2001). The spinal cords of this species indeed have steroid receptors. It will be interesting to see if other suboscines that don't wing-snap lack these receptors and whether similar brain mechanisms have been co-opted to control wing-snap sound production to those used for courtship vocalizations in suboscines (Schlinger et al. 2001).

Female singing of learned songs occurs in birds, but because of uncertainties in the phylogenies of birds it is not always possible to tell whether this is the ancestral or derived state. Chapter 5 discussed hormonal mechanism hypotheses for evolutionary changes in song system dimorphism among closely related species. This is not the same problem as the initial origins of song systems in birds, although it would certainly be interesting to know whether song systems began as monomorphic or dimorphic systems. At both macro and micro scales, however, there might be something about the developmental or other mechanisms underlying singing that make evolutionary changes in song dimorphism relatively easy to achieve as soon as the sexual selection pressures change. These mechanisms could be different in oscines (because of the telencephalic involvement and song learning) than other birds whose songs are unlearned. Male quail crow, as do males of other species of Phasianidae, whereas females don't; in buttonquail (Turnicidae), however, sex roles are reversed and females give loud advertising "booms" (Madge and McGowan 2002). It will be interesting to find out whether the vocalization systems are homologous or homoplasic in true quail and buttonquail, and what has produced the sex role reversal in buttonquail.

In frogs, another famous group of singers, mating vocalizations by females (not just by males) have evolved several times (Emerson and Boyd 1999). Something mechanistic has made this an easy adaptive peak to reach. During mating female androgen levels are higher than male levels. Emerson and Boyd proposed that this has allowed androgen-sensitive neural pathways typically used only by males but present in both sexes to be repeatedly co-opted by females. This is an activational hormone hypothesis, one that could be tested by (1) seeing if androgen antagonists block calling in females that normally call, and (2) seeing if increasing androgens further stimulates calling in females of species that don't normally call. If an activational hypothesis is correct, the question becomes why all female frogs don't sing, that is, what is downregulated in the vocal circuits of most species so that they don't respond to female androgens. The research with *Xenopus* (chapter 4) suggests a version of an organizational hypothesis, but it is not yet known whether it will generalize to other frog lineages.

Mating Behavior

There seems little doubt that mating behavior as loosely defined (sexes come close together or in physical contact and males release sperm) is highly conserved across vertebrates, because all species have such behavior with only a tiny number of self-fertilizing exceptions that are not in basal lineages. Some of the hormonal and neural mechanisms appear to be conserved as well. An obvious candidate is the involvement of the gonadal sex steroids in some aspect of male courtship or mating in many species in all major vertebrate clades that have been studied. Another is that across vertebrates, the preoptic area is critical for male mating behavior, the ventromedial hypothalamus is critical for female mating behavior, and the ventral tegmentum of the midbrain is important in both sexes.

The sensory and motor requirements of mating are different for male vertebrates that mate by making genital contact with the female, especially for those that have an intromittent organ. This organ can take the form of a phallus (as in ratite birds, ducks and geese, crocodiles, turtles), penis (as in mammals), a pair of claspers (modified pelvic fins) (as in sharks), or two hemipenes (lizards, snakes). In mammals the role of the steroid-sensitive portions of the preoptic area in receiving sensory information from the penis and modulating its reflexes is well established. Does it have a similar function in other vertebrates with intromittent organs? The structure and tissue origins of the organs themselves suggest that they are not homologous, and that the mammalian penis is derived (Dixson 1998). Yet an intromittent organ of some kind is the ancestral amniote state in a reconstructed phylogeny, one of those things that living on land apparently required. Given the conserved nature of the preoptic area, it is possible that it serves other amniote intromittent organs and not just the mammalian version. That is, the brain level mechanisms could be homologous even if the exact organ at the other end isn't. It also makes sense that when intromittent organs did evolve in amniotes, it was the hormone-sensitive preoptic area that was co-opted for their management. As an important center for male mating behavior for the preceding many millions of years, the preoptic area would be the logical place to handle new equipment for transferring sperm. It will be interesting to see if similar brain mechanisms control the gonopodium of poeciliid fish (guppies and their relatives) and the claspers of sharks. Here the organs themselves are derived and clearly not remotely homologous to amniote intromittent organs (they are modified fins). If the preoptic area was co-opted once (in mammals or amniotes) for the task, we can hypothesize that it was also co-opted for the poeciliid gonopodium and the shark claspers.

Diversity in mechanisms of mating behavior is layered on top of these conserved features. Chapter 2 presented two notable examples. In one males with dissociated reproductive strategies are mating at a time different from when sperm and high androgens are produced. In the other the hormonal control of female mating in externally fertilizing species (whose receptivity needs to be coordinated with oviposition) differs from the hormonal control in internally fertilizing vertebrates (where coordination with ovulation is needed). This diversity has occasionally resulted in homoplasy. A good example of this is the similarity in the estrous cycles of mammals and of green anoles (*Anolis carolinensis*). The green anole is an internally fertilizing oviparous species with a two-week-long estrous cycle. Females ovulate and lay one egg every two weeks. They mate only around the time of ovulation (are polyestrous), and their mating behavior is controlled by the same ovarian sex steroids (estradiol and progesterone) as in female rats and a number of other mammals (Crews 2002). Other lineages of diapsida do not have an estrous cycle; many lay a whole batch of eggs at once and some lay just once a year (at best they could be considered monoestrous). This is a striking example of convergent evolution in the hormonal regulation of a social behavior.

Parental Behavior

Parental behavior is a appealing case study in conservation and diversity in vertebrate evolution. Parental care by females (either alone or with the male) is characteristic of two large clades, mammals and birds. In mammals it is the rule, even in the most basal lineages. In birds it is absent in much of the most basal lineage (Sibley and Ahlquist's [1990] parvclass Ratitae) and has been lost (a derived state) in a few sex-role-reversed species and in the brood (interspecific) parasites. The recognition that birds are the living descendants of one of the dinosaur clades and the sister group to the Crocodylia has led to a new insight about just how far back in the tree female parental care might go (Greene 1999; Tullberg et al. 2002). Crocodylia have biparental or female parental care, and there is fossil evidence that some dinosaurs had parental care as well (Meng et al. 2004). Female parental care may be the ancestral state in amniotes and in archosaurs, with biparental care evolving later, in the ancestors of the modern birds. Its absence in turtles is derived. If this phylogeny is correct, then maternal care in crocodiles, birds, and mammals has a single origin and its current instantiations are homologous. If male dinosaurs were the parents, however, this scenario would need revision.

Why does it matter whether female care in these animals is homologous? It matters because if there is a single origin, it is more plausible to hypothesize that the endocrinology of parental behavior can be generalized. If there are separate origins, there could still be similar (homoplasic) hormonal mechanisms, but hypothesizing such similarity would be based on a guess about the likelihood of the same mechanisms being co-opted.

Male care has a different distribution in vertebrates (Clutton-Brock 1991; Reynolds et al. 2002; Tullberg et al. 2002). In teleosts, care by the male alone is at least as common (probably more so) than female care and has a distinct distribution, as if it has evolved multiple times independently. Within the avian and mammalian clades biparental and (in birds) male-only care have come and gone several times (Tullberg et al. 2002). With respect to the male's role, there are significant phylogenetic effects in some clades, for example, the marmosets and tamarins, and not others, for example, the diversity within the vole genus *Microtus*. Some paternal care is facultative rather than obligatory and occurs under certain social and ecological conditions, for example, occasional care of chicks by male galliform birds (Madge and McGowan 2002) or by subordinates in cooperatively breeding groups. As chapter 2 pointed out, this plasticity and evolutionary lability reflects the importance of induction by stimuli from the young and the presence of the parental behavior repertoire in the brains of males (Wynne-Edwards and Reburn 2000). From a mechanistic perspective we can see how parental care has evolved so many times. Why there has been repeated evolution of parental care of some kind in vertebrates may not at first glance seem like a huge mystery given the obvious way in which it benefits the parent's fitness to have the offspring survive instead of die. That doesn't mean, however, that it is easy to understand when it is and is not advantageous to forgo additional immediate reproduction to care for young, or why teleost species A has parental care and teleost species B doesn't (Reynolds et al. 2002).

Parenting in some clades is associated with special secretions for feeding the young that have a hormonal basis. In three unrelated groups—mammals, pigeons and doves, and discus fish (*Symphysodon aequifasciata*)—these secretions (mammary gland milk, crop sac "milk," and epidermal mucus, respectively) are stimulated by prolactin or a prolactin-like peptide, a remarkable example of convergence in which a homologous molecule is co-opted for non-homologous parental innovations (van Tienhoven 1983). The traditional explanation for why it is a prolactin family peptide that was co-opted in all three cases is that throughout vertebrates these peptides have an important role in osmoregulation and fluid absorption. Prolactin has been shown to stimulate parental behavior in these three groups as well (chapter 2; Blüm and Fiedler 1965). What differs across these three is whether the innovation is sexually dimorphic. Mammals are the oddballs, because male pigeons and discus fish produce "milk."

Viviparity is a remarkable parental investment innovation with important connections to hormones and behavior. It has arisen at least 90 times independently in vertebrates (Reynolds et al. 2002). It occurs in more than half of the sharks and rays, in a few teleosts, in the coelacanth, in a few amphibians, in all mammals except for monotremes, and in some lizards and snakes. It shows strong phylogenetic effects in some clades, as when all marsupial and eutherian mammals are viviparous but no birds are, but in other clades such as snakes and scincomorph lizards it is extraordinarily labile, varying even within species (Cree 1994; Shine 1995). In these labile clades viviparity is associated with cold climate. Explaining its distribution elsewhere in the vertebrate tree has been difficult. For example, why hasn't viviparity evolved in flightless birds, given that all birds have internal fertilization and most care for the eggs and chicks (Blackburn and Evans 1986)? However mysterious the repeated homoplasy of viviparity, here is another opportunity to see whether similar hormonal mechanisms have been co-opted. In mammals the set of mechanisms includes the formation of corpora lutea by the postovulatory follicles in response to prolactin that then secrete large quantities of progesterone that help maintain the pregnancy. Most other vertebrates don't have a corpus luteum, but viviparous snakes, lizards, sharks, and rays do.

Primates have a unique hormonal innovation associated with pregnancy: chorionic gonadotrophic hormone (CG), a peptide produced by the placenta. Genetically speaking the placenta is fetal tissue. CG is the pregnancy signal that tells the maternal system that there is a fetus. The gene that produces one of its two subunits arose by duplication of the gene for the same subunit of luteinizing hormone (Maston and Ruvolo 2002). The duplication first arose in the lineage that became the New World monkeys and Old World monkeys and apes after it diverged from the prosimians. There is evidence in New World monkeys that this gene family has been subjected to strong positive selection. What the selective pressures were is unclear. Because CG can cross the blood–brain barrier, it is tempting to hypothesize that it might have some brain-mediated behavioral effects on the mother intended to serve the fetus' needs, with corresponding maternal mechanisms designed to tone that down to meet her needs (Haig 1993). We can also wonder if analogous hormones have evolved in any other viviparous lineages. Placenta-like structures are not unique to mammals. What kinds of signals they might produce, and which party they are designed to benefit, would depend on whether they are genetically maternal.

Sex Determination and Sexual Differentiation

The phylogeny of genetic and chromosomal sex-determining systems is a similarly fascinating mix of conservation and diversity. Chapter 4 pointed out that the molecular developmental cascades leading to the formation of ovaries or testes are promising to be relatively conserved. In different kinds of vertebrates these cascades seem to have been captured by different genes on different chromosomes, such as temperature-sensitive genes in animals without genetic sex determination or genes on an autosome. Such an autosome then may become a sex chromosome (Graves and Shetty 2001). Chapter 5 explored some of the consequences of having one system or another. Except in sex-changing teleosts, sex determination itself is not social behavior, but its consequences for the development and evolution of sexually dimorphic social behavior are sufficiently profound to merit a closer look at its phylogeny. Who has what system? Which came first? What sent different clades down different paths? To make these questions more manageable, sex-determining systems will be lumped into four categories: environmental sex determination (ESD), which is often by temperature, genetic sex determination (GSD) of the ZZ/ZW type (males have the two similar sex alleles), genetic sex determination of the XX/XY type (females have the two similar sex alleles), and unknown (?GSD?) but presumably genetic because sex ratios are relatively invariant and close to 50:50.

The coarse-scale distribution of these vertebrate sex-determining systems is summarized in fig. 7.5. The tree includes the nonvertebrate chordates and invertebrates whose sex determining mechanisms will be essential for eventually reconstructing the ancestry of the systems. Some vertebrate clades, most notably snakes, birds, and mammals, are very homogeneous. All sex determination is genetic, and within each clade the system is highly conserved. The majority of the species in these clades also have heteromorphic sex chromosomes, that is, a pair of markedly dissimilar sized chromosomes in one sex, with the W or Y smaller than the Z or X. These were originally autosomes that became heteromorphic after genes on them captured the gonad-determination cascade. The autosomes of origin are not the same across clades; the avian Z and W are not homologous to the mammalian X and Y (Ellegren 2000; Graves and Shetty 2001). Evolutionary biologists have been quite successful at figuring out why a reduction in size occurs in one member of the pair once genes on a chromosome become sex-determining alleles, and why the evolution of hetermorphic chromosomes then makes it less likely that the sex-determining system will change in the future (Bull 1983; Charlesworth 1991; Rice 1994).

Other parts of the vertebrate tree contain a mixture of two or three sex-determining systems, as if there have been repeated changes in the genes and their chromosomal locations that have captured the cascades. In amphibians, scincomorph and gekkomorph lizards, and teleost fishes, variation is seen within families, genera, and even species. In *Rana rugosa*, some populations are XX/XY (as in other ranid frogs) and some are ZZ/ZW (Ohtani et al. 2000). Some populations of *Xiphophorus maculatus* are currently evolving a ZZ/ZW system (Kallman 1984). Evidently, relatively few genetic changes are required to change an XX/XY system to a ZZ/ZW system, a conclusion supported by

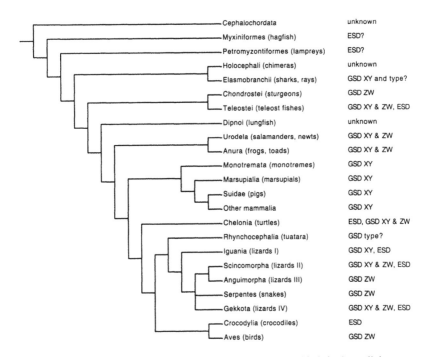

FIGURE 7.5. Simplified cladogram of the living vertebrates along with their closest living ancestors, the cephalochordates, showing the sex-determining systems of the clades. ESD, environmental sex determination; GSD, genetic sex determination; XY, XX/XY system (male heterogamety); ZW, ZZ/ZW system (female heterogamety). Information about sex determining systems is based on Hillis and Green (1990), Janzen and Paukstis (1991), Schmid and Steinlein (2001), and Devlin and Nagahama (2002).

population genetics models of changes in sex-determining systems (Bull 1983; Graves 1995).

Environmental sex determination has often been proposed to be the ancestral system in vertebrates, but does the phylogeny support this hypothesis? It is tempting to assume that ESD is ancestral to GSD in turtles because nearly all living turtles that have been studied have ESD (Janzen and Paukstis 1991). However, if GSD turtles are placed correctly on a more detailed turtle tree, the reconstructed ancestral state for turtles is equivocal (equally likely to be either ESD or GSD). ESD in the occasional lizard is probably derived. Adult sex determination by social environment is derived in teleosts (it occurs only in an apical order, Perciformes). The situation for other forms of ESD in teleosts is unclear, however, because of insufficient information about both sex determination and phylogeny and because some species seem to have a combination of ESD (by the physical environment) and GSD. Sex-determining systems are poorly understood for most basal vertebrate lineages and for those inverte-brates (including chordates) that are relevant to reconstructing the ancestors of vertebrates. Therefore, it is still an open question whether environmental sex determination is ancestral in the broad scale of vertebrate evolution.

The first use of phylogenetic reconstruction plus a parsimony algorithm to reach a conclusion about an ancestral genetic sex-determining system was Hillis and Green's (1990) analysis of the amphibian clade, which contains a mixture of ZZ/ZW and XX/XY GSD. Their analysis showed that ZZ/ZW is ancestral in amphibians, a conclusion that remains unchanged by the addition since their review of some new species (Schmid and Steinlein 2001; Adkins-Regan, unpublished).

Regardless of what the ancestral states are at the key nodes where major clades diverged, it is clear that in the vertebrate clade overall there are multiple instances of convergent evolution toward male heterogamety and female heter-ogamety—multiple pivotal origins. Is there any rhyme or reason to when this happens or are they random events? Few genetic changes are required so long as there aren't highly heteromorphic sex chromosomes, but are there any selec-tive pressures that might have favored a change to one system over another? The phylogenetic distribution of male vs. female heterogamety doesn't imme-diately suggest any promising hypotheses. Claims that viviparity requires ge-netic sex determination don't stand up to either reason or evidence.

Genes (DNA sequences) similar to the mammalian *Sry* are present in other vertebrates (Tiersch et al. 1991; Spotila et al. 1994). Outside of the mammals they are on autosomes, their expression is not sex-specific, and they are not sex determiners. The evolution of a sex-determining function was an important mammalian innovation that was likely responsible for the loss of genes on the Y that began 300 million years ago (Quintana-Murci et al. 2001). The mole voles (*Ellobius*) don't have either a sex-determining *Sry* or a Y chromosome, as if there has been a recent innovation in this rodent lineage in which some other gene has captured the gonad-determining cascade (Just et al. 1995; Graves and Shetty 2001). The avian W is probably more recent than the mam-malian Y (Nanda et al. 2002); no reversals in which it has been subsequently lost have been detected.

Sex steroids, especially estrogens, seem to contribute to gonadal sex induc-tion in some vertebrates (some teleost species, turtles, lizards, alligators, birds) and not others (amphibians, mammals) (chapter 4). The aromatase gene is part of the cascade in the former but not the latter, as if its role has come and gone in evolution. At some point in chordate evolution, organization and/or activation of the extragonadal sexual phenotype by sex steroids appeared. As chapter 4 explained, the development of sexual phenotypes is well understood in only a few groups of vertebrates, and so it is a mystery when this innovation occurred. There are several reasons to think that the ancestral version of the innovation might have been organization by estrogens from the ovary. First, where sexual differentiation is hormonally organized, estrogens always act (either to feminize or to masculinize), whereas androgens are not always effective. When they are, it is usually because of aromatization. Second, the molecular phylogeny of steroid receptors suggests that the estrogen receptor was the first to evolve (Thornton 2001). Third, except in mammals there is a tendency toward ovarian primacy in development (Mittwoch 1998). In teleosts and frogs, testes form later than ovaries, and gonads of genetic males sometimes differentiate into ovaries first and later into testes (Devlin and Nagahama 2002). In oviparous animals the close relationship between ovary size and fecundity has meant strong selection for ovaries with rapid growth rates that get off to a fast start beginning in the embryo.

Brain-directed (top-down) sexual differentiation, discovered in the sex-reversing teleost fishes, is an important derived innovation in the Perciformes (chapter 4). Sexual differentiation of the brain through gonadal steroids (bottom-up) of the kind seen today in mammals and some birds began at some unknown point (points?) in evolution. Who knows how many other new and different ways of developing sex differences in social behavior exist among the thousands of unknown vertebrates?

Conclusions

This chapter has emphasized the mixture of conserved and innovative features of vertebrates related to steroids, peptides, and social behavior. Telling which is which requires a phylogenetic context. As Tinbergen acknowledged with his four aims, the present cannot be understood without the past. Because so much is conserved, it is important to understand the developmental and genetic unpinnings that enable innovations to originate. The same developmental mechanisms can be repeatedly co-opted for nonhomologous behavioral outcomes, as in the evolution of vocalization systems for both learned and unlearned social signals. Changes in the structure of peptides, steroid receptors, and steroidogenic enzymes have occurred in vertebrate evolution. Some of these might be neutral but others have clearly had long-term consequences. Gene duplication is thought to have provided some important raw material for vertebrate evolution, and it has been hypothesized that two large-scale genome expansions were pivotal events (Chiang et al. 2001). These expansions plus changes in gene expression may have produced diversity in the brain distributions of steroid receptors, peptides, and peptide receptors across species and also within individual species. Alternative splicing is another way that such diversity could result from a rather conserved genome, and it will be interesting to see to what extent this has created multiple receptor subtypes that might account for different steroid actions in different tissues or brain regions (Thornton et al. 2003).

Steroids, peptides, and their mechanisms of action have been essential components of the engine of social behavior for many millions of years. They have the flexible properties that have enabled a great diversity of social behavior with relatively few changes in the mechanisms themselves. They promise to help bridge the genome–behavior gap for large-scale vertebrate evolution, the subject of this chapter, as well as for smaller-scale evolution, the focus of chapter 5. In a recent essay, Baker (2003) proposed that system-wide gene regulation by gonadal and adrenal steroids is the key to the success of vertebrates, accounting for the evolution of overall complexity in spite of an unremarkable genome size. Add to this the appearance of neurosteroids (which began who knows when in evolution) and steroids take their rightful place alongside social complexity in the grand scheme of the history of life on earth.

Afterword

The Preface explained that the goal of this book is to encourage crossing the bridge between behavioral endocrinology, on one side, and behavioral ecology (and other versions of animal behavior), on the other side. The hope was that by using the book as a road map, the traveler can reach a more informed and conceptually elevated vantage point, one providing integrated views of whole forests and not just single trees. As in any science, a good view also reveals the gullies and canyons where lie the hypotheses that failed their tests, and reveals the saplings of those not yet adequately tested.

A wide-angle view shows that some kind of important link to steroids and peptides is characteristic of many forms of social behavior at all levels of biological organization. Hormonal actions on the nervous system during development and adulthood underpin the fitness-impacting behavior of individuals, their dyadic social interactions, their life in groups or more spatially dispersed forms of social organization such as territoriality, and their life course. Hormones, and their receptors and other downstream mechanisms, promise to be important for understanding the evolution of social behavior in populations over relatively short timescales and the evolution of species differences. Hormones and their receptors reflect the deeper evolutionary divergences, convergences, and continuities of the tree of life.

Advances in methods for measuring steroids and studying their receptors have been instrumental in the substantial progress that has occurred, along with advances in theory and increased sophistication in the use of integrative and comparative approaches. The experimental study of hormones and social behavior increasingly has been extended to free-living or naturalistically housed animals. The "real-world" ecological context is messy with respect to experimental control and measurement precision, but it has compensating virtues, especially for behavior that is sensitive to current social circumstances (is tactically flexible). Field and laboratory studies provide complementary insights.

There is a dark cloud hanging over the landscape, however. Just as there are dozens of ways for animals to use hormonal mechanisms to achieve success in social life, there are dozens of ways for the residues of modern human life to interfere with that success. The more we learn about the fitness impact of steroids and their receptors for the development and function of the nervous system, the more we realize how many routes there are for anthropogenic chemicals in the environment to act as endocrine disrupters, producing maladaptive outcomes. Steroid receptors are to some extent blind to whether the ligand is an endogenous steroid or an industrial waste steroid mimic. Particularly disturbing is the prospect of steroid mimics with organizing actions on the brain, because such effects are permanent. There is already worrisome evidence for serious consequences for the bodies of aquatic species. Effects on the brain, the organ of adaptive social behavior, are likely to be equally devastating for the future of an animal population.

The book has celebrated the diversity of uses for steroids and peptides along with the ways the varied lifestyles of animals have co-opted the same mechanisms repeatedly. At the level of the genome, all organisms seem to be cousins under the skin, but at the level of the organism, no one species or group of species is a model for all the others. As we head toward the sixth major extinction event on earth, we realize that many of the twigs on the living tree of life could be lost. It cannot be promised that further advances in the field of hormones and behavior will stop this process. It is hoped, however, that such advances, like those from other domains of organismal biology and biological psychology, can serve to increase appreciation of the variety of animal social life and strengthen the resolve to slow the rate of extinction to a more prehuman level. Without conservation and a drastic change in priorities, there will be no more diversity, no more discoveries of new hormonal mechanisms, no more new uses for old hormones.

References

Abbott, D. H., Barrett, J., Faulkes, C. G. and George, L. M. 1989. Social contraception in naked mole-rats and marmoset monkeys. *J. Zool. (Lond.)* 219: 703–710.

Abbott, D. H., Saltzman, W., Schultz-Darken, N. J. and Tannenbaum, P. L. 1998. Adaptations to subordinate status in female marmoset monkeys. *Comp. Biochem. Physiol. C* 119: 261–274.

Abbott, D. H., Keverne, E. B., Bercovitch, F. B., Shively, C. A., Mendoza, S. P., Saltzman, W., Snowdon, C. T., Ziegler, T. E., Banjevic, M., Garland, T. and Sapolsky, R. M. 2003. Are subordinates always stressed? A comparative analysis of rank differences in cortisol levels among primates. *Horm. Behav.* 43: 67–82.

Ábrahám, I. M., Han, S.-K., Todman, M. G., Korach, K. S. and Herbison, A. E. 2003. Estrogen receptor β mediates rapid estrogen actions on gonadotropin-releasing hormone neurons in vivo. *J. Neurosci.* 23: 5771–5777.

Acher, R. 1986. Common patterns of neuroendocrine integration in vertebrates and invertebrates. *Gen. Comp. Endocrinol.* 61: 452–458.

Ader, R. and Cohen, N. 1992. Conditioned immunopharmacologic effects on cell-mediated immunity. *Int. J. Immunopharmacol.* 14: 323–327.

Adkins, E. K. 1975. Hormonal basis of sexual differentiation in the Japanese quail. *J. Comp. Physiol. Psychol.* 89: 61–71.

Adkins, E. K. 1981. Hormone specificity, androgen metabolism, and social behavior. *Am. Zool.* 21: 257–271.

Adkins, E. K. and Adler, N. T. 1972. Hormonal control of behavior in the Japanese quail. *J. Comp. Physiol. Psychol.* 81: 27–36.

Adkins, E. K. and Nock, B. 1976. Behavioral responses to sex steroids of gonadectomized and sexually regressed quail. *J. Endocrinol.* 68: 49–55.

Adkins, E. K. and Pniewski, E. E. 1978. Control of reproductive behavior by sex steroids in male quail. *J. Comp. Physiol. Psychol.* 92: 1169–1178.

Adkins, E. K. and Schlesinger, L. 1979. Androgens and the social behavior of male and female lizards, *Anolis carolinensis*. *Horm. Behav.* 13: 139–152.

Adkins-Regan, E. 1981a. Effect of sex steroids on the reproductive behavior of castrated male ring doves (*Streptopelia* sp.). *Physiol. Behav.* 26: 561–565.

Adkins-Regan, E. K. 1981b. Early organizational effects of hormones: an evolutionary perspective. In N. T. Adler, ed., *Neuroendocrinology of Reproduction: Physiology and Behavior*, 159–228. New York: Plenum Press.

Adkins-Regan, E. 1985. Nonmammalian psychosexual differentiation. In N. T. Adler, R. Goy, and D. Pfaff, eds., *Handbook of Behavioral Neurobiology*, Vol. 7: *Reproduction*, 43–76. New York: Plenum Press.

Adkins-Regan, E. 1987. Hormones and sexual differentiation. In D. O. Norris and R. E. Jones eds., *Hormones and Reproduction in Fishes, Amphibians, and Reptiles*, 1–29. New York: Plenum Press.

Adkins-Regan, E. 1988. Sex hormones and sexual orientation in animals. *Psychobiology.* 16: 335–347.

Adkins-Regan, E. 1996. Neural and hormonal mechanisms of behavior: physiological causes and consequences. In L. Houck and L. Drickamer, eds., *Foundations of Animal Behavior*, 389–405. Chicago: University of Chicago Press.

Adkins-Regan, E. 1998. Hormonal mechanisms of mate choice. *Am. Zool.* 38: 166–178.

Adkins-Regan, E. 1999a. Foam produced by male *Coturnix* quail: what is its function? *Auk* 116: 184–193.

Adkins-Regan, E. 1999b. Testosterone increases singing and aggression but not male-typical sexual partner preference in early estrogen treated female zebra finches. *Horm. Behav.* 35: 63–70.

Adkins-Regan, E. and Ascenzi, M. 1987. Social and sexual behavior of male and female zebra finches treated with oestradiol during the nestling period. *Anim. Behav.* 35: 1100–1112.

Adkins-Regan, E. and Ascenzi, M. 1990. Sexual differentiation of behavior in the zebra finch: effect of early gonadectomy or androgen treatment. *Horm. Behav.* 24: 114–127.

Adkins-Regan, E. and MacKillop, E. 2003. Japanese quail inseminations are more likely to fertilise eggs in a context predicting mating opportunities. *Proc. R. Soc. Lond. B* 270: 1685–1689.

Adkins-Regan, E. and Wade, J. 2001. Masculinized sexual partner preference in female zebra finches with sex-reversed gonads. *Horm. Behav.* 39: 22–28.

Adkins-Regan, E. and Watson, J. T. 1990. Sexual dimorphism in the avian brain is not limited to the song system of songbirds: a morphometric analysis of the brain of the quail (*Coturnix japonica*). *Brain Res.* 514: 320–326.

Adkins-Regan, E., Signoret, J.-P. and Orgeur, P. 1989. Sexual differentiation of reproductive behavior in pigs: defeminizing effects of prepubertal estradiol. *Horm. Behav.* 23: 290–303.

Adkins-Regan, E., Abdelnabi, M., Mobarak, M. and Ottinger, M. A. 1990. Sex steroid levels in developing and adult male and female zebra finches (*Poephila guttata*). *Gen. Comp. Endocr.* 78: 93–109.

Adkins-Regan, E., Mansukhani, V., Seiwert, C. and Thompson, R. 1994. Sexual differentiation of brain and behavior in the zebra finch: critical periods for effects of early estrogen treatment. *J. Neurobiol.* 25: 865–877.

Adkins-Regan, E., Ottinger, M. A. and Park, J. 1995. Maternal transfer of estradiol to egg yolks alters sexual differentiation of avian offspring. *J. Exp. Zool.* 271: 466–470.

Adkins-Regan, E., Yang, S. and Mansukhani, V. 1996. Behavior of male and female zebra finches treated with an estrogen synthesis inhibitor as nestlings. *Behaviour* 133: 847–862.

Adler, N. T. 1978. On the mechanisms of sexual behaviour and their evolutionary constraints. In J. B. Hutchison, ed., *Biological Determinants of Sexual Behavior*, 657–695. Chichester, UK: Wiley.

Afonso, L.O.B., Wassermann, G. J. and de Oliveira, R. T. 2001. Sex reversal in Nile tilapia (*Oreochromis niloticus*) using a nonsteroidal aromatase inhibitor. *J. Exp. Zool.* 290: 177–181.

Ågmo, A. and Kihlström, J. E. 1974. Sexual behaviour in castrated rabbits treated with varying doses of testosterone. *Anim. Behav.* 22: 705–710.

Aguilera, G. 1998. Corticotropin releasing hormone, receptor regulation and the stress response. *Trends Endocrinol. Metab.* 9: 329–336.

Aida, K., Moore, R. and Negishi, M. 1994. Lack of the steroid 15 alpha-hydroxylase gene (Cyp2a-4) in wild mouse strain *Mus spretus*: rapid evolution of the P450 gene superfamily. *Genomics* 19: 564–566.

Aikey, J. L., Nyby, J. G., Anmuth, D. M. and James, P. J. 2002. Testosterone rapidly reduces anxiety in male house mice (*Mus musculus*). *Horm. Behav.* 42: 448–460.

Airey, D. C., Castillo-Juarez, H., Casella, G., Pollak, E. J. and DeVoogd, T. J. 2000. Variation in the volume of zebra finch song control nuclei is heritable: developmental and evolutionary implications. *Proc. R. Soc. Lond. B* 267: 2099–2104.

Alatalo, R., Hoglund, J., Lundberg, A., Rintamaki, P. T. and Silverin, B. 1996. Testosterone and male mating success on the black grouse leks. *Proc. R. Soc. Lond. B* 263: 1697–1702.

Albers, H. E. and Bamshad, M. 1998. Role of vasopressin and oxytocin in the control of social behavior in Syrian hamsters (*Mesocricetus auratus*). *Prog. Brain Res.* 119: 395–408.

Albert, D. J., Jonik, R. H., Watson, N. V., Gorzalka, B. B. and Walsh, M. L. 1990. Hormone-dependent aggression in male rats is proportional to serum testosterone concentration but sexual behavior is not. *Physiol. Behav.* 48: 409–416.

Albert, D. J., Jonik, R. H., Tanco, S. A. and Walsh, M. L. 1992. Cohabitation with a sterile male facilitates the development of retrieval behavior in nulliparous female rats exposed to pups. *Physiol Behav.* 52: 727–729.

Alexander, G. M. and Sherwin, B. B. 1991. The association between testosterone, sexual arousal, and selective attention for erotic stimuli in men. *Horm. Behav.* 25: 367–381.

Alexander, G. M. and Sherwin, B. B. 1993. Sex steroids, sexual behavior, and selection attention for erotic stimuli in women using oral contraceptives. *Psychoneuroendocrinol.* 18: 91–102.

Alexander, G., Packard, M. G. and Hines, M. 1994. Testosterone has rewarding affective properties in male rats: implications for the biological basis of sexual motivation. *Behav. Neurosci.* 108: 424–428.

Allen, T. O. and Adler, N. T. 1985. Neuroendocrine consequences of sexual behavior. In N. Adler, D. Pfaff and R. W. Goy, eds., *Handbook of Behavioral Neurobiology*, vol. 7, 725–766. New York: Plenum Press.

al Sadoon, M. K., el Banna, A. A., Ibrahim, M. M., Abdo, N. M. and al Rasheid, K. A. 1990. Effect of gonadal steroid hormones on the metabolic rate of the cold-acclimated gonadectomized male and female *Chalcides ocellatus* (Forskal). *Gen. Comp. Endocrinol.* 80: 345–348.

Amateau, S. K. and McCarthy, M. M. 2004. *Nature Neurosci.* 7: 643–650.

Andersson, M. 1994. *Sexual Selection*. Princeton, NJ: Princeton University Press.

Andrew, R. J. 1966. Precocious adult behaviour in the young chick. *Anim. Behav.* 14: 485–500.

Andrew, R. J. 1991. Testosterone, attention and memory. In P. Bateson, ed., *The Development and Integration of Behaviour: Essays in Honor of Robert Hinde*, 171–190. Cambridge, UK: Cambridge University Press.

Andrew, R. J. and Jones, R. B. 1992. Increased distractability in capons: an adult parallel to androgen-induced effects in the domestic chick. *Behav. Processes* 26: 201–210.

Anglade, I., Pakdel, F., Bailhache, T., Petit, F., Salbert, G., Jego, P., Valotaire, Y. and Kah, O. 1994. Distribution of estrogen receptor-immunoreactive cells in the brain of the rainbow trout (*Oncorhynchus mykiss*). *J. Neuroendocrinol.* 6: 573–583.

Aragona, B. J., Liu, Y., Curtis, T., Stephan, F. K. and Wang, Z. 2003. A critical role for nucleus accumbens dopamine in partner-preference formation in male prairie voles. *J. Neurosci.* 23: 3483–3490.

Arai, O., Taniguchi, I. and Saito, N. 1989. Correlation between the size of song control nuclei and plumage color change in orange bishop birds. *Neurosci. Lett.* 98: 144–148.

Archer, G. S., Friend, T. H., Piedrahita, J., Nevill, C. H., Walker, S. 2003. Behavioral variation among cloned pigs. *Appl. Anim. Behav. Sci.* 81: 321–331.

Archer, J. 1988. *The Behavioural Biology of Aggression.* Cambridge, UK: Cambridge University Press.

Ardia, D. R., Schat, K. A. and Winkler, D. W. 2003. Reproductive effort reduces long-term immune function in breeding tree swallows (*Tachycineta bicolor*). *Proc. R. Soc. Lond. B* 270: 1679–1683.

Arnold, A. P. 1975. The effects of castration and androgen replacement on song, court-ship, and aggression in zebra finches (*Poephila guttata*). *J. Exp. Zool.* 191: 309–325.

Arnold, A. P. 1984. Androgen regulation of motor neuron size and number. *Trends Neurosci.* 7: 239–242.

Arnold, A. P. 1992. Developmental plasticity in neural circuits controlling birdsong: sexual differentiation and the neural basis of learning. *J. Neurobiol.* 23: 1506–1528.

Arnold, A. P. 2002. Concepts of genetic and hormonal induction of vertebrate sexual differentiation in the twentieth century, with special reference to the brain. In D. W. Pfaff, A. P. Arnold, A. M. Etgen, S. E. Fahrbach and R. T. Rubin, eds., *Hormones, Brain and Behavior*, vol. 4, 105–136. Amsterdam: Academic Press (Elsevier).

Arnold, A. P. and Breedlove, S. M. 1985. Organizational and activational effects of sex steroids on brain and behavior: a reanalysis. *Horm. Behav.* 19: 469–498.

Arnold, A. P. and Gorski, R. A. 1984. Gonadal steroid induction of structural sex differences in the central nervous system. *Annu. Rev. Neurosci.* 7: 413–442.

Arnold, A. P. and Saltiel, A. 1979. Sexual difference in pattern of hormone accumulation in the brain of a songbird. *Science* 205: 702–705.

Arnold A. P., Nottebohm, F. and Pfaff, D. W. 1976. Hormone concentrating cells in vocal control and other areas of the brain of the zebra finch (*Poephila guttata*). *J. Comp. Neurol.* 165: 487–511.

Arnold, S. J. 1987. Genetic correlation and the evolution of physiology. In M. E. Feder, A. F. Bennett, W. W. Burggren and R. B. Huey, eds., *New Directions in Ecological Physiology*, 189–212. Cambridge, UK: Cambridge University Press.

Aronson, L. 1959. Hormones and reproductive behavior: some phylogenetic considerations. In A. Gorbman, ed., *Comparative Endocrinology*, 98–120. New York: Wiley.

Asa, C. S. 1997. Hormonal and experiential factors in the expression of social and parental behavior in canids. In N. G. Solomon and J. A. French, eds., *Cooperative Breeding in Mammals*, 129–149. Cambridge, UK: Cambridge University Press.

Asa, C. S., Goldfoot, D. A., Garcia, M. C. and Ginther, O. J. 1980. Dexamethasone suppression of sexual behavior in the ovariectomized mare. *Horm. Behav.* 14: 55–64.

Astiningsih, K. and Rogers, L. J. 1996. Sensitivity to testosterone varies with strain, sex, and site of action in chickens. *Physiol. Behav.* 59: 1085–1091.

Aubret, F., Bonnet, X., Shine, R. and Lourdais, O. 2002. Fat is sexy for females but not males: the influence of body reserves on reproduction in snakes (*Vipera aspis*). *Horm. Behav.* 42: 135–147.

Auger, A. P. 2001. Ligand-independent activation of progestin receptors: relevance for female sexual behavior. *Reproduction* 122: 847–855.

Auger, A. P. 2004. Steroid receptor control of reproductive behavior. *Horm. Behav.* 45: 168–172.

Auger, A. P., Tetel, M. J. and McCarthy, M. M. 2000. Steroid receptor coactivator-1 (SRC-1) mediates the development of sex-specific brain morphology and behavior. *Proc. Natl. Acad. Sci. U S A* 97: 7551–7555.

Auger, A. P., Perrot-Sinal, T. S. and McCarthy, M. M. 2001. Excitatory versus inhibitory GABA as a divergence point in steroid-mediated sexual differentiation of the brain. *Proc. Natl. Acad. Sci. U S A* 98: 8059–8064.

Bacon, W. L. 2001. Secretory patterns of luteinizing hormone and gonadal steroids in relationship to reproductive status in male and female turkeys. In A. Dawson and C. M. Chaturvedi, eds., *Avian Endocrinology*, 167–179. New Delhi: Narosa Publishing House.

Badura, L. L. and Friedman, H. 1988. Sex reversal in female *Betta splendens* as a function of testosterone manipulation and social influence. *J. Comp. Psychol.* 102: 262–268.

Badyaev, A. V., Hill, G. E. and Weckworth, B. V. 2002. Species divergence in sexually selected traits: increase in song elaboration is related to decrease in plumage ornamentation in finches. *Evolution* 56: 412–419.

Bagatell, C. J., Heiman, J. R., Rivier, J. E. and Bremner, W. J. 1994. Effects of endogenous testosterone and estradiol on sexual behavior in normal young men. *J. Clin. Endocrinol. Metab.* 78: 711–716.

Baker, B. S. 1989. Sex in flies: the splice of life. *Nature* 340: 521–524.

Baker, M. E. 1994. Sequence analysis of steroid- and prostaglandin-metabolizing enzymes: application to understanding catalysis. *Steroids* 59: 248–258.

Baker, M. E. 1997. Steroid receptor phylogeny and vertebrate origins. *Mol. Cell. Endocrinol.* 135: 101–107.

Baker, M. E. 2002. Recent insights into the origins of adrenal and sex steroid receptors. *J. Mol. Endocrinol.* 28: 149–152.

Baker, M. E. 2003. Evolution of adrenal and sex steroid action in vertebrates: a ligand-based mechanism for complexity. *Bioessays* 25: 396–400.

Bakker, J. 2003. Sexual differentiation of the neuroendocrine mechanisms regulating mate recognition in mammals. *J. Neuroendocrinol.* 15: 615–621.

Bakker, J., Honda, S.-I., Harada, N. and Balthazart, J. 2002a. The aromatase knock-out mouse provides new evidence that estradiol is required during development in the female for the expression of sociosexual behaviors in adulthood. *J. Neurosci.* 22: 9104–9112.

Bakker, J., Honda, S., Harada, N. and Balthazart, J. 2002b. Sexual partner preference requires a functional aromatase (cyp19) gene in male mice. *Horm. Behav.* 42: 158–171.

Bakker, T.C.M. 1994. Genetic correlation and the control of behavior, exemplified by aggressiveness in sticklebacks. *Adv. Study Behav.* 23:135–171.

Bakker, T.C.M. and Pomiankowski, A. 1995. The genetics of female mate preferences. *J. Evol. Biol.* 8: 129–171.

Ball, G. F. 1993. The neural integration of environmental information by seasonally breeding birds. *Am. Zool.* 33: 185–199.

Ball, G. F. and Silver, R. J. 1983. Timing of incubation bouts by ring doves (*Streptopelia risoria*). *J. Comp. Psychol.* 97: 213–225.

Ball, G. F., Riters, L. V. and Balthazart, J. 2002. Neuroendocrinology of song behavior and avian brain plasticity: multiple sites of action of sex steroid hormones. *Front. Neuroendocrinol.* 23: 137–178.

Balthazart, J. 1983. Hormonal correlates of behavior. In D. S. Farner, J. R. King and K. C. Parkes, eds., *Avian Biology*, vol. VII, 221–366. New York: Academic Press.

Balthazart, J. 1989. Steroid metabolism and the activation of social behavior. In J. Balthazart, ed., *Advances in Comparative and Environmental Physiology*, vol. 3, 105–159. Berlin: Springer-Verlag.

Balthazart, J. and Adkins-Regan, E. 2002. In D. W. Pfaff, A. P. Arnold, A. M. Etgen, S. E. Fahrbach and R. T. Rubin, eds., *Hormones, Brain and Behavior*, vol. 4, 223–302. Amsterdam: Academic Press (Elsevier).

Balthazart, J. and Ball, G. F. 1993. Neurochemical differences in two steroid-sensitive areas mediating reproductive behaviors. In *Advances in Comparative and Environmental Physiology*, vol. 15, 133–161. Berlin: Springer-Verlag.

Balthazart, J. and Ball, G. F. 1998. New insights into the regulation and function of brain estrogen synthase (aromatase). *Trends Neurosci.* 21: 243–249.

Balthazart, J., Turek, R. and Ottinger, M. A. 1984. Altered brain metabolism of testosterone is correlated with reproductive decline in aging quail. *Horm. Behav.* 18: 330–345.

Balthazart, J., Absil, P., Fiasse, V. and Ball, G. F. 1994. Effects of the aromatase inhibitor R76713 on sexual differentiation of brain and behavior in zebra finches. *Behaviour* 131: 225–260.

Balthazart, J., Reid, J., Absil, P., Foidart, A. and Ball, G. F. 1995. Appetitive as well as consummatory aspects of male sexual behavior in quail are activated by androgens and estrogens. *Behav. Neurosci.* 109: 485–501.

Balthazart, J., Tlemçani, O. and Ball, G. F. 1996. Do sex differences in the brain explain sex differences in the hormonal induction of reproductive behavior? What 25 years of research on the Japanese quail tells us. *Horm. Behav.* 30: 627–661.

Balthazart, J., Castagna, C. and Ball, G. F.1997. Aromatase inhibition blocks the activation and sexual differentiation of appetitive male sexual behavior in Japanese quail. *Behav. Neurosci.* 111: 381–397.

Balthazart, J., Baillien, M., Cornil, C. A. and Ball, G. F. 2004. Preoptic aromatase modulates male sexual behavior: slow and fast mechanisms of action. *Physiol. Behav.* 83: 247–270.

Barfield, R. J. 1969. Activation of copulatory behavior by androgen implanted into the preoptic area of the male fowl. *Horm. Behav.* 1: 37–52.

Barfield, R. J. 1984. Reproductive hormones and aggressive behavior. In K. J. Flannelly, R. J. Blanchard and D. C. Blanchard, eds., *Biological Perspectives on Aggression*, 105–134. New York: A.R. Liss.

Barfield, R. J. and Krieger, M. S. 1977. Ejaculatory and postejaculatory behavior of male and female rats: effects of sex hormones and electric shock. *Physiol. Behav.* 19: 203–208.

Barki, A., Karplus, I., Khalaila, I., Manor, R. and Sagi, A. 2003. Male-like behavioral patterns and physiological alterations induced by androgenic gland implantation in female crayfish. *J. Exp. Biol.* 206: 1791–1797.

Barrington, E. J. W. 1986. The phylogeny of the endocrine system. *Experientia* 42: 775–781.

Barry, T. P., Unwin, M. J., Malison, J. A. and Quinn, T. P. 2001. Free and total cortisol levels in semelparous and iteroparous chinook salmon. *J. Fish Biol.* 59: 1673–1676.

Bartels, M., Van den Berg, M., Sluyter, F., Boomsma, D. I. and de Geus, E. J. C. 2003. Heritability of cortisol levels: review and simultaneous analysis of twin studies. *Psychoneuroendocrinology.* 28: 121–137.

Barth, R. H. 1968. The comparative physiology of reproductive processes in cockroaches, Part I: mating behaviour and its endocrine control. In A. McLaren, ed., *Advances in Reproductive Physiology*, vol. 3, 167–207. New York: Academic Press.

Barth, R. H., Lester, L. J., Sroka, P., Kessler, T. and Hearn, B. 1975. Juvenile hormone promotes dominance behavior and ovarian development in social wasps (*Polistes annularis*). *Experientia* 15: 691–692.

Bass, A. H. 1989. Evolution of vertebrate motor systems for acoustic and electric communication: peripheral and central elements. *Brain Behav. Evol.* 33: 237–247.

Bass, A. 1992. Dimorphic male brains and alternative reproductive tactics in a vocalizing fish. *Trends Neurosci.* 15: 139–145.

Bass, A. H. and Baker, R. 1997. Phenotypic specification of hindbrain rhombomeres and the origins of rhythmic circuits in vertebrates. *Brain Behav. Evol.* 50 (suppl. 1): 3–16.

Bass, A. H. and Grober, M. S. 2001. Social and neural modulation of sexual plasticity in teleost fish. *Brain Behav. Evol.* 57: 293–300.

Bates, R. O., Buchanan, D. S., Johnson, R. K., Wettemann, R. P., Fent R. W. and Hutchens, L. K. 1986. Genetic parameter estimates for reproductive traits of male and female littermate swine. *J. Anim. Sci.* 63: 377–385.

Batty, J. 1978. Plasma levels of testosterone and male sexual behavior in strains of the house mouse (*Mus musculus*). *Anim. Behav.* 26: 339–348.

Baulieu, E.-E. and Robel, P. 1990. Neurosteroids: a new brain function? *J. Steroid Biochem. Mol. Biol.* 37: 395–403.

Baum, M. J. and Erskine, M. S. 1984. Effect of neonatal gonadectomy and administration of testosterone on coital masculinization in the ferret. *Endocrinology* 115: 2440–2444.

Baum, M. J., Erskine, M. S., Kornberg, E. and Weaver, C. E. 1990. Prenatal and neonatal testosterone exposure interact to affect differentiation of sexual behavior and partner preference in female ferrets. *Behav. Neurosci.* 104: 183–198.

Beach, F. A. 1947. Evolutionary changes in the physiological control of mating behavior in mammals. *Psychol. Rev.* 54: 297–315.

Beach, F. A. and Inman, N. G. 1965. Effects of castration and androgen replacement on mating in male quail. *Proc. Natl. Acad. Sci. U S A* 54: 1426–1431.

Beani, L. and Dessì-Fulgheri, F. 1995. Mate choice in the grey partridge, *Perdix perdix*: role of physical and behavioural male traits. *Anim. Behav.* 49: 347–356.

Beani, L., Panzica, G., Briganti, F., Persichella, P. and Dessì-Fulgheri, F. 1995. Testosterone-induced changes of call structure, midbrain and syrinx anatomy in partridges. *Physiol. Behav.* 58: 1149–1157.

Beatty, W. W. 1992. Gonadal hormones and sex differences in nonreproductive behaviors. In A. A. Gerall, H. Moltz and I. L. Ward, eds., *Handbook of Behavioral Neurobiology*, vol. 11: *Sexual Differentiation*, 85–128. New York: Plenum.

Becker, J. B., Breedlove, S. M., Crews, D. and McCarthy, M. M., eds. 2002. *Behavioral Endocrinology*, 2nd ed. Cambridge, MA: MIT Press.

Beletsky, L. D., Orians, G. H. and Wingfield, J. C. 1989. Relationships of steroid hormones and polygyny to territorial status breeding experience and reproductive success in male red-winged blackbirds. *Auk* 106: 107–117.

Beletsky, L. D., Orians, G. H. and Wingfield, J. C. 1990. Steroid hormones in relation to territoriality breeding density and parental behavior in male yellow-headed blackbirds. *Auk* 107: 60–68.

Beletsky, L. D., Gori, D. F., Freeman, S. and Wingfield, J. C. 1995. Testosterone and polygyny in birds. In D. M. Power, ed., *Current Ornithology*, vol. 12, 1–42. New York: Plenum Press.

Belthoff, J. R. and Dufty, A. M. 1998. Corticosterone, body condition and locomotor activity: a model for dispersal in screech-owls. *Anim. Behav.* 55: 405–415.

Bennett, P. M. and Harvey, P. H. 1987. Active and resting metabolism in birds: allometry, phylogeny and ecology. *J. Zool. (Lond.)* 213: 327–363.

Bennett, P. M. and Owens, I.P.F. 2002. *Evolutionary Ecology of Birds: Life Histories, Mating Systems, and Extinction.* Oxford, UK: Oxford University Press.

Benoff, F. H. and Siegel, P. B. 1977. Crowing and mating behaviors in lines of chickens selected for mating frequency. *Appl. Anim. Ethol.* 3: 247–254.

Bentley, G. E. and Ball, G. F. 2000. Photoperiod-dependent and -independent regulation of melatonin receptors in the forebrain of songbirds. *J. Neuroendocrinol.* 12: 745–752.

Bentley, G. E., Moore, I. T., Sower, S. A. and Wingfield, J. C. 2004. Evidence for a novel gonadotropin-releasing hormone in hypothalamic and forebrain areas in songbirds. *Brain Behav. Evol.* 63: 34–46.

Bentley, P. J. 1998. *Comparative Vertebrate Endocrinology.* Cambridge, UK: Cambridge University Press.

Benus, R. F., Bohus, B., Koolhaas, J. M. and van Oortmerssen, G. A. 1991. Heritable variation for aggression as a reflection of individual coping strategies. *Experientia* 47: 1008–1019.

Bern, H. A. 1990. The "new" endocrinology: its scope and its impact. *Am. Zool.* 30: 877–885.

Berglund, A. and Rosenqvist, G. 2003. Sex role reversal in pipefish. In P.J.B. Slater, J. S. Rosenblatt, C. T. Snowdon and T. J. Roper, eds., *Advances in the Study of Behavior*, vol. 32, 131–167. Amsterdam: Academic Press (Elsevier).

Bernard, D. J., Bentley, G. E., Balthazart, J., Turek, F. W. and Ball, G. F. 1999. Androgen receptor, estrogen receptor alpha, and estrogen receptor beta show distinct patterns of expression in forebrain song control nuclei of European starlings. *Endocrinology* 140: 4633–4643.

Bernhardt, P. C., Dabbs, J. M., Fielden, J. A. and Lutter, C. D. 1998. Testosterone changes during vicarious experiences of winning and losing among fans at sporting events. *Physiol. Behav.* 65: 59–62.

Bernon, D. E. and Siegel, P. B. 1981. Fertility of chickens from lines divergently selected for mating frequency. *Poultry Sci.* 60: 45–48.

Berry, M. and Signoret, J.-P. 1984. Sex play and behavioral sexualization in the pig. *Reprod. Nutr. Dev.* 24: 507–513.

Berthold, P. 1996. *Control of Bird Migration.* London: Chapman & Hall.

Bester-Meredith, J. K., Young, L. J. and Marler, C. A. 1999. Species differences in paternal behavior and aggression in *Peromyscus* and their associations with vasopressin immunoreactivity and receptors. *Horm. Behav.* 35: 25–38.

Bhasin, S., Woodhouse, L. and Storer, T. W. 2001a. Proof of the effect of testosterone on skeletal muscle. *J. Endocrinol.* 170: 27–38.

Bhasin, S., Woodhouse, L., Casaburi, R., Singh, A. B., Bhasin, D., Berman, N., Chen, X., Yarasheski, K. E., Magliano, L., Dzekov, C., Dzekov, J., Bross, R., Phillips, J., Sinha-Hikim, I., Shen, R. and Storer, T. W. 2001b. Testosterone dose–response relationships in healthy young men. *Am. J. Physiol. Endocrinol. Metab.* 281: E1172–E1181.

Bicker, G. and Menzel, R. 1989. Chemical codes for the control of behaviour in arthropods. *Nature* 337: 33–39.

Billings, H. J. and Katz, L. S. 1999. Male influence on proceptivity in ovariectomized French-Alpine goats (*Capra hircus*). *Appl. Anim. Behav. Sci.* 64: 181–191.

Birkhead, T. R. and Møller, A. P. 1992. *Sperm Competition in Birds: Evolutionary Causes and Consequences.* London: Academic Press.

Birkhead, T. R., Fletcher, F. and Pellatt, E. J. 1999. Nestling diet, secondary sexual traits and fitness in the zebra finch. *Proc. R. Soc. Lond. B* 266: 385–390.

Bischof, H.-J. 1994. Sexual imprinting as a two-stage process. In J. A. Hogan and J. J. Bolhuis, eds., *Causal Mechanisms of Behavioural Development*, 82–97. Cambridge, UK: Cambridge University Press.

Bitterbaum, E. and Baptista, L. F. (1979). Geographical variation in songs of California House Finches (*Carpodacus mexicanus*). *Auk* 96: 462–474.

Black, J. M., ed. 1996. *Partnerships in Birds: The Study of Monogamy.* Oxford, UK: Oxford University Press.

Blackburn, D. G. and Evans, H. E. 1986. Why are there no viviparous birds? *Am. Nat.* 128: 165–190.

Blanchard, D. C., McKittrick, C. R., Hardy, M. P. and Blanchard, R. J. 2002. Effects of social stress on hormones, brain, and behavior. In D. W. Pfaff, A. P. Arnold, A. M. Etgen, S. E. Fahrbach and R. T. Rubin, eds., *Hormones, Brain and Behavior*, vol. 1, 735–772. Amsterdam: Academic Press (Elsevier).

Blaustein, J. D. and Erskine, M. S. 2002. Feminine sexual behavior: cellular integration of hormonal and afferent information in the rodent brain. In D. W. Pfaff, A. P. Arnold, A. M. Etgen, S. E. Fahrbach and R. T. Rubin, eds., *Hormones, Brain and Behavior*, vol. 1, 139–214. Amsterdam: Academic Press (Elsevier).

Blomberg, S. P., Garland, T., Ives, A. R. 2003. Testing for phylogenetic signal in comparative data: Behavioral traits are more labile. *Evolution* 57: 717–745.

Blount, J. D., Metcalfe, N. B., Birkhead, T. R. and Surai, P. F. 2003. Carotenoid modulation of immune function and sexual attractiveness in zebra finches. *Science* 300: 125–127.

Bluhm, C. K. 1988. Temporal patterns of pair formation and reproduction in annual cycles and associated endocrinology in waterfowl. In R. F. Johnston, ed., *Current Ornithology*, vol. 5, 123–186. New York: Plenum.

Bluhm, C. K., Phillips, R. E., Burke, W. H. and Gupta, G. N. 1984. Effects of male courtship and gonadal steroids on pair formation, egg-laying, and serum LH in Canvasback ducks (*Aythya valisineria*). *J. Zool. (Lond.)* 204: 185–200.

Blüm, V. and Fiedler, K. 1965. Hormonal control of reproductive behavior in some cichlid fish. *Gen. Comp. Endocrinol.* 5: 186–196.

Boag, D. A. 1982. How dominance status of adult Japanese quail influences the viability and dominance status of their offspring. *Can. J. Zool.* 60: 1885–1891.

Böhning-Gaese, K. and Oberrath, R. 1999. Phylogenetic effects on morphological, life-history, behavioural and ecological traits of birds. *Evol. Ecol. Res.* 1: 347–364.

Bolhuis, J. J. 1991. Mechanisms of avian imprinting: a review. *Biol. Rev.* 66: 303–345.

Bonga, S.E.W. 1997. The stress response in fish. *Physiol. Rev.* 77: 591–625.

Boorse, G. C. and Denver, R. J. 2002. Acceleration of *Ambystoma tigrinum* metamorphosis by corticotropin-releasing hormone. *J. Exp. Zool.* 293: 94–98.

Borg, B. 1994. Androgens in teleost fish. *Comp. Biochem. Physiol.* 109C: 219–245.

Borgia, G. and Presgraves, D. C. 1998. Coevolution of elaborated male display traits in the spotted bowerbird: an experimental test of the threat reduction hypothesis. *Anim. Behav.* 56: 1121–1128.

Borgia, G. and Wingfield, J. C. 1991. Hormonal correlates of bower decoration and sexual display in the satin bowerbird (*Ptilonorhynchus violaceus*). *Condor* 93: 935–942.

Borowski, R. 1978. Social inhibition of maturation in natural populations of *Xiphophorus variatus* (Pisces: Poeciliidae). *Science* 201: 933–935.

Boswell, T., Hall, M. R. and Goldsmith, A. R. 1993. Annual cycles of migratory fattening, reproduction and moult in European quail (*Coturnix coturnix*). *J. Zool. (Lond.)* 231: 627–644.

Boswell, T., Hall, M. R. and Goldsmith, A. R. 1995. Testosterone is secreted extragonadally by European quail maintained on short days. *Physiol. Zool.* 68: 967–984.

Bottjer, S. W. and Hewer, S. J. 1992. Castration and antisteroid treatment impair vocal learning in male zebra finches. *J. Neurobiol.* 23: 337–353.

Bouissou, M.-F. 1983. Hormonal influences on aggressive behavior in ungulates. In B. B. Svare, ed., *Hormones and Aggressive Behavior*, 507–534. New York: Plenum Press.

Bouloux, P.-M.G., Munroe, P., Kirk, J. and Besser, G. M. 1992. Sex and smell—an enigma resolved. *J. Endocrinol.* 133: 323–326.

Boyd, S. K. 1994. Arginine vasotocin facilitation of advertisement calling and call phonotaxis in bullfrogs. *Horm. Behav.* 28: 232–240.

Boyd, S. K. 1997. Brain vasotocin pathways and the control of sexual behaviors in the bullfrog. *Brain Res. Bull.* 44: 345–350.

Bradbury, J. W. and Vehrencamp, S. L. 1998. *Principles of Animal Communication.* Sunderland, MA: Sinauer.

Bradley, A. J., McDonald, I. R., and Lee, A. K. 1980. Stress and mortality in a small marsupial (*Antechinus stuartii*, Macleay). *Gen. Comp. Endocrinol.* 40: 188–200.

Brahmakshatriya, R. D., Snetsinger, D. C. and Waibel, P. E. 1969. Effects of exogenous estrogen and/or androgen on performance, egg shell characteristics and blood plasma changes in laying hens. *Poultry Sci.* 48: 444–451.

Brantley, R. K., Wingfield, J. C. and Bass, A. H. 1993. Sex steroid levels in *Porichthys notatus*, a fish with alternative reproductive tactics, and a review of the hormonal bases for male dimorphism among teleost fishes. *Horm. Behav.* 27: 332–347.

Braude, S., Tang-Martinez, Z. and Taylor, G. T. 1999. Stress, testosterone, and the immunoredistribution hypothesis. *Behav. Ecol.* 10: 345–350.

Breedlove, S. M. 1984. Steroid influences on the development and function of a neuromuscular system. *Prog. Brain Res.* 61: 147–170.

Brenowitz, E. A. 1991. Evolution of the vocal control system in the avian brain. *Semin. Neurosci.* 3: 399–407.

Brenowitz, E. A. and Arnold, A. P. 1986. Interspecific comparisons of the size of neural song control regions and song complexity in duetting birds: evolutionary implications. *J. Neurosci.* 6: 2875–2879.

Brenowitz, E. A. and Kroodsma, D. E. 1996. The neuroethology of birdsong. In D. E. Kroodsma and E. H. Miller, eds., *Ecology and Evolution of Acoustic Communication in Birds*, 285–304. Ithaca, NY: Cornell University Press.

Breuner, C. W. and Orchinik, M. 2001a. Downstream from corticosterone: seasonality of binding globulins, receptors and behavior in the avian stress response. In A. Dawson and C. M. Chaturvedi, eds., *Avian Endocrinology*, 385–399. New Dehli: Narosa Publishing House.

Breuner, C. W. and Orchinik, M. 2001b. Seasonal regulation of membrane and intracellular corticosteroid receptors in the house sparrow brain. *J. Neuroendocrinol.* 13: 412–420.

Breuner, C. W. and Orchinik, M. 2002. Plasma binding proteins as mediators of corticosteroid action in vertebrates. *J. Endocrinol.* 175: 99–112.

Breuner, C. W., Greenberg, A. L. and Wingfield, J. C. 1998. Noninvasive corticosterone treatment rapidly increases activity in Gambel's white-crowned sparrows (*Zonotrichia leucophrys gambelii*). *Gen. Comp. Endocrinol.* 111: 386–394.

Bribiescas, R. G. 2001. Reproductive ecology and life history of the human male. *Yearb. Phys. Anthropol.* 44: 148–176.

Bridges, R. S. 1996. Biochemical basis of parental behavior in the rat. In J. S. Rosenblatt and C. T. Snowdon, eds., *Parental Care: Evolution, Mechanisms, and Adaptive Significance (Advances in the Study of Behavior*, vol. 25), 215–242. San Diego: Academic Press.

Brien, P. 1962. Induction gamétique chez les Hydres d'eau douce par la méthode des greffes en parabiose. *C. R. Acad. Sci.* [D] (Paris) 255: 1431.

Brockmann, H. J. 2001. The evolution of alternative strategies and tactics. *Advances in the Study of Behavior* 30: 1–52. San Diego: Academic Press.

Bronson, F. H. 1989. *Mammalian Reproductive Biology.* Chicago: University of Chicago Press.

Bronson, F. H. 2000. Puberty and energy reserves: a walk on the wild side. In K. Wallen and J. E. Schneider, eds., *Reproduction in Context: Social and Environmental Influences on Reproductive Physiology and Behavior*, 15–34. Cambridge, MA: MIT Press.

Bronson, F. H. and Desjardins, C. 1982. Endocrine responses to sexual arousal in male mice. *Endocrinology* 111: 1286–1291.

Bronson, F. H. and Rissman, E. F. 1986. The biology of puberty. *Biol. Rev.* 61: 157–195.

Brooks, R. and Endler, J. A. 2001. Direct and indirect sexual selection and quantitative genetics of male traits in guppies (*Poecilia reticulata*). *Evolution* 55: 1002–1015.

Brosens, J. J. and Parker, M. G. 2003. Oestrogen receptor hijacked. *Nature* 423: 487–488.

Brown, D. D., Wang, Z., Kanamori, A., Eliceiri, B., Furlow, J. D. and Schwartz-man, R. 1995. Amphibian metamorphosis: a complex program of gene expression changes controlled by the thyroid hormone. *Recent Progr. Horm. Res.* 50: 309–315.

Brown, J. L. and Vleck, C. M. 1998. Prolactin and helping in birds: has natural selection strengthened helping behavior? *Behav. Ecol.* 9: 541–545.

Brown, S. D. and Bottjer, S. W. 1993. Testosterone-induced changes in adult canary brain are reversible. *J. Neurobiol.* 24: 627–640.

Buchanan, K. L. and Goldsmith, A. R. 2004. Noninvasive endocrine data for behavioural studies: the importance of validation. *Anim. Behav.* 67: 183–185.

Buchanan, K. L., Evans, M. R., Goldsmith, A. R., Bryant, D. M. and Rowe, L. V. 2001. Testosterone influences basal metabolic rate in male house sparrows: a new cost of dominance signalling? *Proc. R. Soc. Lond. B* 268: 1337–1344.

Buchanan, K. L., Spencer, K. A., Goldsmith, A. R. and Catchpole, C. K. 2003. Song as an honest signal of past developmental stress in the European starling (*Sturnus vulgaris*). *Proc. R. Soc. Lond. B* 270: 1149–1156.

Bull, J. J. 1983. *Evolution of Sex Determining Mechanisms.* Menlo Park, CA: Benjamin/Cummings.

Buntin, J. D. 1996. Neural and hormonal control of parental behavior in birds. In J. S. Rosenblatt and C. T. Snowdon, eds., *Parental Care: Evolution, Mechanisms, and Adaptive Significance (Advances in the Study of Behavior*, vol. 25), 161–214. San Diego: Academic Press.

Burley, N. 1988. The differential-allocation hypothesis: an experimental test. *Am. Nat.* 132: 611–628.

Burley, N. and Coopersmith, C. B. 1987. Bill color preferences of zebra finches. *Ethology* 76: 133–151.

Burley, N. T. and Parker, P. G. 1997. Emerging themes and questions in the study of avian reproductive tactics. *Ornithol. Monogr.* 49: 1–20.

Burns, M., Domjan, M. and Mills, A. D. 1998. Effects of genetic selection for fearfulness or social reinstatement behavior on adult social and sexual behavior in domestic quail (*Coturnix japonica*). *Psychobiology* 26: 249–257.

Butler, A. B. and Hodos, W. 1996. *Comparative Vertebrate Neuroanatomy: Evolution and Adaptation.* New York: Wiley.

Butterfield, P. A. 1970. The pair bond in the zebra finch. In J. H. Crook, ed., *Social Behaviour in Birds and Mammals*, 249–278. London: Academic Press.

Butterstein, G. M., Lance, V. A. and Elsey, R. M. 2003. Steroid patterns in juvenile and mature male alligators. *Biol. Reprod.* 68 (suppl.): 127.

Cahan, S. H. and Keller, L. 2003. Complex hybrid origin of genetic caste determination in harvester ants. *Nature* 424: 306–308.

Cahill, L. P., Saumande, J., Ravault, J. P., Blanc, M., Thimonier, J., Mariana, J. C. and Mauleon, P. 1981. Hormonal and follicular relationships in ewes of high and low ovulation rates. *J Reprod Fertil.* 62: 141–150.

Cairns, R.B., Gariépy, J.-L. and Hood, K. E. 1990. Development, microevolution and social behavior. *Psychol. Rev.* 97: 49–65.

Caldwell, G. S., Glickman, S. E. and Smith, E. R. 1984. Seasonal aggression independent of seasonal testosterone in wood rats. *Proc. Natl. Acad. Sci. U S A* 81: 5255–5257.

Calkins, J. D. and Burley, N. T. 2003. Mate choice for multiple ornaments in the California quail, *Callipepla californica. Anim. Behav.* 65: 69–81.

Callard, G. V. 1984. Aromatization in brain and pituitary: an evolutionary perspective. In F. Celotti and L. Martini, eds., *Metabolism of Hormonal Steroids in the Neuroendocrine Structures*, 79–102. New York: Raven Press.

Callard, G. V., Kunz, T. H. and Petro, Z. 1983. Identification of androgen metabolic pathways in the brain of little brown bats (*Myotis lucifugus*): Sex and seasonal differences. *Biol. Reprod.* 28: 1155–1161.

Callery, E. M. and Elinson, R. P. 2000. Thyroid hormone-dependent metamorphosis in a direct developing frog. *Proc. Natl. Acad. Sci. U S A* 97: 2615–2620.

Cameron, C. B., Mackie, G. O., Powell, J.F.F., Lescheid, D. W. and Sherwood, N. M. 1999. Gonadotropin-releasing hormone in mulberry cells of *Saccoglossus* and *Ptytchordera* (Hemichordata: Enteropneusta). *Gen. Comp. Endocrinol.* 114: 2–10.

Canoine, V. and Gwinner, E. 2002. Seasonal differences in the hormonal control of territorial aggression in free-living European stonechats. *Horm. Behav.* 41: 1–8.

Cant, M. A. and Johnstone, R. A. 2000. Power struggles, dominance testing, and reproductive skew. *Am. Nat.* 155: 406–417.

Cardwell, J. R. and Liley, N. R. 1991. Hormonal control of sex and color change in the stoplight parrotfish, *Sparisoma viride. Gen. Comp. Endocrinol.* 81: 7–20.

Cardwell, J. R., Stacey, N. E., Tan, E.S.P., McAdam, D.S.O. and Lang, S.L.C. 1995. Androgen increases olfactory receptor response to a vertebrate sex pheromone. *J. Comp. Physiol. A* 176: 55–61.

Carere, C., Groothuis, T. G., Mostl, E., Daan, S. and Koolhaas, J. M. 2003. Fecal corticosteroids in a territorial bird selected for different personalities: daily rhythm and the response to social stress. *Horm. Behav.* 43: 540–548.

Carlisle, D. B. 1960. Sexual differentiation in Crustacea Malacostraca. *Mem. Soc. Endocrinol.* 7: 9.

Carlson, A. A., Ziegler, T. E. and Snowdon, C. T. 1997. Ovarian function of pygmy marmoset daughters (*Cebuella pygmaea*) in intact and motherless families. *Am. J. Primatol.* 43: 347–355.

Carroll, S. B. 2001. Chance and necessity: the evolution of morphological complexity and diversity. *Nature* 409: 1102–1109.

Carsia, R. V. and Harvey, S. 2000. Adrenals. In G. C. Whittow, ed., *Sturkie's Avian Physiology*, 5th ed., 489–537. San Diego: Academic Press.

Carsia, R. V., Morin, M. E., Rosen, H. D. and Weber, H. 1987. Ontogenic corticosteroidogenesis of the domestic fowl: response of isolated adrenocortical cells. *Proc. Soc. Exp. Biol. Med.* 184: 436–445.

Carter, C. S., DeVries, A. C. and Getz, L. L. 1995. Physiological substrates of mammalian monogamy: the prairie vole model. *Neurosci. Biobehav. Rev.* 19: 303–314.

Cash, W. B. and Holberton, R. L. 1999. Effects of exogenous corticosterone on locomotor activity in the red-eared slider turtle, *Trachemys scripta elegans. J. Exp. Zool.* 284: 637–644.

Casto, J. M. and Ball, G. F. 1996. Early administration of 17beta-estradiol partially masculinizes song control regions and alpha2-adrenergic receptor distribution in European starlings (*Sturnus vulgaris*). *Horm. Behav.* 30: 387–406.

Casto, J. M., Nolan, V. and Ketterson, E. D. 2001. Steroid hormones and immune function: experimental studies in wild and captive dark-eyed juncos (*Junco hyemalis*). *Am. Nat.* 157: 408–420.

Catchpole, C. K., Dittami, J. and Leisler, B. 1984. Differential responses to male song repertoires in female songbirds implanted with oestradiol. *Nature* 312: 563–564.

Chadwick, D. and Goode, J., eds. 2002. *The Genetics and Biology of Sex Determination* (Novartis Foundation Symposium 244). Chichester, UK: Wiley.

Chan, S.T.H. and Yeung, W.S.B. 1983. Sex control and sex reversal in fish under natural conditions. In W. S. Hoar, D. J. Randall and E. M. Donaldson, eds., *Fish Physiology*, vol. IXB, 171–222. New York: Academic Press.

Chandler, C. R., Ketterson, E. D. and Nolan, V. 1994. Effects of testosterone on spatial activity in free-living male dark-eyed juncos when their mates are fertile. *Anim. Behav.* 54: 543–549.

Chardard, D. and Dournon, C. 1999. Sex reversal by aromatase inhibitor treatment in the newt *Pleurodeles waltl*. *J. Exp. Zool.* 283: 43–50.

Charlesworth, B. 1991. The evolution of sex chromosomes. *Science* 251: 1030–1033.

Charlier, T. D., Lakaye, B., Ball, G. F. and Balthazart, J. 2002. Steroid receptor coactivator SRC-1 exhibits high expression in steroid-sensitive brain areas regulating reproductive behaviors in the quail brain. *Neuroendocrinology* 76: 297–315.

Charniaux-Cotton, H. 1965. Contrôle endocrinien de la différenciation sexuelle chez les crustacés supérieurs. *Arch. Anat. Microsc. Morphol. Exp.* 54: 405.

Charnov, E. L. and Bull, J. 1977. When is sex environmentally determined? *Nature* 266: 828–830.

Chase, I. D., Tovey, C., Spangler-Martin, D. and Manfredonia, M. 2002. Individual differences versus social dynamics in the formation of animal dominance hierarchies. *Proc. Natl. Acad. Sci. U S A* 99: 5744–5749.

Chaudhuri, M. and Ginsberg, J. R. 1990. Urinary androgen concentrations and social status in two species of free ranging zebra (*Equus burchelli* and *E. grevyi*). *J. Reprod. Fertil.* 88: 127–133.

Cheng, K. M., McIntyre, R. F., and Hickman, A. R. 1989. Proctodeal gland foam enhances competitive fertilization in domestic Japanese quail. *Auk* 106: 286–291.

Cheng, M.-F. 1993. Vocal, auditory and endocrine systems: three-way connectivities and implications. *Poultry Sci. Rev.* 5: 37–47.

Cheng, M.-F., Peng, J. P., and Johnson, P. 1998. Hypothalamic neurons preferentially respond to female nest coo stimulation: demonstration of direct acoustic stimulation of luteinizing hormone release. *J. Neurosci.* 18: 5477–5489.

Chenn, A. and Walsh, C. A. 2002. Regulation of cerebral cortical size by control of cell cycle exit in neural precursors. *Science* 297: 365–369.

Chiang, E.F.L., Yan, Y. L., Tong, S. K., Hsiao, P. H., Guiguen, Y., Postlethwait, J. and Chung, B. C. 2001. Characterization of duplicated zebrafish cyp19 genes. *J. Exp. Zool.* 290: 709–714.

Christians, J. K. and Williams, T. D. 1999. Effects of exogenous 17beta-estradiol on the reproductive physiology and reproductive performance of European starlings (*Sturnus vulgaris*). *J. Exp. Biol.* 202: 2679–2685.

Civetta, A. and Singh, R. S. 1998. Sex-related genes, directional sexual selection, and speciation. *Mol. Biol. Evol.* 15: 901–909.

Clark, M. M. and Galef, B. G. 1998. Perinatal influences on the reproductive behavior of adult rodents. In T. A. Mousseau and C. W. Fox, eds., *Maternal Effects as Adaptations*, 261–272. New York: Oxford University Press.

Clark, M. M. and Galef, B. G. 1999. A testosterone-mediated trade-off between parental and sexual effort in male Mongolian gerbils (*Meriones unguiculatus*). *J. Comp. Psychol.* 113: 388–395.

Clarke, F. M., Miethe, G. H., Bennett, N. C. 2001. Reproductive suppression in female Damaraland mole-rats *Cryptomys damarensis*: dominant control or self-restraint? *Proc. R. Soc. Lond. B* 268: 899–909.

Clemens, L. G., Wee, B. E., Weaver, D. R., Roy, E. J., Goldman, B. D., and Rakerd, B. 1988. Retention of masculine sexual behavior following castration in male B6D2F1 mice. *Physiol. Behav.* 42: 69–76.

Clobert, J., Danchin, E., Dhondt, A. A. and Nichols, J. D., eds. 2001. *Dispersal.* Oxford, UK: Oxford University Press.

Clotfelter, E. D., O'Neal, D. M., Gaudioso, J. M., Casto, J. M., Parker-Renga, I. M., Snajdr, E. A., Duffy, D. L., Nolan, V. and Ketterson, E. D. 2004. Consequences of elevating plasma testosterone in females of a socially monogamous songbird: evidence of constraints on male evolution? *Horm. Behav.* 46: 171–178.

Clutton-Brock T. H. 1991. *The Evolution of Parental Care.* Princeton, NJ: Princeton University Press.

Clutton-Brock, T. H. and Parker, G. A. 1992. Potential reproductive rates and the operation of sexual selection. *Q. Rev. Biol.* 67: 437–455.

Cockrem, J. F. and Silverin, B. 2002. Variation within and between birds in corticosterone responses of great tits (*Parus major*). *Gen. Comp. Endocrinol.* 125: 197–206.

Coddington, E. and Moore, F. L. 2003. Neuroendocrinology of context-dependent stress responses: vasotocin alters the effect of corticosterone on amphibian behaviors. *Horm. Behav.* 43: 222–228.

Coe, C. L., Savage, A. and Bromley, L. J. 1992. Phylogenetic influences on hormone levels across the primate order. *Am. J. Primatol.* 28: 81–100.

Cohen-Parsons, M., Van Krey, H. P. and Siegel, P. B. 1983. In vivo aromatization of [^{3}H]testosterone in high and low mating lines of Japanese quail. *Horm. Behav.* 17: 316–323.

Collaer, M. L. and Hines, M. 1995. Human behavioral sex differences: a role for gonadal hormones during early development? *Psychol. Bull.* 118: 55–107.

Collias, N. E., Barfield, R. J. and Tarvyd, E. S. 2002. Testosterone versus psychological castration in the expression of dominance, territoriality and breeding behavior by male village weavers (*Ploceus cucullatus*). *Behaviour* 139: 801–824.

Compaan, J. C., De Ruiter, A.J.H., Koolhaas, J. M., van Oortmerssen, G. A. and Bohus, B. 1992. Differential effects of neonatal testosterone treatment on aggression in two selection lines of mice. *Physiol. Behav.* 51: 7–10.

Conaway, C. H. 1971. Ecological adaptation and mammalian reproduction. *Biol. Reprod.* 4: 239–247.

Congdon, J. D., Nagle, R. D., Kinney, O. M., van Loben Sels, R. C., Quinter, T. and Tinkle, D. W. 2003. Testing hypotheses of aging in long-lived painted turtles (*Chrysemys picta*). *Exp. Gerontol.* 38: 765–772.

Conroy, C. J. and Cook, J. A. 2000. Molecular systematics of a Holarctic rodent (*Microtus*: Muridae). *J. Mammal.* 81: 344–359.

Cooke, B., Hegstrom, C. D., Velleneuve, L. S., and Breedlove, S. M. 1998. Sexual differentiation of the vertebrate brain: principles and mechanisms. *Front. Neuroendocrinol.* 19: 323–362.

Cooper, E. L. and Faisal, M. 1990. Phylogenetic approach to endocrine—immune system interactions. *J. Exp. Zool.* 4 (suppl.): 46–52.

Cooper, R. L. 1978. Sexual receptivity in aged female rats. Behavioral evidence for increased sensitivity to estrogen. *Horm. Behav.* 9: 321–333.

Corbier, P., Dehennin, L., Auchere, D. and Roffi, J. 1992a. Changes in plasma testosterone levels during the peri-hatching period in the chicken. *J. Steroid Biochem. Mol. Biol.* 42: 773–776.

Corbier, P., Edwards, D. A. and Roffi, J. 1992b. The neonatal testosterone surge: a comparative study. *Arch. Internat. Physiol. Biochim. Biophys.* 100: 127–131.

Corley-Smith, G. E., Lim, C. J. and Brandhorst, B. P. 1996. Production of androgenetic zebrafish (*Danio rerio*). *Genetics* 142: 1265–1276.

Cotton, S., Fowler, K. and Pomiankowski, A. 2004. Do sexual ornaments demonstrate heightened condition-dependent expression as predicted by the handicap hypothesis? *Proc. R. Soc. Lond. B* 271: 771–783.

Craig, J. V., al Rawi, B. and Kratzer, D. D. 1977. Social status and sex ratio effects on mating frequency of cockerels. *Poultry Sci.* 56: 767–772.

Cree, A. 1994. Low annual reproductive output in female reptiles from New Zealand. *New Zealand J. Zool.* 21: 351–372.

Creel, S. 2001. Social dominance and stress hormones. *Trends Ecol. Evol.* 16: 491–497.

Creel, S. and Creel, N. M. 1991. *Behav. Ecol. Sociobiol.* 28: 263–270.

Creel, S. R. and Waser, P. M. 1997. Variation in reproductive suppression among dwarf mongooses: interplay between mechanisms and evolution. In N. G. Solomon and J. A. French, eds., *Cooperative Breeding in Mammals*, 150–170. Cambridge, UK: Cambridge University Press.

Creel, S. R., Monfort, S. L., Wildt, D. E. and Waser, P. M. 1991. Spontaneous lactation is an adaptive result of pseudopregnancy. *Nature* 351: 660–662.

Creel, S., Wildt, D. E., and Monfort, S. L. 1993. Aggression, reproduction, and androgens in wild dwarf mongooses: a test of the challenge hypothesis. *Am. Nat.* 141: 816–825.

Creel, S., Creel, N. M. and Monfort, S. L. 1996. Social stress and dominance. *Nature* 379: 212.

Crespi, B. J. and Teo, R. 2002. Comparative phylogenetic analysis of the evolution of semelparity and life history in salmonid fishes. *Evolution* 56: 1008–1020.

Crews, D. 1993. The organizational concept and vertebrates without sex chromosomes. *Brain Behav. Evol.* 42: 202–214.

Crews, D. 1998. On the organization of individual differences in sexual behavior. *Am. Zool.* 38: 118–132.

Crews, D. 2000. Sexuality: The environmental organization of phenotypic plasticity. In K. Wallen and J. E. Schneider, eds., *Reproduction in Context: Social and Environmental Influences on Reproductive Physiology and Behavior*, 473–500. Cambridge, MA: MIT Press.

Crews, D. 2002. Diversity and evolution of hormone-behavior relations in reproductive behavior. In J. B. Becker, S. M. Breedlove, D. Crews and M. M McCarthy, eds., *Behavioral Endocrinology*, 223–288. Cambridge, MA: MIT Press.

Crews, D. and Moore, M. C. 1986. Evolution of mechanisms controlling mating behavior. *Science* 231: 121–126.

Crews, D., Camazine, B., Diamond, M., Mason, R., Tokarz, R. R. and Garstka, W. R. 1984. Hormonal independence of courtship behavior in the male garter snake. *Horm. Behav.* 18: 29–41.

Crews, D., Grassman, M. and Lindzey, J. 1986. Behavioral facilitation of reproduction in sexual and unisexual whiptail lizards. *Proc. Natl. Acad. Sci. U S A* 83: 9547–9550.

Crook, J. H. 1970. Social organization and the environment: aspects of contemporary social ethology. *Anim. Behav.* 18: 197–209.

Cross, E. and Roselli, C. E. 1999. 17β-Estradiol rapidly facilitates chemoinvestigation and mounting in castrated male rats. *Am. J. Physiol.* 276: R1346—R1350.

Cunningham, D. L., Siegel, P. B. and Van Krey, H. P. 1977. Androgen influence on mating behavior in selected lines of Japanese quail. *Horm. Behav.* 8: 166–174.

Cushing, B. S. 1985. Estrous mice and vulnerability to weasel predation. *Ecology* 66: 1976–1978.

Cushing, B. S., Martin, J. O., Young, L. J., Carter, C. S. 2001. The effects of peptides on partner preference formation are predicted by habitat in prairie voles. *Horm Behav.* 39: 48–58.

Cushing, B. S., Okorie, U. and Young, L. J. 2003. The effects of neonatal castration on the subsequent behavioural response to centrally administered arginine vasopressin and the expression of V_{1a} receptors in adult male prairie voles. *J. Neuroendocrinol.* 15: 1021–1026.

Daan, S. and Tinbergen, J. M. 1997. Adaptation of life histories. In J. R. Krebs and N. B. Davies, eds., *Behavioural Ecology: An Evolutionary Approach*, 4th ed., 311–333. Oxford, UK: Blackwell Science.

Da Costa, A.P.C., Guevara-Guzman, R. G., Ohkura, S., Goode, J. A. and Kendrick, K. M. 1996. The role of oxytocin release in the paraventricular nucleus in the control of maternal behaviour in the sheep. *J. Neuroendocrinol.* 8: 163–177.

Dagg, A. I. 1984. Homosexual behaviour and female-male mounting in mammals—a first survey. *Mammal Rev.* 14: 155–185.

Daly, R. C., Su, T.-P., Schmidt, P. J., Pagliaro, M., Pickar, D., and Rubinow, D. R. 2003. Neuroendocrine and behavioral effects of high-dose anabolic steroid administration in male normal volunteers. *Psychoneuroendocrinology* 28: 317–331.

Darlison, M. G. and Richter, D. 1999. Multiple genes for neuropeptides and their receptors: co-evolution and physiology. *Trends Neurosci.* 22: 81–88.

Darwin, C. 1859. *On the Origin of Species by Means of Natural Selection, or the Preservation of Favoured Races in the Struggle of Life.* London: Murray.

Daunt, F., Wanless, S. Harris, M. P. and Monaghan, P. 1999. Experimental evidence that age-specific reproductive success is independent of environmental effects. *Proc. R. Soc. Lond. B* 266: 1489–1493.

Davidson, J. M., Chen, J. J., Crapo, L., Gray, G. D., Greenleaf, W. J. and Catania, J. A. 1983. Hormonal changes and sexual function in aging men. *J. Clin. Endocrinol. Metab.* 57: 71–77.

Davis, G. P. 1993. Genetic parameters for tropical beef cattle in northern Australia: a review. *Austral. J. Agric. Res.* 44: 179–198.

Dawson, A., Williams, T. D. and Nicholls, T. J. 1987. Thyroidectomy of nestling starlings appears to cause neotenous sexual maturation. *J. Endocr.* 112: R5–R6.

Dawson, A., King, V. M., Bentley, G. E. and Ball, G. F. 2001. Photoperiodic control of seasonality in birds. *J. Biol. Rhyth.* 16: 365–380.

de Beun, R., Geerts, N. E., Jansen, E., Slangen, J. L. and van de Poll, N. E. 1991. Luteinizing hormone releasing hormone-induced conditioned place-preference in male rats. *Pharmacol. Biochem. Behav.* 39: 143–147.

DeBold, J. F. and Miczek, K. A. 1984. Aggression persists after ovariectomy in female rats. *Horm. Behav.* 18: 177–190.

de Bruijn, M., Broekman, M. and van der Schoot, P. 1988. Sexual interactions between estrous female rats and castrated male rats treated with testosterone propionate or estradiol benzoate. *Physiol. Behav.* 43: 35–39.

de Catanzaro, D., Muir, C., Sullivan, C. and Boissy, A. 1999. Pheromones and novel male-induced pregnancy disruptions in mice: exposure to conspecifics is necessary for urine alone to induce an effect. *Physiol. Behav.* 66: 153–157.

De Fraipont, M., Clobert, J., John-Alder, H. and Meylan, S. 2000. Increased pre-natal maternal corticosterone promotes philopatry of offspring in common lizards *Lacerta vivipara. J. Anim. Ecol.* 69: 404–413.

De Jonge, F. H., Oldenburger, W. P., Louwerse, A. L. and van de Poll, N. E. 1992. Changes in male copulatory behavior after sexual exciting stimuli: effects of medial amygdala lesions. *Physiol. Behav.* 52: 327–332.

de Kloet, E. R. 1995. Steroids, stability and stress. *Front. Neuroendocrinol.* 16: 416–425.

de la Cruz, C., Solís, E., Valencia, J., Chastel, O. and Sorci, G. 2003. Testosterone and helping behavior in the azure-winged magpie (*Cyanopica cyanus*): natural covariation and an experimental test. *Behav. Ecol. Sociobiol.* 55: 103–111.

Demski, L. S. 1984. The evolution of neuroanatomical substrates of reproductive behavior: sex steroid and LHRH-specific pathways including the terminal nerve. *Am. Zool.* 24: 809–830.

Denver, R. J. 1997. Environmental stress as a developmental cue: corticotropin-releasing hormone is a proximate mediator of adaptive phenotypic plasticity in amphibian metamorphosis. *Horm. Behav.* 31: 169–179.

De Ridder, E., Pinxten, R. and Eens, M. 2000. Experimental evidence of a testosterone-induced shift from paternal to mating behaviour in a facultatively polygynous songbird. *Behav. Ecol. Sociobiol.* 49: 24–30.

de Ruiter, A. J., Wendelaar Bonga, S. E., Slijkhuis, H. and Baggerman, B. 1986. The effect of prolactin on fanning behavior in the male three-spined stickleback, *Gasterosteus aculeatus* L. *Gen. Comp. Endocrinol.* 64: 273–283.

De Sa, W. F., Pleumsamran, P., Morcom, C. B. and Dukelow, W. R. 1981. Exogenous steroid effects on litter size and early embryonic survival in swine. *Theriogenology* 15: 245–256.

De Santo, T. L., Van Krey, H. P., Siegel, P. B. and Gwazdauskas, F. G. 1983. Developmental profile of plasma androgens in cockerels genetically selected for mating frequency. *Poultry Sci.* 62: 2249–2254.

Desjardins, C. , Bronson, F. H. and Blank, J. L. 1986. Genetic selection for reproductive photoresponsiveness in deer mice. *Nature* 322: 172–173.

Deviche, P. and Schumacher, M. 1982. Behavioural and morphological dose-responses to testosterone and to 5α-dihydrotestosterone in the castrated male Japanese quail. *Behav. Processes* 7: 107–119.

Deviche, P., Breuner, C. and Orchinik, M. 2001. Testosterone, corticosterone, and photoperiod interact to regulate plasma levels of binding globulin and free steroid hormone in dark-eyed juncos, *Junco hyemalis. Gen Comp. Endocrinol.* 122: 67–77.

Devlin, R. H. and Nagahama, Y. 2002. Sex determination and sex differentiation in fish: an overview of genetic, physiological, and environmental influences. *Aquaculture* 208: 191–364.

DeVoogd, T. J. 1991. Endocrine modulation of the development and adult function of the avian song system. *Psychoneuroendocrinology* 16: 41–66.

DeVoogd, T. and Nottebohm, F. 1981. Gonadal hormones induce dendritic growth in the adult avian brain. *Science* 214: 202–204.

DeVoogd, T. J., Krebs, J. R., Healy, S. D. and Purvis, A. 1993. Relations between song repertoire size and the volume of brain nuclei related to song: comparative evolutionary analyses amongst oscine birds. *Proc. R. Soc. Lond. B* 254: 75–82.

DeVoogd, T. J., Houtman, A. M. and Falls, J. B. 1995. White-throated sparrow morphs that differ in song production rate also differ in the anatomy of some song-related areas. *J. Neurobiol.* 28: 202–213.

DeVries, A. C. 2002. Interaction among social environment, the hypothalamic–pituitary–adrenal axis, and behavior. *Horm. Behav.* 41: 405–413.

de Vries, G. J. 1995. Studying neurotransmitter systems to understand the development and function of sex differences in the brain: the case of vasopressin. In P. E. Micevych and R. P. Hammer, eds., *Neurobiological Effects of Sex Steroid Hormones*, 254–280. Cambridge, UK: Cambridge University Press.

de Vries, G. J. and Miller, M. A. 1998. Anatomy and function of extrahypothalamic vasopressin systems in the brain. *Prog. Brain Res.* 119: 3–20.

de Vries, G. J., Rissman, E. F., Simerly, R. B., Yang, L.-Y., Scordalakes, E. M., Auger, C. J., Swain, A., Lovell-Badge, R., Burgoyne, P. S. and Arnold, A. P. 2002. A model system for study of sex chromosome effects on sexually dimorphic neural and behavioral traits. *J. Neurosci.* 22: 9005–9014.

Diakow, C. 1978. Hormonal basis for breeding behavior in female frogs: vasotocin inhibits the release call of *Rana pipiens. Science* 199: 1456–1457.

Diakow, C., Wilcox, J. N. and Woltmann, R. 1978. Female frog reproductive behavior elicited in the absence of the ovaries. *Horm. Behav.* 11: 183–189.

Diamond, M. 1970. Intromission pattern and species vaginal code in relation to induction of pseudopregnancy. *Science* 169: 995–997.

Dickhoff, W. W., Brown, C. L., Sullivan, C. V. and Bern, H. A. 1990. Fish and amphibian models for developmental endocrinology. *J. Exp. Zool.* suppl. 4: 90–97.

Di Fiore, A. and Rendall, D. 1994. Evolution of social organization: a reappraisal for primates by using phylogenetic methods. *Proc. Natl. Acad. Sci. U S A* 91: 9941–9945.

Dittami, J. P. and Gwinner, E. 1990. Endocrine correlates of seasonal reproduction and territorial behavior in some tropical passerines. In M. Wada, S. Ishii and C. G. Scanes, eds., *Endocrinology of Birds: Molecular to Behavioral*. Tokyo: Japan Scientific Societies Press and Berlin: Springer-Verlag.

Dittami, J., Hoi, H. and Sageder, G. 1991. Parental investment and territorial/sexual behavior in male and female reed warbers: are they mutually exclusive? *Ethology* 88: 249–255.

Dixson, A. F. 1997. Evolutionary perspectives on primate mating systems and behavior. In C. S. Carter, I. I. Lederhendler and B. Kirkpatrick, eds., *The Integrative Neurobiology of Affiliation (Annals of the New York Academy of Sciences*, vol. 807). New York: New York Academy of Sciences.

Dixson, A. F. 1998. *Primate Sexuality: Comparative Studies of the Prosimians, Monkeys, Apes, and Human Beings*. Oxford, UK: Oxford University Press.

Dixson, A. F. and Herbert, J. 1977. Testosterone, aggressive behavior and dominance rank in captive adult male talapoin monkeys. *Physiol. Behav.* 18: 539–544.

Dixson, A. F., Brown, G. R. and Nevison, C. M. 1998. Developmental significance of the postnatal testosterone "surge" in male primates. In L. Ellis and L. Ebertz, eds., *Males, Females, and Behavior: Toward Biological Understanding*, 129–145. Westport, CT: Praeger.

D'Occhio, M. J. and Brooks, D. E. 1982. Threshold of plasma testosterone required for normal mating activity in male sheep. *Horm. Behav.* 16:383–394.

D'Occhio, M. J. and Ford, J. J. 1988. Sexual differentiation and adult sexual behavior in cattle, sheep and swine: the role of gonadal hormones. In J.M.A. Sitsen, ed., *Handbook of Sexology*, vol. VI: *The Pharmacology and Endocrinology of Sexual Function*. Amsterdam: Elsevier.

Domjan, M. and Hall, S. 1986. Sexual dimorphism in the social proximity behavior of Japanese quail (*Coturnix coturnix japonica*). *J. Comp. Psychol.* 100: 68–71.

Domjan, M., Blesbois, E. and Williams, J. 1998. The adaptive significance of sexual conditioning: Pavlovian control of sperm release. *Psychol. Sci.* 9: 411–415.

Domjan, M., Cusato, B. and Villarreal, R. 2000. Pavlovian feed-forward mechanisms in the control of social behavior. *Behav. Brain Sci.* 23: 235–249.

Drea, C. M., Weldele, M. L., Forger, N. G., Coscia, E. M., Frank, L. G., Licht, P. and Glickman, S. E. 1998. Androgens and masculinization of genitalia in the spotted hyaena (*Crocuta crocuta*), 2: effects of prenatal anti-androgens. *J Reprod. Fertil.* 113: 117–127.

Drent, P. J., van Oers, K. and van Noordwijk, A. J. 2003. Realized heritability of personalities in the great tit (*Parus major*). *Proc. R. Soc. Lond. B* 270: 45–51.

Drickamer, L. C. 1977. Delay of sexual maturation in female house mice by exposure to grouped females or urine from grouped females. *J. Reprod. Fertil.* 51: 77–81.

Drickamer, L. C. 1981. Selection for age of sexual maturation in mice and the consequences for population regulation. *Behav. Neural Biol.* 31: 82–89.

Drickamer, L. C. 1996. Intra-uterine position and anogenital distance in house mice: consequences under field conditions. *Anim. Behav.* 51: 925–934.

Drummond, H., Torres, R. and Krishnan, V. V. 2003. Buffered development: resilience after aggressive subordination in infancy. *Am. Nat.* 161: 794–807.

Dudley, C. A. and Moss, R. L. 1991. Facilitation of sexual receptivity in the female rat by C-terminal fragments of LHRH. *Physiol. Behav.* 50: 1205–1208.

Dufaure, J.-P. 1966. Recherches descriptives et expérimentales sur les modalités et facteurs du développement de l'appareil génital chez le lézard vivipare (*Lacerta vivipara* Jacquin). *Arch. Anat. Microsc. Morphol. Exp.* 55: 437.

Dulka, J. G. and Maler, L. 1994. Testosterone modulates female chirping behavior in the weakly electric fish, *Apteronotus leptorhynchus. J. Comp. Physiol. A* 174: 331–343.

Dunlap, K. D. and Schall, J. J. 1995. Hormonal alterations and reproductive inhibition in male fence lizards (*Sceloporus occidentalis*) infected with the malarial parasite *Plasmodium mexicanum. Physiol. Zool.* 68: 608–621.

Dunlap, K. D., Thomas, P. and Zakon, H. H. 1998. Diversity of sexual dimorphism in electrocommunication signals and its androgen regulation in a genus of electric fish, *Apteronotus. J. Comp. Physiol. A* 183: 77–86.

Durand, S. E., Heaton, J. T., Amateau, S. K. and Brauth, S. E. 1997. Vocal control pathways through the anterior forebrain of a parrot (*Melopsittacus undulatus*). *J. Comp. Neurol.* 377: 179–206.

Easter, S. S., Rusoff, A. C. and Kish, P. E. 1981. The growth and organization of the optic nerve and tract in juvenile and adult goldfish. *J. Neurosci.* 1: 793–811.

Eens, M. and Pinxten, R. 2000. Sex-role reversal in vertebrates: behavioural and endocrinological accounts. *Behav. Processes* 51: 135–147.

Eichmann, F. and Holst, D. V. 1999. Organization of territorial marking behavior by testosterone during puberty in male tree shrews. *Physiol. Behav.* 65: 785–791.

Eising, C. M. and Groothuis, T.G.G. 2003. Yolk androgens and begging behaviour in black-headed gull chicks: an experimental field study. *Anim. Behav.* 66:1027–1034.

Eising, C. M., Eikenaar, C., Schwabl, H. and Groothuis, T.G.G. 2001. Maternal androgens in black-headed gull (*Larus ridibundus*) eggs: consequences for chick development. *Proc. R. Soc. Lond. B* 268: 839–846.

Eising, C. M., Muller, W., Dijkstra, C., and Groothuis, T.G.G. 2003. Maternal androgens in egg yolks: relation with sex, incubation time and embryonic growth. *Gen. Comp. Endocrinol.* 132: 241–247.

Eisthen, H. L. and Nishikawa, K. C. 2002. Convergence: obstacle or opportunity? *Brain Behav. Evol.* 59: 235–239.

Elbrecht, A. and Smith, R. G. 1992. Aromatase enzyme activity and sex determination in chickens. *Science* 255: 467–470.

Elekonich, M. M. and Robinson, G. E. 2000. Organizational and activational effects of hormones on insect behavior. *J. Insect Physiol.* 46: 1509–1515.

Elekonich, M. M. and Wingfield, J. C. 2000. Seasonality and hormonal control of territorial aggression in female song sparrows (Passeriformes: Emberizidae: *Melospiza melodia*). *Ethology* 106: 493–510.

El Halawani, M. E., Burke, W. H., Millam, J. R., Fehrer, S. C. and Hargis, B. M. 1984. Regulation of prolactin and its role in gallinaceous bird reproduction. *J. Exp. Zool.* 232: 521–529.

El Halawani, M. E., Silsby, J. L. and Mauro, L. J. 1990. Vasoactive intestinal peptide is a hypothalamic prolactin-releasing neuropeptide in the turkey (*Meleagris gallopavo*). *Gen. Comp. Endocrinol.* 78: 66–73.

Eliasson, M. and Meyerson, B. J. 1975. Sexual preference in female rats during estrous cycle, pregnancy, and lactation. *Physiol. Behav.* 14: 705–710.

Ellegren, H. 2000. Evolution of the avian sex chromosomes and their role in sex determination. *Trends Ecol. Evol.* 15: 188–192.

Ellis, L. and Ebertz, L., eds. 1997. *Sexual Orientation: Toward Biological Understanding.* Westport, CT: Praeger.

Ellison, P. T. and Panter-Brick, C. 1996. Salivary testosterone levels among Tamang and Kami males of central Nepal. *Hum. Biol.* 68: 955–965.

Emerson, S. B. 2000. Vertebrate secondary sexual characteristics—physiological mechanisms and evolutionary patterns. *Am. Nat.* 156: 84–91.

Emerson, S. B. and Boyd, S. K. 1999. Mating vocalizations of female frogs: control and evolutionary mechanisms. *Brain Behav. Evol.* 53: 187–197.

Emerson, S. B. and Hess, D. L. 1996. The role of androgens in opportunistic breeding, tropical frogs. *Gen. Comp. Endocrinol.* 103: 220–230.

Emerson, S. B. and Hess, D. L. 2001. Glucocorticoids, androgens, testis mass, and the energetics of vocalization in breeding male frogs. *Horm. Behav.* 39: 59–69.

Emerson, S. B., Carroll, L. and Hess, D. L. 1997. Hormonal induction of thumb pads and the evolution of secondary sexual characteristics of the southeast Asian fanged frog, *Rana blythii. J. Exp. Zool.* 279: 587–596.

Emery, N. J., Capitanio, J. P., Mason, W. A., Machado, C. J., Mendoza, S. P. and Amaral, D. G. 2001. The effects of bilateral lesions of the amygdala on dyadic social interactions in rhesus monkeys (*Macaca mulatta*). *Behav. Neurosci.* 115: 515–544.

Emlen, J. T. and Lorenz, F. W. 1942. Pairing responses of free-living Valley Quail to sex-hormone pellet implants. *Auk* 59: 369–378.

Emlen, S. 1997. Predicting family dynamics in social vertebrates. In J. R. Krebs and N. B. Davies, eds., *Behavioural Ecology: An Evolutionary Approach*, 4th ed., 228–253. Oxford, UK: Blackwell Science.

Endler, A., Liebig, J., Schmitt, T., Parker, J. E., Jones, G. R., Schreier, P. and Holldobler, B. 2004. Surface hydrocarbons of queen eggs regulate worker reproduction in a social insect. *Proc. Natl. Acad. Sci. U S A* 101: 2945–2950.

Endler, J. A. 1986. *Natural Selection in the Wild.* Princeton, NJ: Princeton University Press.

Enggist-Dueblin, P. and Pfister, U. 2002. Cultural transmission of vocalizations in ravens, *Corvus corax. Anim. Behav.* 64: 831–841.

Enstrom, D. A., Ketterson, E. D. and Nolan, V. 1997. Testosterone and mate choice in the dark-eyed junco. *Anim. Behav.* 54: 1135–1146.

Erickson, C. J. 1985. Mrs. Harvey's parrot and some problems of socioendocrine response. In P.P.G. Bateson and P. H. Klopfer, eds., *Perspectives in Ethology*, 261–285. New York: Plenum.

Erskine, M. S. 1995. Prolactin release after mating and genitosensory stimulation in females. *Endocrine Rev.* 16: 508–528.

Etches, R. J. 1996. *Reproduction in Poultry.* Oxon, UK: CAB International.

Evans, M. R., Goldsmith, A. R., and Norris, S.R.A. 2000. The effects of testosterone on antibody production and plumage coloration in male house sparrows (*Passer domesticus*). *Behav. Ecol. Sociobiol.* 47: 156–163.

Everitt, B. J. 1990. Sexual motivation: a neural and behavioural analysis of the mechanisms underlying appetitive and copulatory responses of male rats. *Neurosci. Biobehav. Rev.* 14: 217–232.

Ewer, J. and Reynolds, S. 2002. Neuropeptide control of molting in insects. In D. W. Pfaff, A. P. Arnold, A. M. Etgen, S. E. Fahrbach and R. T. Rubin, eds., *Hormones, Brain and Behavior*, vol. 3, 1–92. Amsterdam: Academic Press (Elsevier).

Fadem, B. H. 2000. Perinatal exposure to estradiol masculinizes aspects of sexually dimorphic behavior and morphology in gray short-tailed opossums (*Monodelphis domestica*). *Horm. Behav.* 37: 79–85.

Fahrbach, S. E. 1997. Regulation of age polyethism in bees and wasps by juvenile hormone. *Adv. Study Behav.* 26: 285–316.

Farabaugh, S. M. 1982. The ecological and social significance of duetting. In D. E. Kroodsma and E. H. Miller, eds., *Acoustic Communication in Birds*, vol. 2: *Song Learning and Its Consequences*, 85–124. New York: Academic Press.

Farrell, S. F. and McGinnis, M. Y. 2004. Long-term effects of pubertal anabolic—androgenic steroid exposure on reproductive and aggressive behaviors in male rats. *Horm. Behav.* 46: 193–203.

Farries, M. A. 2001. The oscine song system considered in the context of the avian brain: lessons learned from comparative neurobiology. *Brain Behav. Evol.* 58: 80–100.

Faulkes, C. G. and Abbott, D. H. 1997. The physiology of a reproductive dictatorship: regulation of male and female reproduction by a single breeding female in colonies of naked mole-rats. In N. G. Solomon and J. A. French, eds., *Cooperative Breeding in Mammals*, 302–334. Cambridge, UK: Cambridge University Press.

Faulkes, C. G. and Bennett, N. C. 2001. Family values: group dynamics and social control of reproduction in African mole-rats. *Trends Ecol. Evol.* 16: 184–190.

Faulkes, C. G., Abbott, D. H. and Jarvis, J.U.M. 1990. Social suppression of ovarian cyclicity in captive and wild colonies of naked mole-rats, *Heterocephalus glaber. J. Reprod. Fertil.* 88: 559–568.

Faure, J. M. and Mills, A. D. 1998. Improving the adaptability of animals by selection. In T. Grandin, ed., *Genetics and the Behavior of Domestic Animals*, 235–265. San Diego: Academic Press.

Feder, H. 1981. Perinatal hormones and their role in the development of sexually dimorphic behaviors. In N. T. Adler, ed., *Neuroendocrinology of Reproduction: Physiology and Behavior*, 127–157. New York: Plenum.

Feder, M. E. 1987. The analysis of physiological diversity: the prospects for pattern documentation and general questions in ecological physiology. In M. E. Feder, A. F. Bennett, W. W. Burggren and R. B. Huey, eds., *New Directions in Ecological Physiology*, 38–75. Cambridge, UK: Cambridge University Press.

Feldman, H. A., Longcope, C., Derby, C. A., Johannes, C. B.,Araujo, A. B., Coviello, A. D., Bremner, W. J. and McKinlay, J. B. 2002. Age trends in the level of serum testosterone and other hormones in middle-aged men: longitudinal results from the Massachusetts Male Aging Study. *J. Clin. Endocrinol. Metab.* 87: 589–598.

Ferguson, J. N., Aldag, J. M., Insel, T. R. and Young, L. J. 2001. Oxytocin in the medial amygdala is essential for social recognition in the mouse. *J. Neurosci.* 21: 8278–8285.

Ferkin, M. H. and Gorman, M. R. 1992. Photoperiod and gonadal hormones influence odor preferences of the male meadow vole, *Microtus pennsylvanicus*. *Physiol. Behav.* 51: 1087–1091.

Ferkin, M. H. and Zucker, I. 1991. Seasonal control of odour preferences of meadow voles (*Microtus pennsylvanicus*) by photoperiod and ovarian hormones. *J. Reprod. Fertil.* 92: 433–441.

Fernald, R. D. and White, R. B. 1999. Gonadotropin-releasing hormone genes: phylogeny, structure, and functions. *Front. Neuroendocrinol.* 20: 224–240.

Fernandez-Guasti, A., Swaab, D, and Rodríguez-Manzo, G. 2003. Sexual behavior reduces hypothalamic androgen receptor immunoreactivity. *Psychoneuroendocrinology* 28: 501–512.

Ferris, C. F., Delville, Y., Irvin, R. W. and Potegal, M. 1994. Septo-hypothalamic organization of a stereotyped behavior controlled by vasopressin in golden hamsters. *Physiol. Behav.* 55: 755–759.

Finch, C. E. 1990. *Longevity, Senescence, and the Genome.* Chicago: University of Chicago Press.

Finch, C. E. 2002. Evolution and plasticity of aging in the reproductive schedules in long-lived animals: the importance of genetic variation in neuroendocrine mechanisms. In D. W. Pfaff, A. P. Arnold, A. M. Etgen, S. E. Fahrbach and R. T. Rubin, eds., *Hormones, Brain and Behavior*, vol. 4, 799–820. Amsterdam: Academic Press (Elsevier).

Finch, C. E. and Kirkwood, T.B.L. 2000. *Chance, Development, and Aging.* Oxford, UK: Oxford University Press.

Finch, C. E. and Rose, M. R. 1995. Hormones and the physiological architecture of life history evolution. *Q. Rev. Biol.* 70: 1–52.

Finlay, B. L. and Darlington, R. B. 1995. Linked regularities in the development and evolution of mammalian brains. *Science* 268: 1578–1584.

Firestein, S. 2001. How the olfactory system makes sense of scents. *Nature* 413: 211–218.

Fisher, G. A. 1958. *The Genetical Theory of Natural Selection.* Oxford, UK: Oxford University Press.

Fisher, H., Aron, A., Mashek, D., Li, H., Strong, G. and Brown, L. L. 2002. The neural mechanisms of mate choice: a hypothesis. *Neuroendocrinol. Lett.* 23 (suppl. 4).

Fivizzani, A. J., Oring, L. W., El Halawani, M. E. and Schlinger, B. A. 1990. Hormonal basis of male parental care and female intersexual competition in sex-role reversed birds. In M. Wada, S. Ishii, and Scanes, C. G., eds., *Endocrinology of Birds: Molecular to Behavioral*, 273–286. Tokyo: Japan Scientific Societies Press and Berlin: Springer-Verlag.

Fleming, A. 1986. Psychobiology of rat maternal behavior: how and where hormones act to promote maternal behavior at parturition. *Ann. N. Y. Acad. Sci.* 474: 234–251.

Fleming, A. S., Morgan, H. D. and Walsh, C. 1996. Experiential factors in postpartum regulation of maternal care. In J. S. Rosenblatt and C. T. Snowdon, eds., *Parental Care: Evolution, Mechanisms, and Adaptive Significance* (*Advances in the Study of Behavior*, vol. 25), 295–332. San Diego: Academic Press.

Floody, O. R. 1983. Hormones and aggression in female mammals. In B. B. Svare, ed., *Hormones and Aggressive Behavior*, 39–89. New York: Plenum Press.

Foerster, K., Poesel, A., Kunc, H. and Kempenaers, B. 2002. The natural plasma testosterone profile of male blue tits during the breeding season and its relation to song output. *J. Avian Biol.* 33: 269–275.

Foidart, A., Silverin, B., Baillien, M., Harada, N. and Balthazart, J. 1998. Neuroanatomical distribution and variations across the reproductive cycle of aromatase activity and aromatase-immunoreactive cells in the pied flycatcher (*Ficedula hypoleuca*). *Horm. Behav.* 33: 180–196.

Follett, B. K. 1973. The neuroendocrine regulation of gonadotropin secretion in avian reproduction. In D. S. Farner, ed., *Breeding Biology of Birds: Proceedings of a Symposium on Breeding Behavior and Reproductive Physiology in Birds*, 209–243. Washington, DC: National Academy of Sciences.

Folstad, I. and Karter, A. J. 1992. Parasites, bright males, and the immunocompetence handicap. *Am. Nat.* 139: 603–622.

Foran, C. M. and Bass, A. H. 1999. Preoptic GnRH and AVT: axes for sexual plasticity in teleost fish. *Gen. Comp. Endocrinol.* 116: 141–152.

Ford, J. J. 1983. Serum estrogen concentrations during postnatal development in male pigs. *Proc. Soc. Exp. Biol. Med.* 174: 160–164.

Ford, J. J. and Christenson, R. K. 1987. Influences of pre- and postnatal testosterone treatment on defeminization of sexual receptivity in pigs. *Biol. Reprod.* 36: 581–587.

Forger, N. G. 2001. Development of sex differences in the nervous system. In E. Blass, ed., *Developmental Psychobiology* (*Handbook of Behavioral Neurobiology*, vol. 13), 143–198. New York: Kluwer/Plenum.

Fortman, M., Dellovade, T. L. and Rissman, E. F. 1992. Adrenal contribution to the induction of sexual behavior in the female musk shrew. *Horm. Behav.* 26: 76–86.

Foster, S. A. and Endler, J. A. 1999. Thoughts on geographic variation in behavior. In S. Foster and J. A. Endler, eds., *Geographic Variation in Behavior: Perspectives on Evolutionary Mechanisms*, 287–307. New York: Oxford University Press.

Fowler, C. D., Freeman, M. E. and Wang, Z. 2003. Newly proliferated cells in the adult male amygdala are affected by gonadal steroid hormones. *J. Neurobiol.* 57: 257–269.

Fox, H. E., White, S. A., Kao, M.H.F. and Fernald, R. D. 1997. Stress and dominance in a social fish. *J. Neurosci.* 17: 6463–6469.

Fox, T. O., Tobet, S. A. and Baum, M. J. 1999. Sex differences in human brain and behavior. In G. Adelman and B. Smith, eds., *Encyclopedia of Neuroscience*, 1845–1849. Amsterdam: Elsevier.

Francis, R. C. 1992. Sexual lability in teleosts: developmental factors. *Q. Rev. Biol.* 67: 1–18.

Francis, R. C., Soma, K. and Fernald, R. D. 1993. Social regulation of the brain–pituitary–gonadal axis. *Proc. Natl. Acad. Sci. U S A* 90: 7794–7798.

Frank, L. G., Glickman, S. E. and Licht, P. 1991. Fatal sibling aggression, precocial development, and androgens in neonatal spotted hyenas. *Science* 252: 702–704.

Freed, L. 1987. The long-term pair bond of tropical house wrens: advantage or constraint? *Am. Nat.* 130: 507–525.

Freeman-Gallant, C. R. 1996. DNA fingerprinting reveals female preference for male parental care in Savannah Sparrows. *Proc. R. Soc. Lond. B* 263: 157–160.

French, J. A. 1997. Proximate regulation of singular breeding in callitrichid primates. In N. G. Solomon and J. A. French, eds., *Cooperative Breeding in Mammals*, 34–74. Cambridge, UK: Cambridge University Press.

French, J. A. and Schaffner, C. M. 2000. Contextual influences on sociosexual behavior in monogamous primates. In K. Wallen and J. E. Schneider, eds., *Reproduction in*

Context: Social and Environmental Influences on Reproductive Physiology and Behavior, 325–354. Cambridge, MA: MIT Press.

Frye, C. A. 2001. The role of neurosteroids and non-genomic effects of progestins and androgens in mediating sexual receptivity of rodents. *Brain Res. Rev.* 37: 201–222.

Frye, C. A., Park, D., Tanaka, M., Rosellini, R. and Svare, B. 2001. The testosterone metabolite and neurosteroid 3α-androstanediol may mediate the effects of testosterone on conditioned place preference. *Psychoneuroendocrinology* 26: 731–750.

Fujino, Y., Nagahama, T., Oumi, T., Ukena, K., Morishita, F., Furukawa, Y., Matsushima, O., Ando, M., Takahama, H., Satake, H., Minakata, H. and Nomoto, K. 1999. Possible functions of oxytocin/vasopressin-superfamily peptides in annelids with special reference to reproduction and osmoregulation. *J. Exp. Zool.* 284: 401–406.

Fusani, L. and Hutchison, J. B. 2003. Lack of changes in the courtship behaviour of male ring doves after testosterone treatment. *Ethol. Ecol. Evol.* 15: 143–157.

Fusani, L., Beani, L., Lupo, C. and Dessì-Fulgheri, F. 1997. Sexually selected vigilance behaviour of the grey partridge is affected by plasma androgen levels. *Anim. Behav.* 54: 1013–1018.

Gadgil, M. 1972. Male dimorphism as a consequence of sexual selection. *Am. Nat.* 106: 574–558.

Gahr, M. 2000. Neural song control system of hummingbirds: comparison to swifts, vocal learning (Songbirds) and nonlearning (Suboscines) passerines, and vocal learning (Budgerigars) and nonlearning (Dove, owl, gull, quail, chicken) nonpasserines. *J. Comp. Neurol.* 426: 182–196.

Gahr, M. and Güttinger, H.-R. 1986. Functional aspects of singing in male and female *Uraeginthus bengalus* (Estrildidae). *Ethology* 72: 123–131.

Gahr, M. and Metzdorf, R. 1997. Distribution and dynamics in the expression of androgen and estrogen receptors in vocal control systems of songbirds. *Brain Res. Bull.* 44: 509–517.

Gahr, M., Güttinger, H. R. and Kroodsma, D. E. 1993. Estrogen receptors in the avian brain: survey reveals general distribution and forebrain areas unique to songbirds. *J. Comp. Neurol.* 327: 112–122.

Galdikas, B. 1985. Subadult male orangutan sociality and reproductive behavior at Tanjung Putting. *Am. J. Primatol.* 8: 87–99.

Gandelman, R. 1983. Gonadal hormones and sensory function. *Neurosci. Biobehav. Rev.* 7:1–17.

Ganesan, R. 1994. The aversive and hypophagic effects of estradiol. *Physiol. Behav.* 55: 279–285.

Garey, J., Goodwillie, A., Frohlich, J., Morgan, M., Gustafsson, J. A., Smithies, O., Korach, K. S., Ogawa, S. and Pfaff, D. W. 2003. Genetic contributions to generalized arousal of brain and behavior. *Proc. Natl. Acad. Sci. U S A* 100: 11019–11022.

Garland, T. and Carter, P. A. 1994. Evolutionary physiology. *Annu. Rev. Physiol.* 56 : 579–621.

Geary, N. 2001. Sex differences in disease anorexia. *Nutrition* 17: 499–507.

Gelinas, D. and Callard, G. V. 1993. Immunocytochemical and biochemical evidence for aromatase in neurons of the retina, optic tectum and retinotectal pathways in goldfish. *J. Neuroendocrinol.* 5: 635–641.

Gerall, A. A., Moltz, H. and Ward, I. L. eds. 1992. *Handbook of Behavioral Neurobiology*, vol. 11: *Sexual Differentiation*. New York: Plenum.

Gerhart, J. and Kirschner, M. 1997. *Cells, Embryos, and Evolution*. Oxford, UK: Blackwell Science.

Ghiselin, M. T. 1969. The evolution of hermaphroditism among animals. *Q. Rev. Biol.* 44: 189–208.

Giedd, J. N., Blumenthal, J., Jeffries, N. O., Castellanos, F. X., Liu, H., Zijdenbos, A., Paus, T., Evans, A. C. and Rapoport, J. L. 1999. Brain development during childhood and adolescence: a longitudinal MRI study. *Nature Neurosci.* 2: 861–863.

Gil, D., Graves, J., Hazon, N. and Wells, A. 1999. Male attractiveness and differential testosterone investment in zebra finch eggs. *Science* 286: 126–128.

Gil, D., Leboucher, G., Lacroix, A., Cue, R. and Kreutzer, M. 2004a. Female canaries produce eggs with greater amounts of testosterone when exposed to preferred male song. *Horm. Behav.* 45: 64–70.

Gil, D., Heim, C., Bulmer, E., Rocha, M., Puerta, M. and Naguib, M. 2004b. Negative effects of early developmental stress on yolk testosterone levels in a passerine bird. *J. Exp. Biol.* 207: 2215–2220.

Gilardi, K. V., Shideler, S. E., Valverde, C. R., Roberts, J. A. and Lasley, B. L. 1997. Characterization of the onset of menopause in the rhesus macaque. *Biol. Reprod.* 57: 335–340.

Gilbert, S. F. 2001. Ecological developmental biology: developmental biology meets the real world. *Dev. Biol.* 233: 1–12.

Ginther, A. J., Ziegler, T. E. and Snowdon, C. T. 2001. Reproductive biology of captive male cottontop tamarin monkeys as a function of social environment. *Anim. Behav.* 61: 65–78.

Gittleman, J. L., Anderson, C. G., Kot, M. and Luh, H. K. 1996. Comparative tests of evolutionary lability and rates using molecular phylogenies. In P. H. Harvey, A.J.L. Brown, J. Maynard Smith and S. Nee, eds., *New Uses for New Phylogenies*, 289–307. Oxford, UK: Oxford University Press.

Godwin, J., Crews, D. and Warner, R. R. 1996. Behavioural sex change in the absence of gonads in a coral reef fish. *Proc. R. Soc. Lond. B* 263: 1683–1688.

Godwin, J., Sawby, R., Warner, R. R., Crews, D. and Grober, M. S. 2000. Hypothalamic arginine vasotocin mRNA abundance variation across sexes and with sex change in a coral reef fish. *Brain Behav. Evol.* 55: 77–84.

Godwin, J., Luckenbach, J. A. and Borski, R. J. 2003. Ecology meets endocrinology: environmental sex determination in fishes. *Evol. Dev.* 5: 40–49.

Goldfoot, D. A. and Neff, D. A. 1985. On measuring behavioral sex differences in social contexts. In N. Adler, D. Pfaff and R. W. Goy, eds., *Handbook of Behavioral Neurobiology*, vol. 7, 767–784. New York: Plenum.

Gomendio, M., Harcourt, A. H. and Roldán, E. R. S. 1998. Sperm competition in mammals. In T. R. Birkhead and A. P. Møller, eds., *Sperm Competition and Sexual Selection*, 667–756. San Diego: Academic Press.

González-Mariscal, G.and Rosenblatt, J. S. 1996. Maternal behavior in rabbits: a historical and multidisciplinary perspective. In J. S. Rosenblatt and C. T. Snowdon, eds., *Parental Care: Evolution, Mechanisms, and Adaptive Significance* (*Advances in the Study of Behavior*, vol. 25), 333–360. San Diego: Academic Press.

Goodale, H. D. 1918. Feminized male birds. *Genetics* 3: 276–295.

Goodson, J. L. 1998a. Territorial aggression and dawn song are modulated by septal vasotocin and vasoactive intestinal polypeptide in male field sparrows (*Spizella pusilla*). *Horm. Behav.* 34: 67–77.

Goodson, J. L. 1998b. Vasotocin and vasoactive intestinal polypeptide modulate aggression in a territorial songbird, the violet-eared waxbill (Estrildidae: *Uraeginthus granatina*). *Gen. Comp. Endocrinol.* 111: 233–244.

Goodson, J. L. and Adkins-Regan, E. 1999. Effect of intraseptal vasotocin and vasoactive intestinal polypeptide infusions on courtship song and aggression in the male zebra finch (*Taeniopygia guttata*). *J. Neuroendocrinol.* 11: 19–25.

Goodson, J. L., and Bass, A. H. 2000. Forebrain peptides modulate sexually polymorphic vocal circuitry. *Nature* 403: 769–772.

Goodson, J. L. and Bass, A. H. 2001. Social behavior functions and related anatomical characteristics of vasotocin/vasopressin systems in vertebrates. *Brain Res. Rev.* 35: 246–265.

Goodson, J. L. and Bass, A. H. 2002. Vocal–acoustic circuitry and descending vocal pathways in teleost fish: convergence with terrestrial vertebrates reveals conserved traits. *J. Comp. Neurol.* 448: 298–322.

Goodwin, B. 2001. *How the Leopard Changed Its Spots: The Evolution of Complexity.* Princeton, NJ: Princeton University Press.

Gooren, L. 1990. Biomedical theories of sexual orientation. In D. P. McWhirter, S. A. Sanders and J. M. Reinisch, eds., *Homosexuality/Heterosexuality: Concepts of Sexual Orientation*, 71–87. New York: Oxford University Press.

Gorbman, A. and Sower, S. A. 2003. Evolution of the role of GnRH in animal (Metazoan) biology. *Gen. Comp. Endocrinol.* 134: 207–213.

Gosling, L. M. and Roberts, S. C. 2001. Scent-marking by male mammals: cheat-proof signals to competitors and mates. In P.J.B. Slater, J. S. Rosenblatt, C. T. Snowdon and T. J. Roper, eds., *Advances in the Study of Behavior*, 169–218. San Diego: Academic Press.

Gould, E., Reeves, A. J., Fallah, M., Tanapat, P., Gross, C. G. and Fuchs, E. 1999. Hippocampal neurogenesis in adult Old World primates. *Proc. Natl. Acad. Sci. U S A* 96: 5263–5267.

Goy, R. W. and McEwen, B. S. 1980. *Sexual Differentiation of the Brain.* Cambridge, MA: MIT Press.

Goy, R. W. and Roy, M. 1991. Heterotypical sexual behaviour in female mammals. In M. Haug, P. F. Brain and C. Aron, eds., *Heterotypical Behaviour in Man and Animals*, 71–97. London: Chapman & Hall.

Goymann, W. and Wingfield, J. C. 2004a. Allostatic load, social status and stress hormones: the costs of social status matter. *Anim. Behav.* 67: 591–602.

Goymann, W. and Wingfield, J. C. 2004b. Competing females and caring males. Sex steroids in African black coucals, *Centropus grillii. Anim. Behav.* 68: 733–740.

Goymann, W., Möstl, E. and Gwinner, E. 2002. Corticosterone metabolites can be measured noninvasively in excreta of European stonechats (*Saxicola torquata rubicola*). *Auk* 119: 1167–1173.

Grafen, A. 1990. Biological signals as handicaps. *J. Theor. Biol.* 144: 517–546.

Graham, J. M. and Desjardins, C. 1980. Classical conditioning: induction of luteinizing hormone and testosterone secretion in anticipation of sexual activity. *Science* 210: 1039–1041.

Gram, W. D., Heath, H. W., Wichman, H. A. and Lynch, G. R. 1982. Geographic variation in *Peromyscus leucopus*: short-day induced reproductive regression and spontaneous recrudescence. *Biol. Reprod.* 27: 369–373.

Graur, D. and Li, W.-H. 2000. *Fundamentals of Molecular Evolution*, 2nd ed. Sunderland, MA: Sinauer.

Graves, J. A. 1995. The origin and function of the mammalian Y chromosome and Y-borne genes—an evolving understanding. *Bioessays* 17: 311–320.

Graves, J. A. 2001. From brain determination to testis determination: evolution of the mammalian sex-determining gene. *Reprod. Fertil. Dev.* 13: 665–672.

Graves, J. A. 2002. Evolution of the testis-determining gene—the rise and fall of SRY. *Novartis Found. Symp.* 244: 86–97.

Graves, J.A.M. and Shetty, S. 2001. Sex from W to Z: Evolution of vertebrate sex chromosomes and sex determining genes. *J. Exp. Zool.* 290: 449–462.

Gray, G. D., Smith, E. R., Dorsa, D. M. and Davidson, J. M. 1981. Sexual behavior and testosterone in middle-aged male rats. *Endocrinology* 109: 1597–1604.

Greene, H. W. 1999. Natural history and behavioural homology. In G. R. Bock and G. Cardew, eds., *Novartis Foundation Symposium 222: Homology*, 173–188. New York: Wiley.

Greenwood, P. J., Harvey, P. H., and Perrins C. M. 1979. The role of dispersal in the great tit (*Parus major*): the causes, consequences, and heritability of natal dispersal. *J. Anim. Ecol.* 48: 123–142.

Grisham, W. and Arnold, A. P. 1995. A direct comparison of the masculinizing effects of testosterone, androstenedione, estrogen, and progesterone on the development of the zebra finch song system. *J. Neurobiol.* 26: 163–170.

Grober, M. S. and Bass, A. H. 2002. Life history, neuroendocrinology, and behavior in fish. In D. W. Pfaff, A. P. Arnold, A. M. Etgen, S. E. Fahrbach and R. T. Rubin, eds., *Hormones, Brain and Behavior*, vol. 2, 331–348. Amsterdam: Academic Press (Elsevier).

Grober, M. S. and Sunobe, T. 1996. Serial adult sex change involves rapid and reversible changes in forebrain neurochemistry. *Neuroreport* 7: 2945–2949.

Gross, W. B. and Siegel, P. B. 1985. Selective breeding of chickens for corticosterone response to social stress. *Poultry Sci.* 64: 2230–2233.

Grossman, C. J. 1985. Interactions between the gonadal steroids and the immune system. *Science* 227: 257–261.

Grunt, J. A. and Young, W. C. 1952. Differential reactivity of individuals and the response of the male guinea pig to testosterone propionate. *Endocrinology* 51: 237–248.

Guhl, A. M. 1961. Gonadal hormones and social behavior in infrahuman vertebrates. In W. C. Young, ed., *Sex and Internal Secretions*, vol. II, 3rd ed., 1240–1267. Baltimore: Williams & Wilkins.

Guhl, A. M., Collias, N. E. and Allee, W. C. 1945. Mating behavior and the social hierarchy in small flocks of White Leghorns. *Physiol. Zool.* 18: 365–390.

Guillette, L. J., Crain, D. A., Rooney, A. A. and Pickford, D. B. 1995. Organization vs. activation: the role of endocrine-disrupting contaminants (EDCs) during embryonic development in wildlife. *Environ. Health Perspect.* 103 (suppl. 7): 157–164.

Gurney, M. E. 1981. Hormonal control of cell form and number in the zebra finch song system. *J. Neurosci.* 1: 658–673.

Gurney, M. E. and Konishi, M. 1980. Hormone-induced sexual differentiation of brain and behavior in zebra finches. *Science* 208: 1380–1383.

Gustafson, A. W. and Shemesh, M. 1976. Changes in plasma testosterone levels during the annual reproductive cycle of the hibernating bat, *Myotis lucifugus lucifugus* with a survey of plasma testosterone levels in adult male vertebrates. *Biol. Reprod.* 15: 9–24.

Gvaryahu, G., Snapir, N., Robinzon, B., Goodman, G., el Halawani, M. E. and Grimm, V. E. 1986. The gonadotropic-axis involvement in the course of the filial following response in the domestic fowl chick. *Physiol. Behav.* 38: 651–656.

Gwinner, E., Rödl, T. and Schwabl, H. 1994. Pair territoriality of wintering stonechats: behaviour, function and hormones. *Behav. Ecol. Sociobiol.* 34: 321–327.

Haase, E. 2000. Comparison of reproductive biological parameters in male wolves and domestic dogs. *Z. Säugetierkunde* 65: 257–270.

Haase, E. and Donham, R. S. 1980. Hormones and domestication. In A. Epple and M. H. Stetson, eds., *Avian Endocrinology*, 549–566. New York: Academic Press.

Habib, K. E., Weld, K. P., Rice, K. C., Pushkas, J., Champoux, M., Listwak, S., Webster, E. L., Atkinson, A. J., Schulkin, J., Contoreggi, C., Chrousos, G. P., McCann, S. M., Suomi, S. J., Higley, J. D. and Gold, P. W. 2000. Oral administration of a corticotropin-releasing hormone receptor antagonist significantly attenuates behavioral, neuroendocrine, and autonomic responses to stress in primates. *Proc. Natl. Acad. Sci. U S A* 97: 6079–6084.

Hackl, R., Bromundt, V., Daisley, J., Kotrschal, K. and Mostl, E. 2003. Distribution and origin of steroid hormones in the yolk of Japanese quail eggs (*Coturnix coturnix japonica*). *J. Comp. Physiol. B* 173: 327–331.

Hagelin, J. C. and Ligon, J. D. 2001. Female quail prefer testosterone-mediated traits, rather than the ornate plumage of males. *Anim. Behav.* 61: 465–476.

Haig, D. 1993. Genetic conflicts in human pregnancy. *Quart. Rev. Biol.* 68: 495–532.

Halem, H. A., Cherry, J. A. and Baum, M. J. 1999. Vomeronasal neuroepithelium and forebrain Fos responses to male pheromones in male and female mice. *J. Neurobiol.* 39: 249–263.

Hall, B. K. 1999. *Evolutionary Developmental Biology*, 2nd ed. Dordrecht: Kluwer.

Haller, J. 1995. Biochemical background for an analysis of cost–benefit interrelations in aggression. *Neurosci. Biobehav. Rev.* 19: 599–604.

Haller, J., Makara, G. B. and Kruk, M. R. 1998. Catecholaminergic involvement in the control of aggression: hormones, the peripheral sympathetic, and central noradrenergic systems. *Neurosci. Biobehav. Rev.* 22: 85–97.

Hamilton, J. B. 1948. The role of testicular secretions as indicated by the effects of castration in man and by studies of pathological conditions and the short lifespan associated with maleness. *Recent Prog. Horm. Res.* 3: 257–322.

Hamilton, W. D. and Zuk, M. 1982. Heritable true fitness and bright birds: a role for parasites? *Science* 218: 384–387.

Han, T. M. and de Vries, G. J. 2003. Organizational effects of testosterone, estradiol, and dihydrotestosterone on vasopressin mRNA expression in the bed nucleus of the stria terminalis. *J. Neurobiol.* 54: 502–510.

Hänssler, I. and Prinzinger, R. 1979. The influence of the sex-hormone testosterone on body temperature and metabolism of the male Japanese quail (*Coturnix coturnix japonica*). *Experientia* 35: 509–510.

Harcourt, A. H., Harvey, P. H., Larson, S. G. and Short, R. V. 1981. Testis weight, body weight and breeding system in primates. *Nature* 293: 55–57.

Harding, C. F. 1981. Social modulation of circulating hormone levels in the male. *Am. Zool.* 21: 223–231.

Harding, C. F. 1983. Hormonal influences on avian aggressive behavior. In B. B. Svare, ed., *Hormones and Aggressive Behavior*, 435–468. New York: Plenum Press.

Harding, C. F. 1991. Neuroendocrine integration of social behavior in male songbirds. In T. Archer and S. Hansen, eds., *Behavioral Biology: Neuroendocrine Axis*, 53–66. Hillsdale, NJ: Lawrence Erlbaum Associates.

Harding, C. F. 1992. Hormonal modulation of neurotransmitter function and behavior in male songbirds. *Poultry Sci. Rev.* 4: 261–273.

Harding, C. F. and Follett, B. K. 1979. Hormone changes triggered by aggression in a natural population of blackbirds. *Science* 203: 918–920.

Harding, C. F. and Rowe, S. A. 2003. Vasotocin treatment inhibits courtship in male zebra finches; concomitant androgen treatment inhibits this effect. *Horm. Behav.* 44: 413–418.

Harding, C. F., Sheridan, K., and Walters, M. J. 1983. Hormonal specificity and activation of sexual behavior in male zebra finches. *Horm. Behav.* 17: 111–133.

Harris, G. W. 1955. *Neural Control of the Pituitary Gland*. London: Edward Arnold.

Hart, B. L. 1974. Gonadal androgen and sociosexual behavior of male mammals: a comparative analysis. *Psychol. Bull.* 7: 383–400.

Hart, B. L., Wallach, S.J.R. and Melese-D'Hospital, P. Y. 1983. Differences in responsiveness to testosterone of penile reflexes and copulatory behavior of male rats. *Horm. Behav.* 17: 274–283.

Harvey, L. A. and Propper, C. R. 1997. Effects of androgens on male sexual behavior and secondary sex characters in the explosively breeding spadefoot toad, *Scaphiopus couchii*. *Horm. Behav.* 31: 89–96.

Harzsch, S., Miller, J., Benton, J. and Beltz, B. 1999. From embryo to adult: persistent neurogenesis and apoptotic cell death shape the lobster deutocerebrum. *J. Neurosci.* 19: 3472–3485.

Hasan, S. A., Brain, P. F. and Castano, D. 1988. Studies on effects of tamoxifen (ICI 46474) on agonistic encounters between pairs of intact mice. *Horm. Behav.* 22: 178–185.

Hasselquist, D., Marsh, J. A., Sherman, P. W. and Wingfield, J. C. 1999. Is avian humoral immunocompetence suppressed by testosterone? *Behav. Ecol. Sociobiol.* 45: 167–175.

Hau, M. 2001. Timing of breeding in variable environments: tropical birds as model systems. *Horm. Behav.* 40: 281–290.

Hawkes, K., O'Connell, J. F., Jones, N.G.B., Alvarez, H. and Charnov, E. L. 1998. Grandmothering, menopause, and the evolution of human life histories. *Proc. Natl. Acad. Sci. U S A* 95: 1336–1339.

Hawkins, C. E., Dallas, J. F., Fowler, P. A., Woodroffe, R. and Racey, P. A. 2002. Transient masculinization in the fossa, *Cryptoprocta ferox* (Carnivora, Viverridae). *Biol. Reprod.* 66: 610–615.

Hawkins, M. B., Thornton, J. W., Crews, D., Skipper, J. K., Dotte, A. and Thomas, P. 2000. Identification of a third distinct estrogen receptor and reclassification of estrogen receptors in teleosts. *Proc. Natl. Acad. Sci. U S A* 97: 10751–10756.

Hayes, T. B. 1997. Hormonal mechanisms as potential constraints on evolution: examples from the Anura. *Am. Zool.* 37: 482–490.

Hedricks, C. A. 1994. Female sexual activity across the human menstrual cycle. *Annu. Rev. Sex Res.* 5: 122–172.

Hegner, R. E. and Wingfield, J. C. 1986. Behavioral and endocrine correlates of multiple brooding in the semicolonial house sparrow *Passer domesticus*, I: males. *Horm. Behav.* 20: 294–312.

Hegner, R. E. and Wingfield, J. C. 1987. Effects of experimental manipulation of testosterone levels on parental investment and breeding success in male house sparrows. *Auk* 104: 462–469.

Heistermann, M., Ziegler, T., van Schaik, C. P., Launhardt, K., Winkler, P. and Hodges, J. K. 2001. Loss of oestrus, concealed ovulation and paternity confusion in free-ranging Hanuman langurs. *Proc. R. Soc. Lond. B* 268: 2445–2451.

Helfman, G. S., Collette, B. B. and Facey, D. E. 1997. *The Diversity of Fishes*. Malden, MA: Blackwell Science.

Hennessy, M. B. 1997. Hypothalamic–pituitary–adrenal responses to brief social separation. *Neurosci. Biobehav. Rev.* 21: 11–29.

Hennessy, M. B., Mendoza, S. P., Coe, C. L., Lowe, E. L. and Levine, S. 1980. Androgen-related behavior in the squirrel monkey: an issue that is nothing to sneeze at. *Behav. Neural Biol.* 30: 103–108.

Henson, S. A. and Warner, R. R. 1997. Male and female alternative reproductive behaviors in fishes: a new approach using intersexual dynamics. *Annu. Rev. Ecol. Systematics* 28: 571–592.

Hess, E. H. 1959. Imprinting, an effect of early experience. *Science* 130: 133–141.

Hews, D. K. and Moore, M. C. 1996. A critical period for the organization of alternative male phenotypes of tree lizards by exogenous testosterone? *Physiol. Behav.* 60: 425–429.

Hews, D. K. and Moore, M. C. 1997. Hormones and sex-specific traits: critical questions. In N. E. Beckage, ed., *Parasites and Pathogens: Effects on Host Hormones and Behavior*, 277–292. New York: Chapman & Hall.

Hews, D. K. and Quinn, V. S. 2003. Endocrinology of species differences in sexually dichromatic signals: using the organization and activation model in a phylogenetic framework. In S. F. Fox, J. K. McCoy and T. A. Baird, eds., *Lizard Social Behavior*, 235–277. Baltimore, MD: Johns Hopkins University Press.

Hews, D. K., Knapp, R. and Moore, M. C. 1994. Early exposure to androgens affects adult expression of alternative male types in tree lizards. *Horm. Behav.* 28: 96–115.

Heyes, C. M. and Galef, B. G., eds. 1996. *Social Learning in Animals: The Roots of Culture*. San Diego: Academic Press.

Hillgarth, N., Ramenofsky, M. and Wingfield, J. 1997. Testosterone and sexual selection. *Behav. Ecol.* 8: 108–112.

Hillis, D. M and Green, D. M. 1990. Evolutionary changes of heterogametic sex in the phylogenetic history of amphibians. *J. Evol. Biol.* 3: 49–64.

Hirschenhauser, K., Möstl, E. and Kotrschal, K. 1999. Within-pair testosterone covariation and reproductive output in Greylag Geese *Anser anser*. *Ibis* 141: 577–586.

Hirschenhauser, K. Winkler, H. and Oliveira, R. F. 2003. Comparative analysis of male androgen responsiveness to social environment in birds: the effects of mating system and paternal incubation. *Horm. Behav.* 43: 508–519.

Hirschenhauser, K., Taborsky, M., Oliveira, T., Canàrio, A.V.M. and Oliveira, R. F. 2004. A test of the "challenge hypothesis" in cichlid fish: simulated partner and territory intruder experiments. *Anim. Behav.* 68: 741–750.

Hobbs, J.-P.A., Munday, P. L. and Jones, G. P. 2004. Social induction of maturation and sex determination in a coral reef fish. *Proc. R. Soc. Lond. B* 271: 2109–2114.

Hochberg, R. B. 1998. Biological esterification of steroids. *Endocrinol. Rev.* 19: 331–348.

Hodgkin, J. 2002. Exploring the envelope: systematic alteration in the sex-determination system of the nematode *Caenorhabditis elegans*. *Genetics* 162: 767–780.

Hoffman, K. A., Mendoza, S. P., Hennessy, M. B. and Mason, W. A. 1995. Responses of infant titi monkeys, *Callicebus moloch*, to removal of one of both parents: evidence for paternal attachment. *Dev. Psychobiol.* 28: 399–407.

Holekamp, K. E. and Smale, L. 1998. Dispersal status influences hormones and behavior in the male spotted hyena. *Horm. Behav.* 33: 205–216.

Holekamp, K. E., Smale, L., Simpson, H. B. and Holekamp, N. A. 1984. Hormonal influences on natal dispersal in free-living Belding's ground squirrels (*Spermophilus beldingi*). *Horm. Behav.* 18: 465–483.

Hollis, K. L. 1997. Contemporary research on Pavlovian conditioning: a "new" functional analysis. *Am. Psychol.* 52: 956–965.

Hollis, K. L., Dumas, M. J., Singh, P. and Fackelman, P. 1995. Pavlovian conditioning of aggressive behavior in blue gourami fish (*Trichogaster trichopterus*): winners become winners and losers stay losers. *J. Comp. Psychol.* 109: 123–133.

Hollis, K. L., Pharr, V. L., Dumas, M. J., Britton, G. B. and Field, J. 1997. Classical conditioning provides paternity advantage for territorial male blue gouramis (*Trichogaster trichopterus*). *J. Comp. Psychol.* 111: 219–225.

Holloway, C. C. and Clayton, D. F. 2001. Estrogen synthesis in the male brain triggers development of the avian song control pathway in vitro. *Nature Neurosci.* 4: 170–175.

Holmes, D. J., Fluckiger, R. and Austad, S. N. 2001. Comparative biology of aging in birds: an update. *Exp. Gerontol.* 36: 869–883.

Houck, L. D. and Woodley, S. K. 1995. Field studies of steroid hormones and male reproductive behaviour in amphibians. In H. Heatwole and B. K. Sullivan, eds., *Amphibian Biology*, vol. 2: *Social Behaviour*, 677–703. New South Wales, Australia: Surrey Beatty & Sons.

Houtman, A. M. 1992. Female zebra finches choose extra-pair copulations with genetically attractive males. *Proc. R. Soc. Lond. B* 249: 3–6.

Huhman, K. L., Solomon, M. B., Janicki, M., Harmon, A. C., Lin, S. M., Israel, J. E. and Jasnow, A. M. 2003. Conditioned defeat in male and female Syrian hamsters. *Horm. Behav.* 44: 293–299.

Hulbert, A. J. and Else, P. L. 2000. Mechanisms underlying the cost of living in animals. *Ann. Rev. Physiol.* 62: 207–235.

Hutchison, J. B. 1990. Androgen action in a changing behavioural environment: the role of brain aromatase. In J. Balthazart, ed., *Hormones, Brain and Behaviour in Vertebrates, 2: Behavioural Activation in Males and Females—Social Interaction and Reproductive Endocrinology. Comp. Physiol.* 9: 27–44. Basel: Karger.

Hutchison, J. B. and Bateson, P. 1982. Sexual imprinting in male Japanese quail: The effects of castration at hatching. *Dev. Psychobiol.* 15: 471–477.

Immelmann, K. 1985. Sexual imprinting in zebra finches: mechanisms and biological significance. *Proc. Int. Ornithol. Congr.* 18: 156–172.

Inglis, G. C., Ingram, M. C., Holloway, C. D., Swan, L., Birnie, D., Hillis, W. S., Davies, E., Fraser, R. and Connell, J.M.C. 1999. Familial pattern of corticosteroids and their metabolism in adult human subjects: The Scottish Adult Twin Study. *J. Clin. Endocrinol. Metab.* 84: 4132–4137.

Insel, T. R. and Young, L. J. 2001. The neurobiology of attachment. *Nature Rev. Neurosci.* 2: 129–136.

Isgor, C. and Sengelaub, D. R. 1998. Prenatal gonadal steroids affect adult spatial behavior, CA1 and CA3 pyramidal cell morphology in rats. *Horm. Behav.* 34: 183–198.

Ishikawa, Y., Yoshimoto, M., Yamamoto, N. and Ito, H. 1999. Different brain morphologies from different genotypes in a single teleost species, the medaka (*Oryzias latipes*). *Brain Behav. Evol.* 53: 2–9.

Itzkowitz, M., Santangelo, N. and Richter, M. 2003. How does a parent respond when its mate emphasizes the wrong role? A test using a monogamous fish. *Anim. Behav.* 66: 863–869.

Jaccarini, V., Agius, L., Schembri, P. J. and Rizzo, M. 1983. Sex determination and larval sexual interaction in *Bonellia viridis* Rolando (Echiura: Bonelliidae). *J. Exp. Marine Biol. Ecol.* 66: 25–40.

Jacobs, G. H., Neitz, J. and Neitz, M. 1993. Genetic basis of polymorphism in the color vision of platyrrhine monkeys. *Vision Res.* 33: 269–274.

Jacobs, L. F. 1996. Sexual selection and the brain. *Trends Ecol. Evol.* 11: 82–86.

Jainudeen, M. R. and Hafez, B. 2000. Reproductive failure in males. In B. Hafez and E.S.E. Hafez, eds., *Reproduction in Farm Animals*, 7th ed., 279–290. Philadelphia: Lippincott Williams & Wilkins.

Jainudeen, M. R., Katongole, C. B. and Short, R. V. 1972. Plasma testosterone levels in relation to musth and sexual activity in the male asiatic elephant, *Elephas maximus*. *J. Reprod. Fertil.* 29: 99–103.

James, P. J. and Nyby, J. G. 2002. Testosterone rapidly affects the expression of copulatory behavior in house mice (*Mus musculus*). *Physiol. Behav.* 75: 287–294.

Janzen, F. J. 1996. Is temperature-dependent sex determination in reptiles adaptive? *Trends Ecol. Evol.* 11: 253.

Janzen, F. J. and Paukstis, G. L. 1991. Environmental sex determination in reptiles: ecology, evolution, and experimental design. *Q. Rev. Biol.* 66: 149–179.

Jarvis, E. D., Ribeiro, S., da Silva, M. L., Ventura, D., Vielliard, J. and Mello, C. V. 2000. Behaviourally driven gene expression reveals song nuclei in hummingbird brain. *Nature* 406: 628–632.

Jasnow, A. M., Huhman, K. L., Bartness, T. J. and Demas, G. E. 2000. Short-day increases in aggression are inversely related to circulating testosterone concentrations in male Siberian hamsters (*Phodopus sungorus*). *Horm. Behav.* 38: 102–110.

Jasnow, A. M., Drazen, D. L., Huhman, K. L., Nelson, R. J. and Demas, G. E. 2001. Acute and chronic social defeat suppresses humoral immunity of male Syrian hamsters (*Mesocricetus auratus*). *Horm. Behav.* 40: 428–433.

Jennings, D. H. and Hanken, J. 1998. Mechanistic basis of life history evolution in anuran amphibians: thyroid gland development in the direct-developing frog, *Eleutherodactylus coqui*. *Gen. Comp. Endocrinol.* 111: 225–232.

Jöchle, W. 1975. Current research in coitus-induced ovulation: a review. *J. Reprod. Fertil.* suppl. 22: 165–207.

Johnstone, R. A. 1997. The evolution of animal signals. In J. R. Krebs and N. B. Davies, eds., *Behavioural Ecology: An Evolutionary Approach*, 4th ed., 155–178. Oxford, UK: Blackwell Science.

Johnstone, R. A. 2000. Models of reproductive skew: a review and synthesis. *Ethology* 106: 5–26.

Jones, J. S. and Wynne-Edwards, K. E. 2001. Paternal behaviour in biparental hamsters, *Phodopus campbelli*, does not require contact with the pregnant female. *Anim. Behav.* 62: 453–464.

Jones, R. B. and Andrew, R. J. 1992. Responses of adult domestic cocks and capons to novel and alarming stimuli. *Behav. Processes* 26: 189–200.

Jordan, C. L. 1999. Glia as mediators of steroid hormone action on the nervous system: an overview. *J. Neurobiol.* 40: 434–445.

Jost, A. 1985. Sexual organogenesis. In N. Adler, D. Pfaff and R. W. Goy, eds., *Handbook of Behavioral Neurobiology*, vol. 7: *Reproduction*, 3–20. New York: Plenum Press.

Just, W., Rau, W., Vogel, W., Akhverdian, M., Fredga, K., Graves, J. A. and Lyapunova, E. 1995. Absence of *Sry* in species of the vole *Ellobius*. *Nature Genet.* 11: 117–118.

Kallman, K. D. 1984. A new look at sex determination in poeciliid fishes. In B. J. Turner, ed., *Evolutionary Genetics of Fishes*, 95–171. New York: Plenum.

Kallman, K. D. and Schreibman, M. P. 1973. A sex-linked gene controlling gonadotrop differentiation and its significance in determining the age of sexual maturation and size of the platyfish, *Xiphophorus maculatus*. *Gen. Comp. Endocrinol.* 21: 287–304.

Kalra, P. S., Edwards, T. G., Xu, B., Jain, M. and Kalra, S. P. 1998. The anti-gonadotropic effects of cytokines: the role of neuropeptides. *Domest. Anim. Endocrinol.* 15: 321–332.

Kappeler, P. M. and van Schaik, C. P. 2002. Evolution of primate social systems. *Internat. J. Primatol.* 23:707–740.

Kapusta, J. 1998. Gonadal hormones and intrasexual aggressive behavior in female bank voles (*Clethrionomys glareolus*). *Aggr. Behav.* 24: 63–70.

Karn, R. C. and Nachman, M. W. 1999. Reduced nucleotide variability at an androgenbinding protein locus (Abpa) in house mice: Evidence for positive natural selection. *Mol. Biol. Evol.* 16: 1192–1197.

Katz, P. S. and Harris-Warrick, R. M. 1999. The evolution of neuronal circuits underlying species-specific behavior. *Curr. Opin. Neurobiol.* 9: 628–633.

Kavaliers, M. and Ossenkopp, K. P. 2001. Corticosterone rapidly reduces male odor preferences in female mice. *Neuroreport* 12: 2999–3002.

Kavumpurath, S. and Pandian, T. J. 1994. Masculinization of fighting fish, *Betta splendens* Regan, using synthetic or natural androgens. *Aquacult. Fish. Managemt.* 25: 373–381.

Kelley, D. B. 1988. Sexually dimorphic behaviors. *Annu. Rev. Neurosci.* 11: 225–251.

Kelley, D. B. 1997. Generating sexually differentiated songs. *Curr. Opinion Neurobiol.* 7: 839–843.

Kelley, D. B. and Tobias, M. L. 1999. The vocal repertoire of *Xenopus laevis*. In M. Hauser and M. Konishi, eds., *The Design of Animal Communication*, 9–35. Cambridge, MA: MIT Press.

Kelliher, K. R., Chang, Y. M., Wersinger, S. R. and Baum, M. J. 1998. Sex difference and testosterone modulation of pheromone-induced neuronal Fos in the ferret's main olfactory bulb and hypothalamus. *Biol. Reprod.* 59: 1454–1463.

Kendrick, K. M. and Dixson, A. F. 1984. Ovariectomy does not abolish proceptive behaviour cyclicity in the common marmoset (*Callithrix jacchus*). *J. Endocrinol.* 101: 155–162.

Kendrick, A. M. and Schlinger, B. A. 1996. Independent differentiation of sexual and social traits. *Horm. Behav.* 30: 600–610.

Kendrick, K. M., Lévy, F. and Keverne, E. B. 1992. Changes in the sensory processing of olfactory signals induced by birth in sleep. *Science* 256: 833–836.

Ketterson, E. D. and Nolan, V. 1992. Hormones and life histories: an integrative approach. *Am. Nat.* 140: S33–S62.

Ketterson, E. D. and Nolan, V. 1994. Hormones and life histories: an integrative approach. In L. A. Real, ed., *Behavioral Mechanisms in Evolutionary Ecology*, 328–353. Chicago: University of Chicago Press.

Ketterson, E. D. and Nolan, V. 1999. Adaptation, exaptation, and constraint: a hormonal perspective. *Am. Nat.* 154: S3–S25.

Ketterson, E. D., Nolan, V., Wolf, L. and Ziegenfus, C. 1992. Testosterone and avian life histories: effects of experimentally elevated testosterone on behavior and correlates of fitness in the Dark-eyed Junco (*Junco hyemalis*). *Am. Nat.* 140: 980–999.

Ketterson, E. D., Nolan, V., Cawthorn, J. J., Parker, P. G. and Zeigenfus, C. 1996. Phenotypic engineering: using hormones to explore the mechanistic and functional bases of phenotypic variation in nature. *Ibis* 138: 70–86.

Ketterson, E. D., Nolan, V., Casto, J. M., Buerkle, C. A., Clotfelter, E., Grindstaff, J. A., Hasselquist, D., Jones, K. J., Lipar, J. L., McNabb, F.M.A., Neudorf, D. L., Parker-Renga, I. and Schoech, S. J. 2000. Testosterone, phenotype and fitness: a research program in evolutionary behavioral endocrinology. In A. Dawson and C. M. Chaturvedi, eds., *Avian Endocrinology*, 19–40. New Delhi: Narosa Publishing House.

Keverne, E. B. 1992. Primate social relationships: their determinants and consequences. *Adv. Study Behav.* 21: 1–38.

Keyser-Marcus, L., Stafisso-Sandoz, G., Gerecke, K., Jasnow, A., Nightingale, L., Lambert, K. G., Gatewood, J. and Kinsley, C. H. 2001. Alterations of medial preoptic area neurons following pregnancy and pregnancy-like steroidal treatment in the rat. *Brain Res. Bull.* 55: 737–745.

Khan, M. Z., McNabb, F.M.A., Walters, J. R. and Sharp, P. J. 2001. Patterns of testosterone and prolactin concentrations and reproductive behavior of helpers and breeders in the cooperatively breeding red-cockaded woodpecker (*Picoides borealis*). *Horm. Behav.* 40: 1–13.

Kikuyama, S. and Toyoda, F. 1999. Sodefrin: a novel sex pheromone in a newt. *Rev. Reprod.* 4: 1–4.

Kim, Y. S., Stumpf, W. E., Sar, M. and Martinez-Vargas, M. C. 1978. Estrogen and androgen target cells in the brain of fishes, reptiles and birds: phylogeny and ontogeny. *Am. Zool.* 18: 425–434.

Kimball, R. T. and Ligon, J. D. 1999. Evolution of avian plumage dichromatism from a proximate perspective. *Am. Nat.* 154: 182–193.

Kingsolver, J. G., Hoekstra, H. E., Hoekstra, J. M., Berrigan, D., Vignieri, S. N., Hill, C. E., Hoang, A., Gibert, P. and Beerli, P. 2001. The strength of phenotypic selection in natural populations. *Am. Nat.* 157: 245–261.

Kinsley, C. H., Madonia, L., Gifford, G. W., Tureski, K., Griffin, G. R., Lowry, C., Williams, J., Collins, J., McLearie, H. and Lambert, K. G. 1999. Motherhood improves learning and memory. *Nature* 402: 137–138.

Kirkwood, T.B.L. 1977. Evolution of ageing. *Nature* 270: 301–304.

Kishida, M. and Callard, G. V. 2001. Distinct cytochrome P450 aromatase isoforms in zebrafish (*Danio rerio*) brain and ovary are differentially programmed and estrogen regulated during early development. *Endocrinology* 142:740–750.

Kitano, T., Takamune, K., Kobayashi, T., Nagahama, Y. and Abe, S.-I. 1999. Suppression of P450 aromatase gene expression in sex-reversed males produced by rearing genetically female larvae at a high water temperature during a period of sex differentiation in the Japanese flounder (*Paralichthys olivaceus*). *J. Mol. Endocrinol.* 23: 167–176.

Kitano, T., Takamune, K., Nagahama, Y. and Abe, S. I. 2000. Aromatase inhibitor and 17alpha-methyltestosterone cause sex-reversal from genetical females to phenotypic males and suppression of P450 aromatase gene expression in Japanese flounder (*Paralichthys olivaceus*). *Mol. Reprod. Dev.* 56: 1–5.

Kleiman, D. G. 1980. The sociobiology of captive propagation. In M. E. Soule and B. A. Wilcox, eds., *Conservation Biology, an Evolutionary—Ecological Perspective*, 243–261. Sunderland, MA: Sinauer.

Klint, T. 1985. Mate choice and male behaviour following castration and replacement of testosterone in mallards (*Anas platyrhynchos*). *Behav. Processes* 11: 419–424.

Knapp, R. and Moore, M. C. 1997. Male morphs in tree lizards have different testosterone responses to elevated levels of corticosterone. *Gen. Comp. Endocrinol.* 107: 273–279.

Knapp, R., Hews, D. K., Thompson, C. W., Ray, L. E., and Moore, M. C. 2003. Environmental and endocrine correlates of tactic switching by nonterritorial male tree lizards (*Urosaurus ornatus*). *Horm. Behav.* 43: 83–92.

Kodric-Brown, A. 1998. Sexual dichromatism and temporary color changes in the reproduction of fishes. *Am. Zool.* 38: 70–81.

Kölliker, M. and Richner, H. 2001. Parent-offspring conflict and the genetics of offspring solicitation and parental response. *Anim. Behav.* 62: 395–407.

Komisaruk, B. R., Adler, N. T. and Hutchison, J. 1972. Genital sensory field: enlargement by estrogen treatment in female rats. *Science* 178: 1295–1298.

Koolhaas, J. M., Korte, S. M., De Boer, S. F., Van Der Vegt, B. J., Van Reenen, C. G., Hopster, H., De Jong, I. C., Ruis, M. A. and Blokhuis, H. J. 1999. Coping styles in animals: current status in behavior and stress-physiology. *Neurosci. Biobehav. Rev.* 23: 925–935.

Korpelainen, H. 1990. Sex ratios and conditions required for environmental sex determination in animals. *Biol. Rev.* 65: 147–184.

Kotiaho, J. S. 2001. Costs of sexual traits: a mismatch between theoretical considerations and empirical evidence. *Biol. Rev.* 76: 365–376.

Kouros-Mehr, H., Pintchovski, S., Melnyk, J., Chen, Y. J., Friedman, C., Trask, B. and Shizuya, H. 2001. Identification of non-functional human VNO receptor genes provides evidence for vestigiality of the human VNO. *Chem. Senses* 26: 1167–1174.

Kow, L. M. and Pfaff, D. W. 1973/1974. Effects of estrogen treatment on the size of receptive field and response threshold of pudendal nerve in the female rat. *Neuroendocrinology* 13: 299–313.

Kramer, C. R. and Imbriano, M. A. 1997. Neuropeptide Y (NPY) induces gonad reversal in the protogynous bluehead wrasse, *Thalassoma bifasciatum* (Teleostei: Labridae). *J. Exp. Zool.* 279: 133–144.

Krebs, J. R. and Davies, N. B., eds. 1997. *Behavioural Ecology: An Evolutionary Approach*, 4th edition. Oxford, UK: Blackwell Science.

Kriner, E. and Schwabl, H. 1991. Control of winter song and territorial aggression of female Robin (*Erithacus rubecula*) by testosterone. *Ethology* 87: 37–44.

Kroodsma, D. E. 1982. Learning and the ontogeny of sound signals in birds. In D. E. Kroodsma and E. H. Miller, eds., *Acoustic Communication in Birds*, vol. 2: *Song Learning and Its Consequences*, 1–23. New York: Academic Press.

Kroodsma, D. E. and Canady, R. A. 1985. Differences in repertoire size, singing behavior, and associated neuroanatomy among marsh wren populations have a genetic basis. *Auk* 102: 439–447.

Krubitzer, L. and Kahn, D. M. 2003. Nature versus nurture revisited: an old idea with a new twist. *Prog. Neurobiol.* 70: 33–52.

Kunelius, P., Lukkarinen, O., Hannuksela, M. L., Itkonen, O., and Tapanainen, J. S. 2002. The effects of transdermal dihydrotestosterone in the aging male: a prospective, randomized, double blind study. *J. Clin. Endocrinol. Metab.* 87: 1467–1472.

Kunkel, P. 1974. Mating systems of tropical birds: the effects of weakness or absence of external reproduction-timing factors, with special reference to prolonged pair bonds. *Z. Tierpsychol.* 34: 265–307.

Künzl, C. and Sachser, N. 1999. The behavioral endocrinology of domestication: a comparison between the domestic guinea pig (*Cavia aperea* f. *porcellus*) and its wild ancestor, the cavy (*Cavia aperea*). *Horm. Behav.* 35: 28–37.

Kwon, J. Y., Haghpanah, V., Kogson-Hurtado, L. M., McAndrew, B. J. and Penman, D. J. 2000. Masculinization of genetic female nile tilapia (*Oreochromis niloticus*) by dietary administration of an aromatase inhibitor during sexual differentiation. *J. Exp. Zool.* 287: 46–53.

Kwon, J. Y., McAndrew, B. J. and Penman, D. J. 2001. Cloning of brain aromatase gene and expression of brain and ovarian aromatase genes during sexual differentiation in genetic male and female Nile tilapia *Oreochromis niloticus*. *Mol. Reprod. Dev.* 59: 359–370.

Lacey, E. A. and P. W. Sherman 1997. Cooperative breeding in naked mole-rats: implications for vertebrate and invertebrate sociality. In N. G. Solomon and J. A. French, eds., *Cooperative Breeding in Mammals*, 267–301. Cambridge, UK: Cambridge University Press.

Lack D. 1968. *Ecological Adaptations for Breeding in Birds*. London: Chapman & Hall.

Lahdenperä, M., Lummaa, V., Helle, S., Tremblay, M. and Russell, A. F. 2004. Fitness benefits of prolonged post-reproductive lifespan in women. *Nature* 428: 178–181.

Lahr, G., Maxson, S. C., Mayer, A., Just, W., Pilgrim, C. and Reisert, I. 1995. Transcription of the Y chromosomal gene, *Sry*, in adult mouse brain. *Mol. Brain Res.* 33: 179–182.

Lance, V. A., Vliet, K. A. and Bolaffi, J. L. 1985. Effect of mammalian luteinizing hormone-releasing hormone on plasma testosterone in male alligators, with observations on the nature of alligator hypothalamic gonadotropin-releasing hormone. *Gen. Comp. Endocrinol.* 60: 138–143.

Land, R. B. 1984. Genetics and reproduction. In C. R. Austin and R. V. Short, eds., *Reproduction in Mammals,* book 4: *Reproductive Fitness.* Cambridge, UK: Cambridge University Press.

Lande, R. 1980. Sexual dimorphism, sexual selection, and adaptation in polygenic characters. *Evolution* 34: 292–305.

Lande, R. 1987. Genetic correlations between the sexes in the evolution of sexual dimorphism and mating preferences. In J. W. Bradbury and M. B. Andersson, eds., *Sexual Selection: Testing the Alternatives* (Report of the Dahlem Workshop), 83–94. Chichester, UK: Wiley.

Lande, R. and S. J. Arnold. 1985. Evolution of mating preference and sexual dimorphism. *J. Theor. Biol.* 117: 651–664.

Langmore, N. E., Cockrem, J. F. and Candy, E. J. 2002. Competition for male reproductive investment elevates testosterone levels in female dunnocks, *Prunella modularis. Proc. R. Soc. Lond. B* 269: 2473–2478.

Lank, D. B., Coupe, M. and Wynne-Edwards, K. E. 1999. Testosterone-induced male traits in female ruffs (*Philomachus pugnax*): autosomal inheritance and gender differentiation. *Proc. R. Soc. Lond. B* 266: 2323–2330.

Larson, E. T., Norris, D. O., Grau, E. G., and Summers, C. H. 2003. Monoamines stimulate sex reversal in the saddleback wrasse. *Gen. Comp. Endocrinol.* 130: 289–298.

Lathe, R. 2001. Hormones and the hippocampus. *J. Endocrinol.* 169: 205–231.

Lauay, C., Gerlach, N. M., Adkins-Regan, E. and DeVoogd, T. J. 2004. Female zebra finches require early song exposure to prefer high quality song as adults. *Anim. Behav.,* 68: 1249–1255.

Laudet, V., Hanni, C., Coll, J., Catzeflis, F. and Stehelin, D. 1992. Evolution of the nuclear receptor gene superfamily. *EMBO J.* 11: 1003–1013.

Laughlin, G. A. and Barrett-Connor, E. 2000. Sexual dimorphism in the influence of advanced aging on adrenal hormone levels: The Rancho Bernardo Study. *J. Clin. Endocrinol. Metab.* 85: 3561–3568.

Lea, R. W., Richard-Yris, M. A. and Sharp, P. J. 1996. The effect of ovariectomy on concentrations of plasma prolactin and LH and parental behavior in the domestic fowl. *Gen. Comp. Endocrinol.* 101: 115–121.

LeDoux, J. E. 1995. Emotion: clues from the brain. *Annu. Rev. Psychol.* 46: 209–235.

Lee, A. K. and Cockburn, A. 1985. *Evolutionary Ecology of Marsupials.* New York: Cambridge University Press.

Lee, K. A. and Klasing, K. C. 2004. A role for immunology in invasion biology. *Trends Ecol. Evol.* 19: 523–528.

Lehrman, D. S. 1965. Interaction between internal and external environments in the regulation of the reproductive cycle of the ring dove. In F. A. Beach, ed., *Sex and Behavior,* 344–380. New York: Wiley.

Leitner, S., Nicholson, J., Leisler, B., DeVoogd, T. J. and Catchpole, C. K. 2002. Song and the song control pathway in the brain can develop independently of exposure to song in the sedge warbler. *Proc. R. Soc. Lond. B* 269; 2519–2524.

Leonard, R. B., Coggeshall, R. E. and Willis, W. D. 1978. A documentation of an age-related increase in neuronal and axon numbers in the stingray, *Dasyatis abina*, Le-seuer. *J. Comp. Neurol.* 179: 13–22.

Leonard, S. L. 1939. Induction of singing in female canaries by injections of male hormone. *Proc. Soc. Exp. Biol. Med.* 41: 229–230.

Leopold, A. S. 1944. The nature of heritable wildness in turkeys. *Condor* 46: 133–197.

Lephart, E. D. 1997. Molecular aspects of brain aromatase cytochrome P450. *J. Steroid Biochem. Mol. Biol.* 61: 375–380.

Lerner, D. T. and Mason, R. T. 2001. The influence of sex steroids on the sexual size dimorphism in the red-spotted garter snake, *Thamnophis sirtalis concinnus*. *Gen. Comp. Endocrinol.* 124: 218–225.

Leshner, A. I. 1983. Pituitary-adrenocortical effects on intermale agonistic behavior. In B. B. Svare, ed., *Hormones and Aggressive Behavior*, 27–38. New York: Plenum Press.

Lethimonier, C., Madigou, T., Muñoz-Cueto, J.-A., Lareyre, J.-J. and Kah, O. 2004. Evolutionary aspects of GnRHs, GnRH neuronal systems and GnRH receptors in teleost fish. *Gen. Comp. Endocrinol.* 135: 1–16.

Levin, R. 1996. Song behaviour and reproductive strategies in a duetting wren. *Anim. Behav.* 52: 1093–1106.

Levin, R. N. and Wingfield, J. C. 1992. The hormonal control of territorial aggression in tropical birds. *Ornis Scandinavica* 23: 284–291.

Levine, S. and Mody T. 2003. The long-term psychobiological consequences of inter-mittent postnatal separation in the squirrel monkey. *Neurosci. Biobehav. Rev.* 27: 83–89.

Levine, S. and Mullins, R. F. 1968. Hormones in infancy. In G. Newton, G. and S. Levine, eds., *Early Experience and Behavior: The Psychobiology of Development*, 168–197. Springfield, IL: Charles C. Thomas.

Lévy, F., Kendrick, K. M., Keverne, E. B., Porter, R. H., and Romeyer, A. 1996. Physio-logical, sensory, and experiential factors of parental care in sheep. In J. S. Rosenblatt and C. T. Snowdon, eds., *Parental Care: Evolution, Mechanisms, and Adaptive Sig-nificance (Advances in the Study of Behavior*, vol. 25), 385–422. San Diego: Aca-demic Press.

Lewis, C. M. and Rose, J. D. 2003. Rapid corticosterone-induced impairment of amplectic clasping occurs in the spinal cord of roughskin newts (*Taricha granulosa*). *Horm. Behav.* 43: 93–98.

Licht, P., Frank, L. G., Pavgi, S., Yalcinkaya, T. M., Siiteri, P. K. and Glickman, S. E. 1992. Hormonal correlates of 'masculinization' in female spotted hyaenas (*Crocuta crocuta*), 2: maternal and fetal steroids. *J. Reprod. Fertil.* 95: 463–474.

Ligon, J. D. 1997. A single functional testis as a unique proximate mechanism promot-ing sex-role reversal in coucals. *Auk* 114: 800–801.

Ligon, J. D. 1999. *The Evolution of Avian Breeding Systems*. Oxford, UK: Oxford University Press.

Liley, N. R. and Stacey, N. E. 1983. Hormones, pheromones, and reproductive behavior in fish. In W. S. Hoar and D. J. Randall, eds., *Fish Physiology*, vol. IXB, 1–63. New York: Academic Press.

Lim, M. M., Wang, Z., Olazábal, D. E., Ren, X., Terwilliger, E. F. and Young, L. J. 2004. Enhanced partner preference in a promiscuous species by manipulating the expression of a single gene. *Nature* 429: 754–757.

Liman, E. R. and Innan, H. 2003. Relaxed selective pressure on an essential component of pheromone transduction in primate evolution. *Proc. Natl. Acad. Sci. U S A* 100: 3328–3332.

Lincoln, G. A. and Tyler, N.J.C. 1999. Role of oestradiol in the regulation of the seasonal antler cycle in female reindeer, *Rangifer tarandus*. *J. Reprod. Fertil.* 115: 167–174.

Lincoln, G. A., Guinness, F. and Short, R. V. 1972. The way in which testosterone controls the social and sexual behavior of the red deer stag (*Cervus elaphus*). *Horm. Behav.* 3: 375–396.

Lincoln, G. A., Racey, P. A., Sharp, P. J. and Klandorf, H. 1980. Endocrine changes associated with spring and autumn sexuality in the rook (*Corvus frugilegus*). *J. Zool.* 190: 137–153.

Lindholm, A.and Breden, F. 2002. Sex chromosomes and sexual selection in poeciliid fishes. *Am. Nat.* 160: suppl. S214—S224.

Lindzey, J. and Crews, D. 1988. Effects of progestins on sexual behaviour in castrated lizards (*Cnemidophorus inornatus*). *J. Endocrinol.* 119: 265–273.

Lipar, J. L. and Ketterson, E. D. 2000. Maternally derived yolk testosterone enhances the development of the hatching muscle in the red-winged blackbird *Agelaius phoeniceus*. *Proc. R. Soc. Lond. B* 267: 2005–2010.

Lisk, R. D. and Nachtigall, M. J. 1988. Estrogen regulation of agonistic and proceptive responses in the golden hamster. *Horm. Behav.* 22: 35–48.

Lisk, R. D. and Reuter, L. A. 1980. Relative contributions of oestradiol and progesterone to the maintenance of sexual receptivity in mated female hamsters. *J. Endocrinol.* 87: 175–183.

Lisk, R. D., Ciaccio, L. A. and Caranzaro, C. 1983. Mating behaviour of the golden hamster under seminatural conditions. *Anim. Behav.* 31: 659–666.

Liu, Y., Curtis, T. and Wang, Z. 2001. Vasopressin in the lateral septum regulates pair bond formation in male prairie voles (*Microtus ochrogaster*). *Behav. Neurosci.* 115: 910–919.

Logan, C. A. and Wingfield, J. C. 1990. Autumnal territorial aggression is independent of plasma testosterone in mockingbirds. *Horm. Behav.* 24: 568–581.

Lonstein, J. S. and de Vries, G. J. 1999. Sex differences in the parental behaviour of adult virgin prairie voles: independence from gonadal hormones and vasopressin. *J. Neuroendocrinol.* 11: 441–449.

Lonstein, J. S. and Gammie, S. C. 2002. Sensory, hormonal, and neural control of maternal aggression in laboratory rodents. *Neurosci. Biobehav. Rev.* 26: 869–888.

Lott, D. F. 1991. *Intraspecific Variation in the Social Systems of Wild Vertebrates*. Cambridge, UK: Cambridge University Press.

Lovejoy, D. A. and Balment, R. J. 1999. Evolution and physiology of the corticotropin-releasing factor (CRF) family of neuropeptides in vertebrates. *Gen. Comp. Endocrinol.* 115: 1–22.

Lovell-Badge, R., Canning, C. and Sekido, R. 2002. Sex-determining genes in mice: building pathways. *Novartis Found. Symp.* 244: 4–18.

Lovern, M. B. and Wade, J. 2003. Yolk testosterone varies with sex in eggs of the lizard, *Anolis carolinensis*. *J. Exp. Zool.* 295: 206–210.

Lovern, M. B., McNabb, F.M.A., and Jenssen, T. A. 2001. Developmental effects of testosterone on behavior in male and female green anoles (*Anolis carolinensis*). *Horm. Behav.* 39: 131–143.

Luine, V. N. and Harding, C. F., eds. 1994. *Hormonal Restructuring of the Adult Brain, Basic and Clinical Perspectives*. New York: New York Academy of Sciences.

Luine, V., Nottebohm, F., Harding, C. and McEwen, B. S. 1980. Androgen affects cholinergic enzymes in syringeal motor neurons and muscle. *Brain Res.* 192: 89–107.

Lunn, S. F., Recio, R., Morris, K. and Fraser, H. M. 1994. Blockade of the neonatal rise in testosterone by a gonadotrophin-releasing hormone antagonist: effects on timing of puberty and sexual behaviour in the male marmoset monkey. *J. Endocrinol.* 141: 439–447.

Lynch, J. W., Khan, M. Z., Altmann, J., Njahira, M. N. and Rubenstein, N. 2003. Concentrations of four fecal steroids in wild baboons: short-term storage conditions and consequences for data interpretation. *Gen. Comp. Endocrinol.* 132: 264–271.

Lynn, S. E., Houtman, A. M., Weathers, W. W., Ketterson, E. D. and Nolan, V. 2000. Testosterone increases activity but not daily energy expenditure in captive male dark-eyed juncos, *Junco hyemalis*. *Anim. Behav.* 60: 581–587.

Lyons, H. A. 1969. Respiratory effects of gonadal hormones. In H. A. Salhanick, D. M. Kipnis and R. L. Vande Wiele, eds., *Metabolic Effects of Gonadal Hormones and Contraceptive Steroids*, 394–402. New York: Plenum.

MacDougall-Shackleton, S. A. and Ball, G. F. 1999. Comparative studies of sex differences in the song-control system of songbirds. *Trends Neurosci.* 22: 432–436.

Maddison, D. R. and Maddison, W. P. 2001. *MacClade 4: Analysis of Phylogeny and Character Evolution*. Version 4.03. Sunderland, MA: Sinauer Associates.

Madge, S. and McGowan, P. 2002. *Pheasants, Partridges, and Grouse*. Princeton, NJ: Princeton University Press.

Maggioncalda, A., Sapolsky, R. and Czekala, N. 1999. Reproductive hormone profiles in captive male orangutans: implications for understanding developmental arrest. *Am. J. Physical Anthropol.* 109: 19–32.

Magrath, M.J.L. and Komdeur, J. 2003. Is male care compromised by additional mating opportunity? *Trends Ecol. Evol.* 18: 424–430.

Mahendroo, M. S., Cala, K. M., Landrum, C. P. and Russell, D. W. 1997. Fetal death in mice lacking 5-alpha-reductase type 1 caused by estrogen excess. *Mol. Endocrinol.* 11: 917–927.

Majewska, M. D., Harrison, N. L., Schwartz, R. D., Barker, J. L., and Paul, S. M. 1986. Steroid hormone metabolites are barbiturate-like modulators of the GABA receptor. *Science* 232: 1004–1007.

Makara, G. B. and Haller, J. 2001. Non-genomic effects of glucocorticoids in the neural system: evidence, mechanisms and implications. *Prog. Neurobiol.* 65: 367–390.

Maldonado, T. A., Jones, R. E. and Norris, D. O. 2000. Distribution of beta-amyloid and amyloid precursor protein in the brain of spawning (senescent) salmon: a natural, brain-aging model. *Brain Res.* 858: 237–251.

Malmnas, C. O. 1977. Short-latency effect of testosterone on copulatory behaviour and ejaculation in sexually experienced intact male rats. *J. Reprod. Fertil.* 51: 351–354.

Maney, D. L. and Wingfield, J. C. 1998. Neuroendocrine suppression of female court-ship in a wild passerine: corticotropin-releasing factor and endogenous opioids. *J. Neuroendocrinol.* 10: 593–599.

Maney, D. L., Goode, C. T. and Wingfield, J. C. 1997a. Intraventricular infusion of arginine vasotocin induces singing in a female songbird. *J. Neuroendocrinol.* 9: 487–491.

Maney, D. L., Richardson, R. D. and Wingfield, J. C. 1997b. Central administration of chicken gonadotropin-releasing-hormone-II enhances courtship behavior in a female sparrow. *Horm. Behav.* 32: 11–18.

Maney, D. L., Schoech, S. J., Sharp, P. J. and Wingfield, J. C. 1999. Effects of vasoactive intestinal peptide on plasma prolactin in passerines. *Gen. Comp. Endocrinol.* 113: 323–330.

Mann, D. R., Akinbami, M. A., Gould, K. G., Paul, K. and Wallen, K. 1998. Sexual maturation in male rhesus monkeys: importance of neonatal testosterone exposure and social rank. *J. Endocrinol.* 156: 493–501.

Mansukhani, V., Adkins-Regan, E. and Yang, S. 1996. Sexual partner preference in female zebra finches: the role of early hormones and social environment. *Horm. Behav.* 30: 506–513.

March, J. B., Sharp, P. J., Wilson, P. W. and Sang, H. M. 1994. Effect of active immuni-zation against recombinant-derived chicken prolactin fusion protein on the onset of broodiness and photoinduced egg laying in bantam hens. *J. Reprod. Fertil.* 101: 227–233.

Marin, R. H. and Satterlee, D. G. 2003. Selection for contrasting adrenocortical respon-siveness in Japanese quail (*Coturnix japonica*) influences sexual behaviour in males. *Appl. Anim. Behav. Sci.* 83: 187–199.

Marler, C. A., Walsberg, G., White, M. L. and Moore, M. 1995. Increased energy expen-diture due to increased territorial defense in male lizards after phenotypic manipula-tion. *Behav. Ecol. Sociobiol.* 37: 225–231.

Marler, C. A., Bester-Meredith, J. K. and Trainor, B. C. 2003. Paternal behavior and aggression: endocrine mechanisms and nongenomic transmission of behavior. In P.J.B. Slater, J. S. Rosenblatt, C. T. Snowdon and T. J. Roper, eds., *Advances in the Study of Behavior* vol. 32, 263–323. Amsterdam: Academic Press (Elsevier).

Marler, P. and Tamura, P. 1962. Song "dialects" in three populations of white-crowned sparrows. *Condor* 64: 368–377.

Marler, P., Peters, S., Ball, G. F., Dufty, A. M. and Wingfield, J. C. 1988. The role of sex steroids in the acquisition and production of birdsong. *Nature* 336: 770–772.

Martin, T. 1981. ACTH and brain mechanisms controlling approach-avoidance and imprinting in birds. In J. L. Martinez and R. A. Jensen, eds., *Endogenous Peptides and Learning and Memory Processes*, 99–116. New York: Academic Press.

Mason, P. and Adkins, E. K. 1976. Hormones and social behavior in the lizard, *Anolis carolinensis*. *Horm. Behav.* 7: 75–86.

Massa, R., Davies, D. T., Bottoni, L. and Martini, L. 1979. Photoperiodic control of testosterone metabolism in the central and peripheral structures of avian species. *J. Ster. Biochem.* 11: 937–944.

Maston, G. A. and Ruvolo, M. 2002. Chorionic gonadotropin has a recent origin within primates and an evolutionary history of selection. *Mol. Biol. Evol.* 19: 320–335.

Mathieu, M., Mensah-Nyagan, A. G., Vallarino, M., Do-Rego, J. L., Beaujean, D., Vaudry, D., Luu-The, V., Pelletier, G. and Vaudry, H. 2001. Immunohistochemical localization of 3 beta-hydroxysteroid dehydrogenase and 5 alpha-reductase in the brain of the African lungfish *Protopterus annectens*. *J. Comp. Neurol.* 438: 123–135.

Matsuda, M., Nagahama, Y., Shinomiya, A., Sato, T., Matsuda, C., Kobayashi, T., Morrey, C. E., Shibata, N., Asakawa, S., Shimizu, N., Hori, H., Hamaguchi, S. and Sakaizumi, M. 2002. *DMY* is a Y-specific DM-domain gene required for male development in the medaka fish. *Nature* 417: 559–563.

Matsumoto, A., ed. 1999. *Sexual Differentiation of the Brain*. Boca Raton, FL: CRC Press.

Mayer, A., Lahr, G., Swaab, D. F., Pilgrim, C. and Reisert, I. 1998. The Y-chromosomal genes SRY and ZFY are transcribed in adult human brain. *Neurogenetics* 1: 281–288.

Maynard Smith, J., Burian, R., Kauffman, S., Alberch, P., Campbell, J., Goodwin, B., Lande, R., Raup, D. and Wolpert, L. 1985. Developmental constraints and evolution. *Q. Rev. Biol.* 60: 265–287.

Mazur, A. and Booth, A. 1998. Testosterone and dominance in men. *Behav. Brain Sci.* 21: 353–362.

McCarthy, M. M. 1994. Molecular aspects of sexual differentiation of the rodent brain. *Psychoneuroendocrinology* 19: 415–427.

McClintock, M. K. 2002. Pheromones, odors, and vasanas: the neuroendocrinology of social chemosignals in humans and animals. In D. W. Pfaff, A. P. Arnold, A. M. Etgen, S. E. Fahrbach and R. T. Rubin, eds., *Hormones, Brain and Behavior*, vol. 1, 797–870. Amsterdam: Academic Press (Elsevier).

McCollom, R. E., Siegel, P. B. and Van Krey, H. P. 1971. Responses to androgen in lines of chickens selected for mating behavior. *Horm. Behav.* 2: 31–42.

McCormick, M. I. 1998. Behaviorally induced maternal stress in a fish influences progeny quality by a hormonal mechanism. *Ecology* 79: 1873–1883.

McDonald, I. R., Lee, A. K., Bradley, A. J. and Than, K. A. 1981. Endocrine changes in dasyurid marsupials with differing mortality patterns. *Gen. Comp. Endocrinol.* 44: 292–301.

McDonald, I. R., Lee, A. K., Than, K. A. and Martin, R. W. 1986. Failure of glucocorticoid feedback in males of a population of small marsupials (*Antechinus swainsonii*) during the period of mating. *J. Endocrinol.* 108: 63–68.

McGary, S., Estevez, I., Bakst, M. R. and Pollock, D. L. 2002. Phenotypic traits as reliable indicators of fertility in male broiler breeders. *Poultry Sci.* 81: 102–111.

McGaugh, J. L. 1989. Involvement of hormonal and neuromodulatory systems in the regulation of memory storage. *Annu. Rev. Neurosci.* 12: 255–287.

McGill, T. E. 1978. Genotype-hormone interactions. In T. E. McGill, D. A. Dewsbury and B. D. Jinks, eds., *Sex and Behavior: Status and Prospectus*, 161–187. New York: Plenum Press.

McGinnis, M. Y., Marcelli, M. and Lamb, D. J. 2002. Consequences of mutations in androgen receptor genes: molecular biology and behavior. In D. W. Pfaff, A. P. Arnold, A. M. Etgen, S. E. Fahrbach and R. T. Rubin, eds., *Hormones, Brain and Behavior*, vol. 5, 347–379. Amsterdam: Academic Press (Elsevier).

McGraw, K. and Ardia, D. R. 2003. Carotenoids, immunocompetence, and the information content of sexual colors: an experimental test. *Am. Nat.* 162: 704–712.

Meaney, M. J. 2001. Maternal care, gene expression, and the transmission of individual differences in stress reactivity across generations. *Annu. Rev. Neurosci.* 24: 1161–1192.

Meaney, M. J., Stewart, J. and Beatty, W. W. 1985. Sex differences in social play: the socialization of sex roles. In *Advances in the Study of Behavior*, vol. 15, 1–58. Orlando, FL: Academic Press.

Medawar, P. 1952. *An Unsolved Problem of Biology.* London: H. K. Lewis.

Meeks, J. J., Weiss, J. and Jameson, J. L. 2003. Dax1 is required for testis determination. *Nature Genet.* 34: 32–33.

Meikle, A. W., Stringham, J. D., Bishop, D. T. and West, D. W. 1988. Quantitating genetic and nongenetic factors influencing androgen production and clearance rates in men. *J. Clin. Endocrinol. Metab.* 67: 104–109.

Meisel, R. L. and Sachs, B. D. 1994. The physiology of male sexual behavior. In E. Knobil and J. D. Neill, eds., *The Physiology of Reproduction*, 2nd ed., vol. 2, 3–106. New York: Raven.

Mellon, S. H. and Griffin, L. D. 2002. Neurosteroids: biochemistry and clinical significance. *Trends Endocrinol. Metab.* 13: 35–43.

Melrose, D. R., Reed, H. C. and Patterson, R. L. 1971. Androgen steroids associated with boar odour as an aid to the detection of oestrus in pig artificial insemination. *Br. Vet. J.* 127: 497–502.

Mendonça, M. T. 1987. Timing of reproductive behaviour in male musk turtles, *Sternotherus odoratus*: effects of photoperiod, temperature, and testosterone. *Anim. Behav.* 35: 1002–1014.

Mendonça, M. T., Chernetsky, S. D., Nester, K. E. and Gardner, G. L. 1996. Effects of gonadal sex steroids on sexual behavior in the big brown bat, *Eptesicus fuscus*, upon arousal from hibernation. *Horm. Behav.* 30: 153–161.

Mendoza, S. P. and Mason, W. A. 1986. Contrasting responses to intruders and to involuntary separation by monogamous and polygynous New World monkeys. *Physiol. Behav.* 38: 795–801.

Meng, Q., Liu, J., Varricchio, D. J., Huang, T. and Gao, C. 2004. Parental care in an ornithischian dinosaur. *Nature* 431: 145–146.

Menuet, A., Anglade, I., Le Guevel, R., Pellegrini, E., Pakdel, F. and Kah, O. 2003. Distribution of aromatase mRNA and protein in the brain and pituitary of female rainbow trout: comparison with estrogen receptor alpha. *J. Comp. Neurol.* 462: 180–193.

Meredith, M. 2001. Human vomeronasal organ function: a critical review of best and worst cases. *Chem. Senses* 26: 433–445.

Merker, G. P. and Nagy, K. A. 1984. Energy utilization by free-ranging *Sceloporus virgatus* lizards. *Ecology* 65: 575–581.

Metzdorf, R., Gahr, M. and Fusani, L. 1999. Distribution of aromatase, estrogen receptor, and androgen receptor mRNA in the forebrain of songbirds and nonsongbirds. *J. Comp. Neurol.* 407: 115–129.

Meyer, A. 1999. Homology and homoplasy: the retention of genetic programmes. *Novartis Found. Symp.* 222: 141–153.

Meyer, A. and Zardoya, R. 2003. Recent advances in the (molecular) phylogeny of vertebrates. *Annu. Rev. Ecol. Evol. Syst.* 34: 311–338.

Meyer-Bahlburg, H.F.L. and Ehrhardt, A. A. 1982. Prenatal sex hormones and human aggression: a review, and new data on progestogen effects. *Aggr. Behav.* 8: 39–62.

Micevych, P. E. and Hammer, R. P., eds. 1995. *Neurobiological Effects of Sex Steroid Hormones.* Cambridge, UK: Cambridge University Press.

Michael, R. P. and Zumpe, D. 1993. A review of hormonal factors influencing the sexual and aggressive behavior of macaques. *Am. J. Primatol.* 30: 213–241.

Miczek, K. A., Maxson, S. C., Fish, E. W. and Faccidomo, S. 2001. Aggressive behavioral phenotypes in mice. *Behav. Brain Res.* 125:167–181.

Midgley, A. R., Niswender, G. D., and Rebar, R. W. 1969. Principles for the assessment of radioimmunoassay methods (precision, accuracy, sensitivity, specificity). In A. Diczfalusy, ed., *Karolinska Symposia on Research Methods in Reproductive Endocrinology*, first symposium, 163–180. Stockholm: Karolinska Institute.

Millar, R. P. 2003. GnRH II and type II GnRH receptors. *Trends Endocrinol. Metab.* 14: 35–43.

Milligan, S. R. 1982. Induced ovulation in mammals. In C. A. Finn, ed., *Oxford Reviews in Reproductive Biology*, 1–46. London: Clarendon Press.

Mindell, D. P. and Meyer, A. 2001. Homology evolving. *Trends Ecol. Evol.* 16: 434–440.

Miranda, J. A., Oliveira, R. F., Carneiro, L. A., Santos, R. S. and Grober, M. S. 2003. Neurochemical correlates of male polymorphism and alternative reproductive tactics in the Azorean rock-pool blenny, *Parablennius parvicornis. Gen. Comp. Endocrinol.* 132: 183–189.

Mittwoch, U. 1998. Phenotypic manifestations during the development of the dominant and default gonads in mammals and birds. *J. Exp. Zool.* 281: 466–471.

Mittwoch, U. 2000. Genetics of sex determination: exceptions that prove the rule. *Mol. Genet. Metab.* 71: 405–410.

Mock, D. W. and Fujioka, M. 1990. Monogamy and long-term pair bonding in vertebrates. *Trends Ecol. Evol.* 5: 39–43.

Moffatt, C. A. 2003. Steroid hormone modulation of olfactory processing in the context of socio-sexual behaviors in rodents and humans. *Brain Res. Rev.* 43: 192–206.

Møller, A. P. and Briskie, J. V. 1995. Extra-pair paternity, sperm competition and the evolution of testis size in birds. *Behav. Ecol. Sociobiol.* 36: 357–365.

Molnár, Z. and Butler, A. B. 2002. The corticostriatal junction: a crucial region for forebrain development and evolution. *Bioessays* 24: 530–541.

Money, J. 1987. Propaedeutics of diecious G-I/R: theoretical foundations for understanding dimorphic gender-identity/role. In J. M. Reinisch, L. A. Rosenblum and S. A. Sanders, eds., *Masculinity/Femininity: Basic Perspectives*, 13–34. New York: Oxford University Press.

Moore, C. L. 1995. Maternal contributions to mammalian reproductive development and the divergence of males and females. In P.J.B. Slater, J. S. Rosenblatt, C. T. Snowdon and M. Milinski, eds., *Advances in the Study of Behavior*, vol. 24, 47–118. San Diego: Academic Press.

Moore, C. L. and Rogers, S. A. 1984. Contribution of self-grooming to onset of puberty in male rats. *Dev. Psychobiol.* 17: 243–253.

Moore, F. L. and Evans, S.J. 1999. Steroid hormones use non-genomic mechanisms to control brain functions and behaviors: a review of evidence. *Brain Behav. Evol.* 54: 41–50.

Moore, F. L. and Lowry, C. A. 1998. Comparative neuroanatomy of vasotocin and vasopressin in amphibians and other vertebrates. *Comp. Biochem. Physiol.* C 119: 251–260.

Moore, F. L. and Orchinik, M. 1991. Multiple molecular actions for steroids in the regulation of reproductive behaviors. *Semin. Neurosci.* 3: 489–496.

Moore, F. L. and Zoeller, R. T. 1979. Endocrine control of amphibian sexual behavior: evidence for a neurohormone–androgen interaction. *Horm. Behav.* 13: 207–213.

Moore, F. L., Roberts, J. and Bevers, J. 1984. Corticotropin-releasing factor (CRF) stimulates locomotor activity in intact and hypophysectomized newts (Amphibia). *J. Exp. Zool.* 213: 331–333.

Moore, I. T. and Jessop, T. S. 2003. Stress, reproduction, and adrenocortical modulation in amphibians and reptiles. *Horm. Behav.* 43: 39–47.

Moore, I. T., Lemaster, M. P. and Mason, R. T. 2000. Behavioural and hormonal responses to capture stress in the male red-sided garter snake, *Thamnophis sirtalis parietalis*. *Anim. Behav.* 59: 529–534.

Moore, I. T., Perfito, N., Wada, H., Sperry, T. S. and Wingfield, J. C. 2002. Latitudinal variation in plasma testosterone levels in birds of the genus *Zonotrichia*. *Gen. Comp. Endocrinol.* 129: 13–19.

Moore, M. C. 1987. Castration affects territorial and sexual behaviour of free-living male lizards, *Sceloporus jarrovi*. *Anim. Behav.* 35: 1193–1199.

Moore, M. C. 1988. Testosterone control of territorial behavior: tonic-release implants fully restore seasonal and short-term aggressive responses in free-living castrated lizards. *Gen. Comp. Endocrinol.* 70: 450–459.

Moore, M. C. 1991. Application of organization–activation theory to alternative male reproductive strategies: a review. *Horm. Behav.* 25: 154–179.

Moore, M. C. and Crews, D. 1986. Sex steroid hormones in natural populations of a sexual whiptail lizard *Cnemidophorus inornatus*, a direct evolutionary ancestor of a unisexual parthenogen. *Gen. Comp. Endocrinol.* 63: 424–430.

Moore, M. C. and Kranz, R. 1983. Evidence for androgen independence of male mounting behavior in white-crowned sparrows (*Zonotrichia leucophrys gambelii*). *Horm. Behav.* 17: 414–423.

Moore, M. C. and Lindzey, J. 1992. The physiological basis of sexual behavior in male reptiles. In C. Gans and D. Crews, eds., *Biology of the Reptilia*, vol. 18, Physiology E: *Hormones, Brain, and Behavior*, 70–113. Chicago: University of Chicago Press.

Moore, M. C. and Marler, C. A. 1988. Hormones, behavior, and the environment: an evolutionary perspective. In M. H. Stetson, ed., *Processing of Environmental Information in Vertebrates*, 71–84. New York: Springer-Verlag.

Moore, M. C., Hews, D. K. and Knapp, R. 1998. Hormonal control and evolution of alternative male phenotypes: generalizations of models for sexual differentiation. *Am. Zool.* 38: 133–151.

Moreau, J., Bertin, A., Caubet, Y. and Rigaud, T. 2001. Sexual selection in an isopod with *Wolbachia*-induced sex reversal: males prefer real females. *J. Evol. Biol.* 14: 388–394.

Moreno, J., Veiga, J. P., Cordero, P. J. and Minguez, E. 1999. Effects of parental care on reproductive success in the polygynous spotless starling *Sturnus unicolor*. *Behav. Ecol. Sociobiol.* 47: 47–53.

Mormède, P., Courvoisier, H., Ramos, A., Marissal-Arvy, N., Ousova, O., Désautés, C., Duclos, M., Chaouloff, F. and Moisan, M.-P. 2002. Molecular genetic approaches to investigate individual variations in behavioral and neuroendocrine stress responses. *Psychoneuroendocrinology* 27: 563–583.

Morrell, J. I. and Pfaff, D. W. 1981. Autoradiographic technique for steroid hormone localization: application to the vertebrate brain. In N. T. Adler, ed., *Neuroendocrinology of Reproduction: Physiology and Behavior*, 519–532. New York: Plenum Press.

Morris, D. J. and van Aarde, R. J. 1985. Sexual behavior of the female porcupine *Hystrix africaeaustralis*. *Horm. Behav.* 19: 400–412.

Morris, N. M. and Udry, J. R. 1982. Epidemiological patterns of sexual behavior in the menstrual cycle. In R. C. Friedman, ed., *Behavior and the Menstrual Cycle*. New York: Marcel Dekker.

Morrish, B. C. and Sinclair, A. H. 2002. Vertebrate sex determination: many means to an end. *Reproduction* 124: 447–457.

Moss, R. L., McCann, S. M. and Dudley, C. A. 1975. Releasing hormones and sexual behavior. *Prog. Brain Res.* 42: 37–46.

Moss, R., Parr, R. and Lambin, X. 1994. Effects of testosterone on breeding density, breeding success and survival of red grouse. *Proc. R. Soc. Lond. B* 258: 175–180.

Mougeot, F., Redpath, S. M., Leckis, F. and Hudson, P. J. 2003. The effect of aggressiveness on the population dynamics of a territorial bird. *Nature* 421: 737–739.

Moyer, K. E. 1976. *The Psychobiology of Aggression*. New York: Harper & Row.

Muller, M. N. and Wrangham, R. 2002. Sexual mimicry in hyenas. *Q. Rev. Biol.* 77: 3–16.

Müller, W., Eising, C. M., Dijkstra, C. and Groothuis, T.G.G. 2002. Sex differences in yolk hormones depend on maternal social status in Leghorn chickens (*Gallus gallus domesticus*). *Proc. R. Soc. Lond. B* 269: 2249–2255.

Mumme, R. L. 1997. A bird's eye view of mammalian cooperative breeding. In N. G. Solomon, J. A. French, eds., *Cooperative Breeding in Mammals*, 364–388. Cambridge, UK: Cambridge University Press.

Muñoz, R. C. and Warner, R. R. 2003. A new version of the size-advantage hypothesis for sex change: incorporating sperm competition and size-fecundity skew. *Am. Nat.* 161: 749–761.

Murphy, W. J., Eizirik, E., Johnson, W. E., Zhang, Y. P., Ryder, O. A. and O'Brien, S. J. 2001. Molecular phylogenetics and the origins of placental mammals. *Nature* 409: 614–618.

Muske, L. E. 1993. Evolution of gonadotropin-releasing hormone (GnRH) neuronal systems. *Brain Behav. Evol.* 42: 215–230.

Myers, L. S. 1995. Methodological review and meta-analysis of sexuality and menopause research. *Neurosci. Biobehav. Rev.* 19: 331–341.

Naftolin, F. and MacLusky, N. 1984. Aromatization hypothesis revisited. In M. Serio, M. Motta, M. Zanisi and L. Martini, eds., *Sexual Differentiation: Basic and Clinical Aspects*, 79–92. New York: Raven.

Naftolin, F., Horvath, T. L. and Balthazart, J. 2001. Estrogen synthetase (aromatase) immunohistochemistry reveals concordance between avian and rodent limbic systems and hypothalami. *Exp. Biol. Med.* 226: 717–725.

Naisse, J. 1966. Contrôle endocrinien de la différenciation sexuelle chez *Lampyris noctiluca* (Coléoptère Lampyride). *Gen. Comp. Endocrinol.* 7: 85.

Nakashima, Y., Kuwamura, T. and Yogo, Y. 1995. Why be a both-ways sex changer? *Ethology* 101: 301–307.

Nanda, I., Haaf, T., Schartl, M., Schmid, M. and Burt, D. W. 2002. Comparative mapping of Z-orthologous genes in vertebrates: implications for the evolution of avian sex chromosomes. *Cytogenet. Genome Res.* 99: 178–184.

Nealen, P. M. and Perkel, D. J. 2000. Sexual dimorphism in the song system of the Carolina wren *Thryothorus ludovicianus*. *J. Comp. Neurol.* 418: 346–360.

Negri-Cesi, P., Poletti, A., Martini, L. and Piva, F. 2000. Steroid metabolism in the brain: role in sexual differentiation. In A. Matsumoto, ed., *Sexual Differentiation of the Brain*, 33–58. Boca Raton, FL: CRC Press.

Nelson, E. E. and Panksepp, J. 1998. Brain substrates of infant-mother attachment: contributions of opioids, oxytocin, and norepinephrine. *Neurosci. Biobehav. Rev.* 22: 437–452.

Nelson, R. J. 1997. The use of genetic "knockout" mice in behavioral endocrinology research. *Horm. Behav.* 31: 188–197.

Nelson, R. J. 2005. *An Introduction to Behavioral Endocrinology*, 3rd ed. Sunderland, MA: Sinauer.

Nelson, R. J. and Demas, G. E. 1996. Seasonal changes in immune function. *Q. Rev. Biol.* 71: 511–548.

Nelson, R. J. and Klein, S. L. 2000. Environmental and social influences on seasonal breeding and immune function. In K. Wallen and J. E. Schneider, eds., *Reproduction in Context: Social and Environmental Influence on Reproductive Physiology and Behavior*, 219–256. Cambridge, MA: MIT Press.

Nelson, R. J., Demas, G. E., Klein, S. L. and Kriegsfeld, L. J. 2002. *Seasonal Patterns of Stress, Immune Function, and Disease*. Cambridge, UK: Cambridge University Press.

Newman, S. W. 2002. Pheromonal signals access the medial extended amygdala: one node in a proposed social behavior network. In D. W. Pfaff, A. P. Arnold, A. M. Etgen, S. E. Fahrbach and R. T. Rubin, eds., *Hormones, Brain and Behavior*, vol. 2, 17–32. Amsterdam: Academic Press (Elsevier).

Newton, I., ed. 1989. *Lifetime Reproduction in Birds*. London: Academic Press.

Nichols, E. H., Burke, T. A. and Birkhead, T. R. 2001. Ejaculate allocation by male sand martins *Riparia riparia*. *Proc. R. Soc. Lond. B* 268: 1265–1270.

Nijhout, H. F. 1994. *Insect Hormones*. Princeton, NJ: Princeton University Press.

Nisbet, I. C., Finch, C. E., Thompson, N., Russek-Cohen, E., Proudman, J. A. and Ottinger, M. A. 1999. Endocrine patterns during aging in the common tern (*Sterna hirundo*). *Gen. Comp. Endocrinol.* 114: 279–286.

Nock, B. L. and Leshner, A. I. 1976. Hormonal mediation of the effects of defeat on agonistic responding in mice. *Physiol. Behav.* 17: 111–119.

Noden, D. M. 1992. Vertebrate craniofacial development: novel approaches and new dilemmas. *Curr. Opin. Genet. Dev.* 2: 576–581.

Nol, E., Cheng, K. and Nichols, C. 1996. Heritability and phenotypic correlations of behaviour and dominance rank of Japanese quail. *Anim. Behav.* 52: 813–820.

Nolan, V., Ketterson, E. D., Ziegenfus, C., Cullen, D. P. and Chandler, C. R. 1992. Testosterone and avian life histories: effects of experimentally elevated testosterone on prebasic molt and survival in male dark-eyed juncos. *Condor* 94: 364–370.

Nordeen, E. J. and Nordeen, K. W. 1989. Estrogen stimulates the incorporation of new neurons into avian song nuclei during adolescence. *Dev. Brain Res.* 49: 27–32.

Nordeen, E. J. and Nordeen, K. W. 1994. Hormonally-regulated neuron death in the avian brain. *Semin. Neurosci.* 6: 299–306.

Norris, D. O. 1996. *Vertebrate Endocrinology*, 3rd ed. San Diego: Academic Press.

Northcutt, R. G. 2002. Understanding vertebrate brain evolution. *Integr. Comp. Biol.* 42: 743–756.

Nottebohm, F. 1980. Testosterone triggers growth of brain vocal control nuclei in adult female canaries. *Brain Res.* 189: 429–436.

Nottebohm, F. 1999. The anatomy and timing of vocal learning in birds. In M. D. Hauser and M. Konishi, eds., *The Design of Animal Communication*. Cambridge, MA: MIT Press.

Nottebohm, F. and Arnold, A. P. 1976. Sexual dimorphism in the vocal control areas in the song bird brain. *Science* 194: 211–213.

Nottebohm, F., Stokes, T. M. and Leonard, C. M. 1976. Central control of song in the canary, *Serinus canarius. J. Comp. Neurol.* 165: 457–486.

Novacek, M. J. 1992. Mammalian phylogeny: shaking the tree. *Nature* 356: 121–125.

Novotny, M. V., Ma, W. D., Wiesler, D. and Zidek, L. 1999. Positive identification of the puberty-accelerating pheromone of the house mouse: the volatile ligands associating with the major urinary protein. *Proc. R. Soc. Lond. B* 266: 2017–2022.

Nowicki, S., Searcy, W. A. and Peters, S. 2002a. Brain development, song learning and mate choice in birds: a review and experimental test of the "nutritional stress hypothesis." *J. Comp. Physiol. A* 188: 1003–1014.

Nowicki, S., Searcy, W. A. and Peters, S. 2002b. Quality of song learning affects female response to male bird song. *Proc. R. Soc. Lond. B* 269: 1949–1954.

Numan, M. 1994. Maternal behavior. In E. Knobil and J. D. Neill, eds., *The Physiology of Reproduction*, 2nd ed., vol. 2, 221–302. New York: Raven.

Nunez, J. L., Huppenbauer, C. B., McAbee, M. D., Juraska, J. M. and DonCarlos, L. L. 2003. Androgen receptor expression in the developing male and female rat visual and prefrontal cortex. *J. Neurobiol.* 56: 293–302.

Nunn, C. L., van Schaik, C. P. and Zinner, D. 2001. Do exaggerated sexual swellings function in female mating competition in primates? A comparative test of the reliable indicator hypothesis. *Behav. Ecol.* 12: 646–654.

Ogata, T. and Matsuo, N. 1993. Sex chromosome aberrations and stature: Deduction of the principal factors involved in the determination of adult height. *Hum. Genet.* 91: 551–562.

Ogawa, S. and Pfaff, D. W. 2000. Genetic contributions to the sexual differentiation of behavior. In A. Matsumoto, ed., *Sexual Differentiation of the Brain*, 11–20. Boca Raton, FL: CRC Press.

Ohno, S. 1970. *Evolution by Gene and Genome Duplication*. Berlin: Springer.

Ohno, S., Wolf, U., and Atkin, N. B. 1968. Evolution from fish to mammals by gene duplication. *Hereditas* 59: 169–187.

Ohtani, H., Miura, I., Hanada, H. and Ichikawa, Y. 2000. Alteration of the sex determining system resulting from structural change of the sex chromosomes in the frog *Rana rugosa. J. Exp. Zool.* 286: 313–319.

Ojeda, S. R. and Terasawa, E. 2002. Neuroendocrine regulation of puberty. In D. W. Pfaff, A. P. Arnold, A. M. Etgen, S. E. Fahrbach and R. T. Rubin, eds., *Hormones, Brain and Behavior*, vol. 4, pp. 589–660. Amsterdam: Academic Press (Elsevier).

Oliveira, R. F., Almada, V. C. and Canario, A. V. 1996. Social modulation of sex steroid concentrations in the urine of male cichlid fish *Oreochromis mossambicus*. *Horm. Behav.* 30: 2–12.

Oliveira, R. F., Canario, A.V.M. and Grober, M. S. 2001a. Male sexual polymorphism, alternative reproductive tactics, and androgens in combtooth blennies (Pisces: Blenniidae). *Horm. Behav.* 40: 266–275.

Oliveira, R. F., Carneiro, L. A., Canario, A. V. and Grober, M. S. 2001b. Effects of androgens on social behavior and morphology of alternative reproductive males of the Azorean rock-pool blenny. *Horm. Behav.* 39: 157–166.

Oliveira, R. F., Carneiro, L. A., Gonçalves, D. M., Canario, A. V. and Grober, M. S. 2001c. 11-Ketotestosterone inhibits the alternative mating tactic in sneaker males of the peacock blenny, *Salaria pavo. Brain Behav. Evol.* 58: 28–37.

Oliveira, R. F., Hirschenhauser, K., Carneiro, L. A. and Canario, A. V. 2002. Social modulation of androgen levels in male teleost fish. *Comp. Biochem. Physiol. B* 132: 203–215.

Olmstead, A. W. and Leblanc, G. A. 2002. Juvenoid hormone methyl farnesoate is a sex determinant in the crustacean *Daphnia magna. J. Exp. Zool.* 293: 736–739.

Olsson, L., Falck, P., Lopez, K., Cobb, J. and Hanken, J. 2001. Cranial neural crest cells contribute to connective tissue in cranial muscles in the anuran amphibian, *Bombina orientalis. Dev. Biol.* 237: 354–367.

Olsson, M. 2001. "Voyeurism" prolongs copulation in the dragon lizard *Ctenophorus fordi. Behav. Ecol. Sociobiol.* 50: 378–381.

Orchard, I., Ramirez, J.-M. and Lange, A. B. 1993. A multifunctional role for octopamine in locust flight. *Annu. Rev. Entomol.* 38: 227–249.

Orchinik, M. 1998. Glucocorticoids, stress, and behavior: shifting the timeframe. *Horm. Behav.* 34: 320–327.

Orchinik, M., Licht, P. and Crews, D. 1988. Plasma steroid concentrations change in response to sexual behavior in *Bufo marinus. Horm. Behav.* 22: 338–350.

Orchinik, M., Gasser, P. and Breuner, C. 2002. Rapid corticosteroid actions on behavior: cellular mechanisms and organismal consequences. In D. W. Pfaff, A. P. Arnold, A. M. Etgen, S. E. Fahrbach and R. T. Rubin, eds., *Hormones, Brain and Behavior*, vol. 3, pp. 567–600. Amsterdam: Academic Press (Elsevier).

Ortman, L. L. and Craig, J. V. 1968. Social dominance in chickens modified by genetic selection–physiological mechanisms. *Anim. Behav.* 16: 33–37.

Oster, G. F., Shubin, N., Murray, J. D. and Alberch, P. 1988. Evolution and morphogenetic rules: the shape of the vertebrate limb in ontogeny and phylogeny. *Evolution* 42: 862–884.

Ottaviani, E. and Franceschi, C. 1996. The neuroimmunology of stress from invertebrates to man. *Progr. Neurobiol.* 48: 421–440.

Ottinger, M. A. 1983. Short term variation in serum luteinizing hormone and testosterone in the male Japanese quail. *Poultry Sci.* 62: 908–913.

Ottinger, M. A. 1992. Altered neuroendocrine mechanisms during reproductive aging. *Poultry Sci. Rev.* 4: 235–248.

Ottinger, M. A. and Bakst, M. R. 1981. Peripheral androgen concentrations and testicular morphology in embryonic and young male Japanese quail. *Gen. Comp. Endocrinol.* 43: 170–177.

Ottinger, M. A. and Balthazart, J. 1986. Altered endocrine and behavioral responses with reproductive aging in the male Japanese quail. *Horm. Behav.* 20: 83–94.

Ottinger, M. A. and vom Saal, F. S. 2002. Impact of environmental endocrine disruptors on sexual differentiation in birds and mammals. In D. W. Pfaff, A. P. Arnold, A. M. Etgen, S. E. Fahrbach and R. T. Rubin, eds., *Hormones, Brain and Behavior*, vol. 4, pp. 325–384. Amsterdam: Academic Press (Elsevier).

Ottinger, M. A., Nisbet, I.C.T. and Finch, C. E. 1995. Aging and reproduction: comparative endocrinology of the common tern and Japanese quail. *Am. Zool.* 35: 299–306.

Ottinger, M. A., Pitts, S. and Abdelnabi, M. A. 2001. Steroid hormones during embryonic development in Japanese quail: plasma, gonadal, an adrenal levels. *Poultry Sci.* 80: 795–799.

Øverli, Ø., Harris, C. A. and Winberg, S. 1999. Short-term effects of fights for social dominance and the establishment of dominant-subordinate relationships on brain monoamines and cortisol in rainbow trout. *Brain Behav. Evol.* 54: 263–275.

Owens, I.P.F. and Short, R. V. 1995. Hormonal basis of sexual dimorphism in birds: implications for new theories of sexual selection. *Trends Ecol. Evol.* 10: 44–47.

Packer, C., Tatar, M. and Collins, A. 1998. Reproductive cessation in female mammals. *Nature* 392: 807–811.

Palanza, P., Morellini, F., Parmigiani, S. and vom Saal, F. S. 1999. Prenatal exposure to endocrine disrupting chemicals: effects on behavioral development. *Neurosci. Biobehav. Rev.* 23: 1011–1027.

Páll, M. K., Mayer, I. and Borg, B. 2002a. Androgen and behavior in the male three-spined stickleback, *Gasterosteus aculeatus*, I: changes in 11-ketotestosterone levels during the nesting cycle. *Horm. Behav.* 41: 377–385.

Páll, M. K., Mayer, I. and Borg, B. 2002b. Androgen and behavior in the male three-spined stickleback, *Gasterosteus aculeatus*, II: castration and 11-ketotestosterone effects on courtship and parental care during the nesting cycle. *Horm. Behav.* 42: 337–344.

Panzica, G. C., García-Ojeda, E., Viglietti-Panzica, C., Thompson, N. E. and Ottinger, M. A. 1996a. Testosterone effects on vasotocinergic innervation of sexually dimorphic medial preoptic nucleus and lateral septum during aging in male quail. *Brain Res.* 712: 190–198.

Panzica, G. C., Viglietti-Panzica, C. and Balthazart, J. 1996b. The sexually dimorphic medial preoptic nucleus of quail: a key brain area mediating steroid action on male sexual behavior. *Front. Neuroendocrinol.* 17: 51–125.

Panzica, G. C., García-Ojeda, E., Viglietti-Panzica, C., Aste, N. and Ottinger, M. A. 1997. Role of testosterone in the activation of sexual behavior and neuronal circuitries in the senescent brain. In G. Filogamo, A. Vernadakis, F. Gremo, A. M. Privat, and P. S. Timiras, eds., *Brain Plasticity: Development and Aging*, 273–287. New York: Plenum Press.

Panzica, G. C., Castagna, C., Viglietti-Panzica, C., Russo, C., Tlemçani, O. and Balthazart, J. 1998. Organizational effects of estrogens on brain vasotocin and sexual behavior in quail. *J. Neurobiol.* 37: 684–699.

Panzica, G. C., Aste, N., Castagna, C., Viglietti-Panzica, C. and Balthazart, J. 2001. Steroid-induced plasticity in the sexually dimorphic vasotocinergic innervation of the avian brain: behavioral implications. *Brain Res. Rev.* 37: 178–200.

Parker, K. J., Phillips, K. M., Kinney, L. F., and Lee, T. M. 2001. Day length and sociosexual cohabitation alter central oxytocin receptor binding in female meadow voles (*Microtus pennsylvanicus*). *Behav. Neurosci.* 115: 1349–1356.

Parker, T. H. and Ligon, J. D. 2003. Female mating preferences in red junglefowl: a meta-analysis. *Ethol. Ecol. Evol.* 15: 63–72.

Parker, T. H., Knapp, R. and Rosenfield, J. A. 2002. Social mediation of sexual selected ornamentation and steroid hormone levels in male junglefowl. *Anim. Behav.* 64: 291–298.

Parmigiani, S., Ferrari, P. F. and Palanza, P. 1998. An evolutionary approach to behavioral pharmacology: using drugs to understand proximate and ultimate mechanisms of different forms of aggression in mice. *Neurosci. Biobehav. Rev.* 23: 143–153.

Partridge, L. 1989. Lifetime reproductive success and life-history evolution. In I. Newton, ed., *Lifetime Reproduction in Birds*, 430–440. London: Academic Press.

Partridge, L. 2001. Evolutionary theories of ageing applied to long-lived organisms. *Exp. Geront.* 36: 641–650.

Pasmanik, M. and Callard, G. V. 1988. A high abundance androgen receptor in goldfish brain: characteristics and seasonal changes. *Endocrinology* 123: 1162–1171.

Paton, J. A., Manogue, K. R. and Nottebohm, F. 1981. Bilateral organization of the vocal control pathway in the budgerigar, *Melopsittacus undulatus*. *J. Neurosci.* 1: 1279–1288.

Pedersen, C. A. 1998. Oxytocin control of maternal behavior: regulation by sex steroids and offspring stimuli. In C. S. Carter, B. Kirkpatrick and I. I. Lederhendler, eds., *The Integrative Neurobiology of Affiliation*, 301–320. Cambridge, MA: MIT Press.

Pedersen, H. C. 1989. Effects of exogenous prolactin on parental behaviour in free-living female willow ptarmigan *Lagopus l. lagopus*. *Anim. Behav.* 38: 926–934.

Pener, M. P. 1991. Locust phase polymorphism and its endocrine relations. *Adv. Insect Physiol.* 23: 1–79.

Penna, M., Capranica, R. R. and Somers, J.1992. Hormone-induced vocal behavior and midbrain auditory sensitivity in the green treefrog, *Hyla cinerea*. *J. Comp. Physiol.* [A] 170: 73–82.

Penton-Voak, I. S. and Perrett, D. I. 2001. Male facial attractiveness: perceived personality and shifting female preferences for male traits across the menstrual cycle. In P.J.B. Slater, J. S. Rosenblatt, C. T. Snowdon and T. J. Roper, eds., *Advances in the Study of Behavior*, 219–260. San Diego: Academic Press.

Perkins, A., Fitzgerald, J. A. and Moss, G. E. 1995. A comparison of LH secretion and brain estradiol receptors in heterosexual and homosexual rams and female sheep. *Horm. Behav.* 29: 31–41.

Perlman, W. R., Ramachandran, B. and Arnold, A. P. 2003. Expression of androgen receptor mRNA in the late embryonic and early posthatch zebra finch brain. *J. Comp. Neurol.* 455: 513–530.

Perrot-Sinal, T. S., Auger, A. P. and McCarthy, M. M. 2003. Excitatory actions of GABA in developing brain are mediated by l-type Ca^{2+} channels and dependent on age, sex, and brain region. *Neuroscience* 116: 995–1003.

Perry, A. N. and Grober, M. S. 2003. A model for social control of sex change: interactions of behavior, neuropeptides, glucocorticoids, and sex steroids. *Horm. Behav.* 43: 31–38.

Peter, R. E. 1983. Evolution of neurohormonal regulation of reproduction in lower vertebrates. *Am. Zool.* 23: 685–695.

Peter, R. E., Yu, K.-L., Marchant, T. A. and Rosenblum, P. M. 1990. Direct neural regulation of the teleost adenohypophysis. *J. Exp. Zool.* suppl. 4: 84–89.

Peters, A. 2000. Testosterone treatment is immunosuppressive in superb fairy-wrens, yet free-living males with high testosterone are more immunocompetent. *Proc. R. Soc. Lond. B* 267, 883–889.

Petersen, S. L. 1986. Age-related changes in plasma oestrogen concentration, behavioural responsiveness to oestrogen, and reproductive success in female gray-tailed voles, *Microtus canicaudus. J. Reprod. Fertil.* 78: 57–64.

Petrie, M., Halliday, T. and Sanders, C. 1991. Peahens prefer peacocks with elaborate trains. *Anim. Behav.* 41: 323–332.

Pfaff, D. W., Schwartz-Giblin, S., McCarthy, M. M. and Kow, L.-M. 1994. Cellular and molecular mechanisms of female reproductive behaviors. In E. Knobil and J. Neill, eds., *The Physiology of Reproduction*, vol. 2, 107–220. New York: Raven Press.

Pfaff, D. W., Arnold, A. P., Etgen, A. M., Fahrbach, S. E. and Rubin, R. T. 2002. *Hormones, Brain and Behavior.* Five volumes. Amsterdam: Academic Press (Elsevier).

Pfaus, J. G., Hippin, T. E. and Centeno, S. 2001. Conditioning and sexual behavior: a review. *Horm. Behav.* 40: 291–321.

Phelps, S. M., Lydon, J. P., O'Malley, B. W., and Crews, D. 1998. Regulation of male sexual behavior by progesterone receptor, sexual experience, and androgen. *Horm. Behav.* 34: 294–302.

Phillips, R. E. and Barfield, R. J. 1977. Effects of testosterone implants in midbrain vocal areas of capons. *Brain Res.* 122: 378–381.

Phoenix, C. H. and Chambers, K. C. 1986. Aging and primate male sexual behavior. *Proc. Soc. Exp. Biol. Med.* 183: 151–162.

Phoenix, C. H., Goy, R. W., Gerall, A. A. and Young, W. C. 1959. Organizing action of prenatally administered testosterone propionate on the tissues mediating mating behavior in the female guinea pig. *Endocrinology* 65: 369–382.

Piazza, P. V. and Le Moal, M. 1997. Glucocorticoids as a biological substrate of reward: physiological and pathophysiological implications. *Brain Res. Rev.* 25: 359–372.

Pieau, C., Dorizzi, M. and Richard-Mercier, N. 1999. Temperature-dependent sex determination and gonadal differentiation in reptiles. *Cell. Mol. Life Sci.* 55: 887–900.

Piferrer, F. and Donaldson, E. M. 1988. Progress in the development of sex control techniques for the culture of Pacific salmon. *Proceedings of the Aquaculture International Congress and Exposition*, Vancouver, BC, 519–530.

Piferrer, F., Zanuy, S., Carrillo, M., Solar, I. I., Devlin, R. H. and Donaldson, E. M. 1994. Brief treatment with an aromatase inhibitor during sex differentiation causes chromosomally female salmon to develop as normal, functional males. *J. Exp. Zool.* 270: 255–262.

Pilz, K. M. and Smith, H. G. 2004. Egg yolk androgen levels increase with breeding density in the European starling, *Sturnus vulgaris. Funct. Ecol.* 18: 58–66.

Pilz, K. M., Smith, H. G., Sandell, M. and Schwabl, H. 2003. Inter-female variation in egg yolk androgen allocation in the European starling: Do high quality females invest more? *Anim. Behav.* 65: 841–850.

Pitkow, L. J., Sharer, C. A., Ren, X., Insel, T. R., Terwilliger, E. F. and Young, L. J. 2001. Facilitation of affiliation and pair-bond formation by vasopressin receptor gene transfer into the ventral forebrain of a monogamous vole. *J. Neurosci.* 21: 7392–7396.

Pizzari, T., Froman, D. P. and Birkhead, T. R. 2002. Pre- and post-insemination episodes of sexual selection in the fowl, *Gallus g. domesticus. Heredity* 88: 112–116.

Plant, T. M. 2001. Neurobiological bases underlying the control of the onset of puberty in the rhesus monkey: a representative higher primate. *Front. Neuroendocrinol.* 22: 107–139.

Platt, N. and Reynolds, S. E. 1988. Invertebrate neuropeptides. In G. G. Lunt and R. W. Olsen, eds., *Comparative Invertebrate Neurochemistry*, 175–226. Ithaca, NY: Cornell University Press.

Plomin, R., DeFries, J. C., McClearn, G. E. and McGuffin, P. 2001. *Behavioral Genetics*, 4th ed. New York: Worth.

Pohl-Apel, G. and Sossinka, R. 1984. Hormonal determination of song capacity in females of the zebra finch. *Z. Tierpsychol.* 64: 330–336.

Pottinger, T. G. 1999. The impact of stress of animal reproductive activities. In P.H.M. Baum, ed., *Stress Physiology in Animals*, 130–177. Boca Raton, FL: CRC Press.

Pottinger, T. G. and Carrick, T. R. 2001. Stress responsiveness affects dominant-subordinate relationships in rainbow trout. *Horm. Behav.* 40: 419–427.

Prendergast, B. J., Nelson, R. J. and Zucker, I. 2002. Mammalian seasonal rhythms: behavior and neuroendocrine substrates. In In D. W. Pfaff, A. P. Arnold, A. M. Etgen, S. E. Fahrbach and R. T. Rubin, eds., *Hormones, Brain and Behavior*, vol. 2, pp. 93–156. Amsterdam: Academic Press (Elsevier).

Price, J. J. 1998. Family- and sex-specific vocal traditions in a cooperatively breeding songbird. *Proc. R. Soc. Lond. B* 265: 497–502.

Propper, C. R. and Dixon, T. B. 1997. Differential effects of arginine vasotocin and gonadotropin-releasing hormone on sexual behaviors in an anuran amphibian. *Horm. Behav.* 32: 99–104.

Prum, R. O. 1998. Sexual selection and the evolution of mechanical sound production in manakins (Aves: Pipridae). *Anim. Behav.* 55: 977–994.

Pusey, A. E. 1987. Sex-biased dispersal and inbreeding avoidance in birds and mammals. *Trends Ecol. Evol.* 2: 295–299.

Pusey, A. and Packer, C. 1997. The ecology of relationships. In J. R. Krebs and N. B. Davies, eds., *Behavioural Ecology: An Evolutionary Approach*, 4th ed., 254–283. Oxford, UK: Blackwell Science.

Quintana-Murci, L., Jamain, S., Fellous, M., 2001. Origin and evolution of mammalian sex chromosomes. *C. R. Acad. Sci. III.* 324: 1–11.

Raberg, L., Grahn, M., Hasselquist, D. and Svensson, E. 1998. On the adaptive significance of stress-induced immunosuppression. *Proc. R. Soc. Lond. B* 265: 1637–1641.

Raisman, G. 1997. An urge to explain the incomprehensible: Geoffrey Harris and the discovery of the neural control of the pituitary gland. *Annu. Rev. Neurosci.* 20: 533–566.

Raisman, G. and Field, P. M. 1971. Sexual dimorphism in the preoptic area of the rat. *Science* 173: 731–733.

Rallu, M., Corbin, J. G. and Fishell, G. 2002. Parsing the prosencephalon. *Nature Rev. Neurosci.* 3: 943–951.

Ramenofsky, M. 1984. Agonistic behaviour and endogenous plasma hormones in male Japanese quail. *Anim. Behav.* 32: 698–708.

Ranz, J. M., Castillo-Davis, C. I., Meiklejohn, C. D. and Hartl, D. L. 2003. Sex-dependent gene expression and evolution of the *Drosophila* transcriptome. *Science* 300: 1742–1745.

Raouf, S. A., Parker, P. G., Ketterson, E. D., Nolan, V., and Ziegenfus, C. 1997. Testosterone affects reproductive success by influencing extra-pair fertilizations in male dark-eyed juncos (Aves: *Junco hyemalis*). *Proc. R. Soc. Lond. B* 264: 1599–1603.

Rasmussen, L. E., Buss, I. O., Hess, D. L. and Schmidt, M. J. 1984. Testosterone and dihydrotestosterone concentrations in elephant serum and temporal gland secretions. *Biol. Reprod.* 30: 352–362.

Reburn, C. J. and Wynne-Edwards, K. E. 1999. Hormonal changes in males of a naturally biparental and a uniparental mammal. *Horm. Behav.* 35: 163–176.

Reed, W. L. and Vleck, C. M. 2001. Functional significance of variation in egg-yolk androgens in the American coot. *Oecologia* 128: 164–171.

Rees, H. H. 1985. Biosynthesis of ecdysone. In G. A. Kerkut and L. I. Gilbert, eds., *Comprehensive Insect Physiology, Biochemistry and Pharmacology*, vol. 7, 249–293. New York: Pergamon.

Reeve, H. K. and Pfennig, D. W. 2003. Genetic biases for showy males: are some genetic systems especially conducive to sexual selection? *Proc. Natl. Acad. Sci. U S A* 100: 1089–1094.

Reeve, H. K. and Shellman-Reeve, J. S. 1997. The general protected invasion theory: sex biases in parental and alloparental care. *Evol. Ecol.* 11: 357–370.

Reeve H. K. and Sherman, P. W. 1993. Adaptation and the goals of evolutionary research. *Q. Rev. Biol.* 68: 1–32.

Reeve, H. K., Emlen, S. T. and Keller, L. 1998. Reproductive sharing in animal societies: reproductive incentives or incomplete control by dominant breeders. *Behav. Ecol.* 9: 267–278.

Reiner, A., Perkel, D. J., Bruce, L., Butler, A. B., Csillag, A., Kuenzel, W., Medina, L., Paxinos, G., Shimizu, T., Striedter, G. F., Wild, M., Ball, G. F., Durand, S., Güntürkün, O., Lee, D. W., Mello, C. V., Powers, A., White, S. A., Hough, G., Kubikova, L., Smulders, T. V., Wada, K., Dugas-Ford, J., Husband, S., Yamamoto, K., Yu, J., Siang, C., and Jarvis, E. D. 2004. Revised nomenclature for avian telencephalon and some related brainstem nuclei. *J. Comp. Neurol.* 473: 377–414.

Reisert, I. and Pilgrim, C. 1991. Sexual differentiation of monoaminergic neurons— genetic or epigenetic? *Trends Neurosci.* 14: 468–473.

Reisert, I., Karolczak, M., Beyer, C., Just, W., Maxson, S. C. and Ehret, G. 2002. *Sry* does not fully sex-reverse female into male behavior towards pups. *Behav. Genet.* 32: 103–111.

Remage-Healey, L., Adkins-Regan, E. and Romero, L. M. 2003. Behavioral and adrenocortical responses to mate separation and reunion in the zebra finch. *Horm. Behav.* 43: 108–114.

Renfree, M. B., Harry, J. L. and Shaw, G. 1995. The marsupial male: a role model for sexual development. *Philos. Trans. R. Soc. Lond. B* 350: 243–251.

Reyer, H. J., Dittami, J. P. and Hall, M. R. 1986. Avian helpers at the nest—are they psychologically castrated? *Ethology* 71: 216–228.

Reynolds, J. D., Goodwin, N. B. and Freckleton, R. P. 2002. Evolutionary transitions in parental care and live bearing in vertebrates. *Proc. R. Soc. Lond. B* 357: 269–281.

Rhen, T. 2000. Sex-limited mutations and the evolution of sexual dimorphism. *Evolution* 54: 37–43.

Rhen, T. and Crews, D. 2000. Organization and activation of sexual and agonistic behavior in the leopard gecko, *Eublepharis macularius. Neuroendocrinology* 71: 252–261.

Rhen, T. and Crews, D. 2002. Variation in reproductive behaviour within a sex: neural systems and endocrine activation. *J. Neuroendocrinol.* 14: 517–531.

Rhen, T., Ross, J. and Crews, D. 1999. Effects of testosterone on sexual behavior and morphology in adult female leopard geckos, *Eublepharis macularius. Horm. Behav.* 36: 119–128.

Rice, W. R. 1994. Degeneration of a nonrecombining chromosome. *Science* 263: 230–232.

Richard-Yris, M. A., Garnier, D. H. and Leboucher, G. 1983. Induction of maternal behavior and some hormonal and physiological correlates in the domestic hen. *Horm. Behav.* 17: 345–355.

Richter, C. P. 1949. Domestication of the Norway rat and its implications for the problem of stress. *Proc. Assoc. Res. Nerv. Mental Dis.* 29: 19–47.

Ricklefs, R. E. 2003. Global diversification rates of passerine birds. *Proc. R. Soc. Lond. B* 270: 2285–2291.

Rigby, R. G. 1972. A study of the behaviour of caged *Antechinus stuartii. Z. Tierpsychol.* 331: 15–25.

Ringo, J. M. 2002. Hormonal regulation of sexual behavior in insects. In D. W. Pfaff, A. P. Arnold, A. M. Etgen, S. E. Fahrbach and R. T. Rubin, eds., *Hormones, Brain and Behavior*, vol. 3, 93–114. Amsterdam: Academic Press (Elsevier).

Rintamaki, P. T., Hoglund, J., Karvonen, E., Alatalo, R. V., Bjorklund, N., Lundberg, A., Ratti, O. and Vouti, J. 2000. Combs and sexual selection in black grouse (*Tetrao tetrix*). *Behav. Ecol.* 11: 465–471.

Rissman, E. F. 1990. The musk shrew, *Suncus murinus*, a unique animal model for the study of female behavioral endocrinology. *J. Exp. Zool. Suppl.* 4: 207–209.

Rissman, E. F. and Adkins-Regan, E. 1984. Androgens and reproductive behavior in ovariectomized ring doves. *Physiol. Behav.* 32: 697–699.

Riters, L. V., Baillien, M., Eens, M., Pinxten, R., Foidart, A., Ball, G. F. and Balthazart, J. 2001. Seasonal variation in androgen-metabolizing enzymes in the diencephalon and telencephalon of the male European starling (*Sturnus vulgaris*). *J. Neuroendocrinol.* 13: 985–997.

Rivest, S. and Rivier, C. 1995. The role of corticotropin-releasing factor and interleukin-1 in the regulation of neurons controlling reproductive functions. *Endocrinol. Rev.* 16: 177–199.

Roberts, M. L., Buchanan, K. L. and Evans, M. R. 2004. Testing the immunocompetence handicap hypothesis: a review of the evidence. *Anim. Behav.* 68: 227–239.

Roberts, R. L., Jenkins, K. T., Lawler, T., Wegner, F. H. and Newman, J. D. 2001. Bromocriptine administration lowers serum prolactin and disrupts parental responsiveness in common marmosets (*Callithrix j. jacchus*). *Horm. Behav.* 39: 106–112.

Robertson, O. H. 1961. Prolongation of the life span of kokanee salmon (*Oncorhynchus nerka kennerlyi*) by castration before beginning gonad development. *Proc. Natl. Acad. Sci. U S A* 47: 609–621.

Robertson, D. R. 1972. Social control of sex-reversal in a coral-reef fish. *Science* 117: 1007–1009.

Robinson, G. E., Fahrbach, S. E. and Winston, M. L. 1997. Insect societies and the molecular biology of social behavior. *Bioessays* 19: 1099–1108.

Roff, D. 2001. *Life History Evolution*. Sunderland, MA: Sinauer.

Rohde Parfet, K. A., Ganjam, V. K., Lamberson, W. R., Rieke, A. R., Vom Saal, F. S. and Day, B. N. 1990. Intrauterine position effects in female swine: subsequent reproductive performance, and social and sexual behavior. *Appl. Anim. Behav. Sci.* 26: 349–362.

Rohwer, S. and Rohwer, F. C. 1978. Status signalling in Harris sparrows: experimental deceptions achieved. *Anim. Behav.* 26: 1012–1022.

Rollmann, S. M., Houck, L. D. and Feldhoff, R. C. 1999. Proteinaceous pheromone affecting female receptivity in a terrestrial salamander. *Science* 285: 1907–1909.

Romeo, R. D., Richardson, H. N. and Sisk, C. L. 2002a. Puberty and the maturation of the male brain and sexual behavior: recasting a behavioral potential. *Neurosci. Biobehav. Rev.* 26: 381–391.

Romeo, R. D., Wagner, C. K., Jansen, H. T., Diedrich, S. L. and Sisk, C. L. 2002b. Estradiol induces hypothalamic progesterone receptors but does not activate mating behavior in male hamsters (*Mesocricetus auratus*) before puberty. *Behav. Neurosci.* 116: 198–205.

Romero, L. M. 2002. Seasonal changes in plasma glucocorticoid concentrations in free-living vertebrates. *Gen. Comp. Endocrinol.* 128: 1–24.

Romero, L. M. and Sapolsky, R. M. 1996. Patterns of ACTH secretagog secretion in response to psychological stimuli. *J. Neuroendocrinol.* 8: 243–258.

Roosenburg, W. M. and Niewiarowski, P. 1998. Maternal effects and the maintenance of environmental sex determination. In T. A. Mousseau and C. W. Fox, eds., *Maternal Effects as Adaptations*, 307–322. New York: Oxford University Press.

Ros, A.F.H., Dieleman, S. J. and Groothuis, T.G.G. 2002. Social stimuli, testosterone, and aggression in gull chicks: support for the challenge hypothesis. *Horm. Behav.* 41: 334–342.

Rosa-Molinar, E., Fritzsch, B. and Hendricks, S. E. 1996. Organizational—activational concept revisited: sexual differentiation in an atherinomorph teleost. *Horm. Behav.* 30: 563–575.

Rosa-Molinar, E., McCaffery, P. J. and Fritzsch, B. 1997. Sex differences in endogenous retinoid release in the post-embryonic spinal cord of the western mosquitofish, *Gambusia affinis affinis*. *Adv. Exp. Med. Biol.* 414: 95–108.

Rose, C. S. 1996. An endocrine-based model for developmental and morphogenetic diversification in metamorphic and paedomorphic urodeles. *J. Zool. (Lond.)* 239: 253–284.

Rose, J. D. and Moore, F. L. 1999. A neurobehavioral model for rapid actions of corticosterone on sensorimotor integration. *Steroids* 64: 92–99.

Rose, J. D. and Moore, F. L. 2002. Behavioral neuroendocrinology of vasotocin and vasopressin and the sensorimotor processing hypothesis. *Front. Neuroendocrinol.* 23: 317–341.

Röseler, P. F. 1985. Endocrine basis of dominance and reproduction in polistine paper wasps Vespidae. In B. Hölldobler and M. Lindauer, eds., *Experimental Behavioral Ecology and Sociobiology*, 259–272. Sunderland, MA: Sinauer.

Roselli, C. E., Resko, J. A. and Stormshak, F. 2002. Hormonal influences on sexual partner preference in rams. *Archiv. Sex. Behav.* 31: 43–49.

Rosenblatt, J. S. and Aronson, L. R. 1958. The decline of sexual behavior in male cats after castration with special reference to the role of prior sexual experience. *Behaviour* 12: 285–338.

Rosenblatt, J. S. and Snowdon, C. T., eds. 1996. *Parental Care: Evolution, Mechanisms, and Adaptive Significance (Advances in the Study of Behavior*, vol. 25). San Diego: Academic Press.

Rosenblatt J. S., Siegel, H. I. and Mayer, A. D. 1979. Progress in the study of maternal behavior in the rat: hormonal, nonhormonal, sensory, and developmental aspects. *Adv. Study Behav.* 10: 226–311.

Ross, M. E. and Walsh, C. A. 2001. Human brain malformations and their lessons for neuronal migration. *Annu. Rev. Neurosci.* 24: 1041–1070.

Roth, G., Nishikawa, K. C. and Wake, D. B. 1997. Genome size, secondary simplification, and the evolution of the brain in salamanders. *Brain Behav. Evol.* 50: 50–59.

Rubenstein, J.L.R., Martinez, S., Shimamura, K. and Puelles, L. 1994. The embryonic vertebrate forebrain: the prosomeric model. *Science* 266: 578–580.

Ruckstuhl, K. E. and Neuhaus, P. 2002. Sexual segregation in ungulates: a comparative test of three hypotheses. *Biol. Rev.* 77: 77–96.

Ryan, B. C. and Vandenbergh, J. G. 2002. Intrauterine position effects. *Neurosci. Biobehav. Rev.* 26: 665–678.

Sachs, B. D. 1969. Photoperiodic control of reproductive behavior and physiology of the male Japanese quail (*Coturnix coturnix japonica*). *Horm. Behav.* 1: 7–24.

Sachser, N., Dürschlag, M. and Hirzel, D. 1998. Social relationships and the management of stress. *Psychoneuroendocrinology* 23: 891–904.

Sagi, A. and Khalaila, I. 2001. The crustacean androgen: a hormone in an isopod and androgenic activity in decapods. *Am. Zool.* 41: 477–484.

Saino, N. and Møller, A. P. 1995. Testosterone-induced depression of male parental behavior in the barn swallow: female compensation and effects on seasonal fitness. *Behav. Ecol. Sociobiol.* 36: 151–157.

Sakai, Y., Karino, K., Kuwamura, T., Nakashima, Y. and Maruo, Y. 2003. Sexually dichromatic protogynous angelfish *Centropyge ferrugata* (Pomacanthidae) males can change back to females. *Zool. Sci.* 20: 627–633.

Sakata, J. T., Gupta, A., Gonzalez-Lima, F. and Crews, D. 2002. Heterosexual housing increases the retention of courtship behavior following castration and elevates metabolic capacity in limbic brain nuclei in male whiptail lizards, *Cnemidophorus inornatus*. *Horm. Behav.* 42: 263–273.

Saldanha, C. J., Popper, P., Micevych, P. E. and Schlinger, B. A. 1998. The passerine hippocampus is a site of high aromatase: inter- and intraspecies comparisons. *Horm. Behav.* 34: 85–97.

Saldanha, C. J., Schultz, J. D., London, S. E. and Schlinger, B. A. 2000. Telencephalic aromatase but not a song circuit in a sub-oscine passerine, the golden collared manakin (*Manacus vitellinus*). *Brain Behav. Evol.* 56: 29–37.

Salhanick, H. A., Kipnis, D. M. and Vande Wiele, R. L., eds. 1969. *Metabolic Effects of Gonadal Hormones and Contraceptive Steroids.* New York: Plenum.

Saligaut, C., Salbert, G., Bailhache, T., Bennani, S. and Jego, P. 1992. Serotonin and dopamine turnover in the female rainbow trout (*Oncorhynchus mykiss*) brain and pituitary: changes during the annual reproductive cycle. *Gen. Comp. Endocrinol.* 85: 261–268.

Salzet, M., Vieau, D. and Day, R. 2000. Crosstalk between nervous and immune systems through the animal kingdom: focus on opioids. *Trends Neurosci.* 23: 550–555.

Sandi, C., Venero, C. and Guaza, C.1996. Nitric oxide synthesis inhibitors prevent rapid behavioral effects of corticosterone in rats. *Neuroendocrinology* 63: 446–453.

Sapolsky, R. M. 1992. Cortisol concentrations and the social significance of rank instability among wild baboons. *Psychoneuroendocrinology* 17: 701–710.

Sapolsky, R. M. 1993. Endocrinology alfresco: psychoendocrine studies of wild baboons. *Recent Progr. Horm. Res.* 48: 437–468.

Sapolsky, R. M. 2002. Endocrinology of the stress response. In J. B. Becker, S. M. Breedlove, D. Crews and McCarthy, M. M., eds., *Behavioral Endocrinology*, 2nd ed., 409–450. Cambridge, MA: MIT Press.

Sapolsky, R. M., Krey, L. C. and McEwen, B. S. 1986. The neuroendocrinology of stress and aging: the glucocorticoid cascade hypothesis. *Endocrinol. Rev.* 7: 284–301.

Sapolsky, R. M., Vogelman, J. H., Orentreich, N. and Altmann, J. 1993. Senescent decline in serum dehydroepiandrosterone sulfate concentrations in a population of wild baboons. *J. Gerontol.* 48: B196–B200.

Satterlee, D. G., Marin, R. H. and Jones, R. B. 2002. Selection of Japanese quail for reduced adrenocortical responsiveness accelerates puberty in males. *Poultry Sci.* 81: 1071–1076.

Sayag, N., Snapir, N., Robinzon, B., Arnon, E., el Halawani, M. E. and Grimm, V. E. 1989. Embryonic sex steroids affect mating behavior and plasma LH in adult chickens. *Physiol. Behav.* 45: 1107–1112.

Scantlebury, M., Russell, A. F., McIlrath, G. M., Speakman, J. R. and Clutton-Brock, T. H. 2002. The energetics of lactation in cooperatively breeding meerkats *Suricata suricatta. Proc. R. Soc. Lond. B* 269: 2147–2153.

Scharrer, B. 1959. The role of neurosecretion in neuroendocrine integration. In A. Gorbman, ed., *Comparative Endocrinology*, 134–148. New York: Wiley.

Scharrer, E. 1959. General and phylogenetic interpretations of neuroendocrine interrelations. In A. Gorbman, ed., *Comparative Endocrinology*, 233–249. New York: Wiley.

Schiml, P. A., Wersinger, S. R. and Rissman, E. F. 2000. Behavioral activation of the female neuroendocrine axis. In K. Wallen and J. E. Schneider, eds., *Reproduction in Context: Social and Environmental Influence on Reproductive Physiology and Behavior*, 445–472. Cambridge, MA: MIT Press.

Schlinger, B. A. and Arnold, A. P. 1991. Brain is the major site of estrogen synthesis in a male songbird. *Proc. Natl. Acad. Sci. U S A* 88: 4191–4194.

Schlinger, B. A. and Arnold, A. P. 1992. Plasma sex steroids and tissue aromatization in hatchling zebra finches: implications for the sexual differentiation of singing behavior. *Endocrinology* 130: 289–299.

Schlinger, B. A. and Brenowitz, E. A. 2002. Neural and hormonal control of birdsong. In D. W. Pfaff, A. P. Arnold, A. M. Etgen, S. E. Fahrbach and R. T. Rubin, eds., *Hormones, Brain and Behavior*, vol. 2, 799–840. Amsterdam: Academic Press (Elsevier).

Schlinger, B. A. and Callard, G. V. 1990. Aromatization mediates aggressive behavior in quail. *Gen. Comp. Endocrinol.* 79: 39–53.

Schlinger, B. A., Fivizzani, A. J. and Callard, G. V. 1989. Aromatase, 5α- and 5β-reductase in brain, pituitary and skin of the sex-role reversed Wilson's phalarope. *J. Endocrinol.* 122: 573–581.

Schlinger, B. A., Greco, C. and Bass, A. H. 1999. Aromatase activity in hindbrain vocal control region of a teleost fish: divergence amoung males with alternative reproductive tactics *Proc. R. Soc. Lond. B* 266: 131–131.

Schlinger, B. A., Schultz, J. D. and Hertel, F. 2001. Neuromuscular and endocrine control of an avian courtship behavior. *Horm. Behav.* 40: 276–280.

Schmid, M. and Steinlein, C. 2001. Sex chromosomes, sex-linked genes, and sex determination in the vertebrate class Amphibia. In G. Scherer and M. Schmid, eds., *Genes and Mechanisms in Vertebrate Sex Determination*, 143–176. Basel: Birkhäuser.

Schmidt, B.M.W., Gerdes, D., Feuring, M., Falkenstein, E., Christ, M. and Wehling, M. 2000. Rapid, nongenomic steroid actions: a new age? *Front. Neuroendocrinol.* 21: 57–94.

Schneider, J. E. and Wade, G. N. 2000. Inhibition of reproduction in service of energy balance. In K. Wallen and J. E. Schneider, eds., *Reproduction in Context: Social and Environmental Influences on Reproductive Physiology and Behavior*, 35–82. Cambridge, MA: MIT Press.

Schneiderman, A. M., Matsumoto, S. G. and Hildebrand, J. G. 1982. Trans-sexually grafted antennae influence development of sexually dimorphic neurones in moth brain. *Nature* 289: 844–846.

Schoech, S. J., Mumme, R. L. and Wingfield, J. C. 1996. Prolactin and helping behaviour in the cooperatively breeding Florida scrub-jay, *Aphelocoma coerulescens*. *Anim. Behav.* 52: 445–456.

Schoech, S. J., Mumme, R. L. and Wingfield, J. C. 1997. Corticosterone, reproductive status, and body mass in a cooperative breeder, the Florida scrub-jay (*Aphelocoma coerulescens*). *Physiol. Zool.* 75: 68–73.

Schoech, S. J., Ketterson, E. D., Nolan, V., Sharp, P. J. and Buntin, J. D. 1998. The effect of exogenous testosterone on parental behavior, plasma prolactin, and prolactin binding sites in dark-eyed juncos. *Horm. Behav.* 34: 1–10.

Schoech, S. J., Reynolds, S. J. and Boughton, R. K. 2004. Endocrinology. In W. D. Koenig and J. Dickenson, eds., *Ecology and Evolution of Cooperative Breeding in Birds*, 128–141. Cambridge, UK: Cambridge University Press.

Schradin, C. and Anzenberger, G. 1999. Prolactin, the hormone of paternity. *News Physiol. Sci.* 14: 223–231.

Schreibman, M. P. and Magliulo-Cepriano, L. 2002. Differentiation/maturation of centers in the brain regulating reproductive function in fishes. In D. W. Pfaff, A. P.

Arnold, A. M. Etgen, S. E. Fahrbach and R. T. Rubin (eds.), *Hormones, Brain and Behavior*, vol. 4, 303–324. Amsterdam: Academic Press (Elsevier).

Schreibman, M. P. and Scanes, C. G., eds. 1989. *Development, Maturation, and Senescence of Neuroendocrine Systems: A Comparative Approach*. San Diego: Academic Press.

Schulte, B. A. and Rasmussen, L.E.L. 1999. Signal–receiver interplay in the communication of male condition by Asian elephants. *Anim. Behav.* 57: 1265–1274.

Schulz, D. J., Sullivan, J. P. and Robinson, G. E. 2002. Juvenile hormone and octopamine in the regulation of division of labor in honey bee colonies. *Horm. Behav.* 42: 222–231.

Schulz, K. M., Richardson, H. N., Zehr, J. L., Osetek, A. J., Menard, T. A. and Sisk, C. L. 2004. Gonadal hormones masculinize and defeminize reproductive behaviors during puberty in the male Syrian hamster. *Horm. Behav.* 45: 242–249.

Schumacher, M., Sulon, J., and Balthazart, J. 1988. Changes in serum concentrations of steroids during embryonic and post-hatching development of male and female Japanese quail (*Coturnix coturnix japonica*). *J. Endocrinol.* 118: 127–134.

Schwabl, H. 1993. Yolk is a source of maternal testosterone for developing birds. *Proc. Natl. Acad. Sci. U S A* 90: 11446–11450.

Schwabl, H. 1996a. Environment modifies the testosterone levels of a female bird and its eggs. *J. Exp. Zool.* 276: 157–163.

Schwabl, H. 1996b. Maternal testosterone in the avian egg enhances postnatal growth. *Comp. Biochem. Physiol. A* 114: 271–276.

Schwabl, H. 1997a. Maternal steroid hormones in the egg. In S. Harvey and R. J. Etches, eds., *Perspectives in Avian Endocrinology*, 3–14. Bristol, UK: *Journal of Endocrinology*.

Schwabl, H. 1997b. The contents of maternal testosterone in house sparrow *Passer domesticus* eggs vary with breeding conditions. *Naturwissenschaften* 84: 406–408.

Schwabl, H., Mock, D. W. and Gieg, J. A. 1997. A hormonal mechanism for parental favouritism. *Nature* 386: 231.

Schwagmeyer, P. L. 1979. The Bruce effect: an evaluation of male/female advantages. *Am. Nat.* 114: 932–938.

Scordalakes, E. M. and Rissman, E. F. 2003. Aggression in male mice lacking functional estrogen receptor α. *Behav. Neurosci.* 117: 38–45.

Searcy, W. A. 1988. Do female red-winged blackbirds limit their own breeding densities? *Ecology* 69: 85–95.

Searcy, W. A. and Wingfield, J. C. 1980. The effects of androgen and anti androgen on dominance and aggressiveness in male red-winged blackbirds *Agelaius phoeniceus*. *Horm. Behav.* 14: 126–135.

Seeley, T. D. 1995. *The Wisdom of the Hive: The Social Physiology of Honey Bee Colonies*. Cambridge, MA: Harvard University Press.

Segovia, S. and Guillamón, A. 1996. Searching for sex differences in the vomeronasal pathway. *Horm. Behav.* 30: 618–626.

Semsar, K. and Godwin, J. 2004. Muliple mechanisms of phenotype development in the bluehead wrasse. *Horm. Behav.* 45: 345–353.

Semsar, K., Klomberg, K. F. and Marler, C. 1998. Arginine vasotocin increases calling-site acquisition by nonresident male grey treefrogs. *Anim. Behav.* 56: 983–989.

Semsar, K., Kandel, F.L.M. and Godwin, J. 2001. Manipulations of the AVT system shift social status and related courtship and aggressive behavior in the bluehead wrasse. *Horm. Behav.* 40: 21–31.

Sengelaub, D. R. and Arnold, A. P. 1989. Hormonal control of neuron number in sexually dimorphic spinal nuclei of the rat, I: testosterone-regulated death in the dorsolateral nucleus. *J. Comp. Neurol.* 280: 622–629.

Setchell, J. M. and Dixson, A. F. 2001. Changes in the secondary sexual ornaments of male mandrills (*Mandrillus sphinx*) are associated with gain and loss of alpha status. *Horm. Behav.* 39: 177–184.

Shaffer, H. B. 1993. Phylogenetics of model organisms: the laboratory axolotl, *Ambystoma mexicanum*. *Syst. Biol.* 42: 508–522.

Shanley, D. P. and Kirkwood, T.B.L. 2001. Evolution of the human menopause. *Bioessays* 23: 282–287.

Shapiro, D. Y. 1983. Distinguishing behavioral interactions from visual cues as causes of adult sex change in a coral reef fish. *Horm. Behav.* 17: 424–432.

Shapiro, D. Y. 1992. Plasticity of gonadal development and protandry in fishes. *J. Exp. Zool.* 261: 194–203.

Shapiro, D. Y. and Boulon, R. H. 1982. The influence of females on the inititation of female-to-male sex change in a coral reef fish. *Horm. Behav.* 16: 66–75.

Sharp, P. J., Macnamee, M. C., Sterling, R. J., Lea, R. W. and Pedersen, H. C. 1988. Relationships between prolactin, LH and broody behaviour in bantam hens. *J. Endocrinol.* 118: 279–286.

Sharp, P. J., Sterling, R. J., Talbot, R. T. and Huskisson, N. S. 1989. The role of hypothalamic vasoactive intestinal polypeptide in the maintenance of prolactin secretion in incubating bantam hens: observations using passive immunization, radioimmunoassay and immunohistochemistry. *J. Endocrinol.* 122: 5–13.

Shaw, B. K. and Kennedy, G. G. 2002. Evidence for species differences in the pattern of androgen receptor distribution in relation to species differences in an androgen-dependent behavior. *J. Neurobiol.* 52: 203–220.

Sheehan, T. and Numan, M. 2000. The septal region and social behavior. In M. Numan, ed., *The Behavioral Neuroscience of the Septal Region*, 175–209. New York: Springer-Verlag.

Sherwin, B. B. 1998. Use of combined estrogen-androgen preparations in the postmenopause: evidence from clinical studies. *Int. J. Fertil. Womens Med.* 43: 98–103.

Sherwin, B. B., Gelfand, M. M. and Brender, W. 1985. Androgen enhances sexual motivation in females: a prospective, crossover study of sex steroid administration in the surgical menopause. *Psychosom. Med.* 47: 339–351.

Sherwood, N. 1987. The GnRH family of peptides. *Trends Neurosci.* 10: 129–132.

Sherwood, N. M. and Parker, D. B. 1990. Neuropeptide families: an evolutionary perspective. *J. Exp. Zool. Suppl.* 4: 63–71.

Shine, R. 1995. A new hypothesis for the evolution of viviparity in reptiles. *Am. Nat.* 145: 809–823.

Shine, R. 1999. Why is sex determined by nest temperature in many reptiles? *Trends Ecol. Evol.* 14: 186–189.

Shire, J.G.M. 1976. The forms, uses and significance of genetic variation in endocrine systems. *Biol. Rev.* 51: 105–141.

Short, R. V. 1979. Sexual selection and its component parts, somatic and genital selection, as illustrated by man and the great apes. *Advances in the Study of Behavior* 9: 131–158. New York: Academic Press.

Shrenker, P. and Maxson, S. C. 1983. The genetics of hormonal influences on male sexual behavior of mice and rats. *Neurosci. Biobehav. Rev.* 7: 349–359.

Shuster, S. M. and Wade, M. J. 2003. *Mating Systems and Strategies.* Princeton, NJ: Princeton University Press.

Sibley, C. G. and Ahlquist, J. E. 1990. *Phylogeny and Classification of Birds: A Study in Molecular Evolution.* New Haven, CT: Yale University Press.

Signoret, J. P. 1967. Attraction de la femelle en oestrus par le mâle chez les porcins. *Rev. Comp. Anim.* 4: 10–22.

Signoret, J. P. 1970. Reproductive behaviour of pigs. *J. Reprod. Fertil.* suppl. 11: 105–117.

Signoret, J. P. 1991. Sexual pheromones in the domestic sheep: importance and limits in the regulation of reproductive physiology. *J. Steroid Biochem. Mol. Biol.* 39: 639–645.

Silver, R. 1984. Prolactin and parenting in the pigeon family. *J. Exp. Zool.* 232: 617–625.

Silver, R., O'Connell, M. and Saad, R. 1979. The effect of androgens on the behavior of birds. In C. Beyer, ed., *Endocrine Control of Sexual Behavior*, 223–278. New York: Raven Press.

Silverin, B. 1980. Effects of long-acting testosterone treatment on free-living pied flycatchers, *Ficedula hypoleuca*, during the breeding period. *Anim. Behav.* 28: 906–912.

Silverin, B. 1997. The stress response and autumn dispersal behaviour in willow tits. *Anim. Behav.* 53: 451–459.

Silverin, B. 1998. Stress responses in birds. *Poultry Avian Biol. Rev.* 9: 153–168.

Simerly, R. B. 2002. Wired for reproduction: organization and development of sexually dimorphic circuits in the mammalian forebrain. *Annu. Rev. Neurosci.* 25: 507–536.

Simon, N. G. and Whalen, R. E. 1986. Hormonal regulation of aggression: evidence for a relationship among genotype, receptor binding, and behavioral sensitivity to androgens and estrogens. *Aggr. Behav.* 12: 255–266.

Simpson, E. R., Michael, M. D., Agarwal, V. R., Hinshelwood, M. M., Bulun, S. E. and Zhao, Y. 1997. Cytochromes P450 11: expression of the CYP19 (aromatase) gene: an unusual case of alternative promoter usage. *FASEB J.* 11: 29–36.

Simpson, H. B. and Vicario, D. S. 1991a. Early estrogen treatment alone causes female zebra finches to produce learned, male-like vocalizations. *J. Neurobiol.* 22: 755–776.

Simpson, H. B. and Vicario, D. S. 1991b. Early estrogen treatment of female zebra finches masculinizes the brain pathway for learned vocalizations. *J. Neurobiol.* 22: 777–793.

Sinervo, B. 2001. Selection in local neighborhoods, the social environment, and ecology of alternative strategies. In L. A. Dugatkin, ed., *Model Systems in Behavioral Ecology: Integrating Conceptual, Theoretical, and Empirical Approaches*, 191–226. Princeton, NJ: Princeton University Press.

Sinervo, B. and Basolo, A. L. 1996. Testing adaptation using phenotypic manipulations. In M. R. Rose and G. V. Lauder, eds., *Adaptation*, 149–185. San Diego: Academic Press.

Sinervo, B. and Zamudio, K. R. 2001. The evolution of alternative reproductive strategies: fitness differential, heritability and genetic correlation between the sexes. *J. Hered.* 92: 198–205.

Sinervo, B., Miles, D. B., Frankino, W. A., Klukowski, M. and DeNardo, D. F. 2000. Testosterone, endurance, and darwinian fitness: natural and sexual selection on the physiological bases of alternative male behaviors in side-blotched lizards. *Horm. Behav.* 38: 222–233.

Sisk, C. L. 2000. Puberty as a period of neural development and behavioral maturation–integration. In J.-P. Bourguignon and T. M. Plant, eds., *The Onset of Puberty in Perspective*, 407–410 Amsterdam: Elsevier.

Sisneros, J. A., Forlano, P. M., Deitcher, D. L. and Bass, A. H. 2004. Steroid-dependent auditory plasticity leads to adaptive coupling of sender and receiver. *Science* 305: 404–407.

Slabbekoorn, H. and Smith, T. B. 2002. Bird song, ecology and speciation. *Philos. Trans. R. Soc. Lond.* 357: 493–503.

Slabbekoorn, D., van Goozen, S.H.M., Megens, J., Gooren, L.J.G. and Cohen-Kettenis, P. T. 1999. Activating effects of cross-sex hormones on cognitive functioning: a study of short-term and long-term hormone effects in transsexuals. *Psychoneuroendocrinology* 24: 423–448.

Sloman, K. A., Desforges, P. R. and Gilmour, K. M. 2001. Evidence for a mineralocorticoid-like receptor linked to branchial chloride cell proliferation in freshwater rainbow trout. *J. Exp. Biol.* 204: 3953–3961.

Smale, L., Nunes, S. and Holekamp, K. E. 1997. Sexually dimorphic dispersal in mammals: patterns, causes, and consequences. *Adv. Study Behav.* 26: 181–251. San Diego: Academic Press.

Smith, E. R., Stefanick, M. L., Clark, J. T. and Davidson, J. M. 1992. Hormones and sexual behavior in relationship to aging in male rats. *Horm. Behav.* 26: 110–135.

Smith, G. T., Wingfield, J. C. and Veit, R. R. 1994. Adrenocortical response to stress in the common diving petrel, *Pelecanoides urinatrix. Physiol. Zool.* 67: 526–537.

Sockman, K. W. and Schwabl, H. 1999. Daily estradiol and progesterone levels relative to laying and onset of incubation in canaries. *Gen. Comp. Endocrinol.* 114: 257–268.

Sockman, K. W. and Schwabl, H. 2000. Yolk androgens reduce offspring survival. *Proc. R. Soc. Lond. B* 267: 1451–1456.

Sockman, K. W., Schwabl, H. and Sharp, P. J. 2004. Removing the confound of time in investigating the regulation of serial behaviours: testosterone, prolactin and the transition from sexual to parental activity in male American Kestrels. *Anim. Behav.* 67: 1151–1161.

Soma, K. K. and Wingfield, J. C. 1999. Endocrinology of aggression in the non-breeding season. In N. J. Adams and R. H. Slotow, eds., *Proceedings of the 22nd International Ornithological Congress*, Durban, 1606–1620. Johannesburg: BirdLife South Africa.

Soma, K. K., Sullivan, K. and Wingfield, J. 1999. Combined aromatase inhibitor and antiandrogen treatment decreases territorial aggression in a wild songbird during the nonbreeding season. *Gen. Comp. Endocrinol.* 115: 442–453.

Soma, K. K., Tramontin, A. D. and Wingfield, J. C. 2000. Oestrogen regulates male aggresion in the non-breeding season. *Proc. R. Soc. Lond. B* 267: 1089–1096.

Soma, K. K., Wissman, A. M., Brenowitz, E. A. and Wingfield, J. C. 2002. Dehydroepiandrosterone (DHEA) increases territorial song and the size of an associated brain region in a male songbird. *Horm. Behav.* 41: 203–212.

Soma, K. K., Schlinger, B. A., Wingfield, J. C. and Saldanha, C. J. 2003. Brain aromatase, 5α-reductase, and 5α-reductase change seasonally in wild male song sparrows: relationship to aggressive and sexual behavior. *J. Neurobiol.* 56: 209–221.

Sorensen, P. W. and Stacey, N. E. 1999. Evolution and specialization of fish hormonal pheromones. In R. E. Johnston, D. Müller-Schwarze and P. W. Sorensen, eds., *Advances in Chemical Signals in Vertebrates*, 15–47. New York: Kluwer/Plenum.

Sorensen, P. W., Stacey, N. E. and Chamberlain, K. J. 1989. Differing behavioral and endocrinological effects of two female sex pheromones on male goldfish. *Horm. Behav.* 23: 317–332.

Sower, S. 1998. Brain and pituitary hormones of lampreys, recent findings and their evolutionary significance. *Am. Zool.* 38: 15–38.

Sperry, T. S. and Thomas, P. 1999a. Characterization of two nuclear androgen receptors in Atlantic croaker: comparison of their biochemical properties and binding specificities. *Endocrinology* 140: 1602–1611.

Sperry, T. S. and Thomas, P. 1999b. Identification of two nuclear androgen receptors in kelp bass (*Paralabrax clathratus*) and their binding affinities for xenobiotics: comparison with Atlantic croaker (*Micropogonias undulatus*) androgen receptors. *Biol. Reprod.* 61:1152–1161.

Spicer, J. I. and Gaston, K. J. 1999. *Physiological Diversity and Its Ecological Implications*. Oxford, UK: Blackwell Science.

Spotila, J. R., Spotila, L. D. and Kaufer, N. F. 1994. Molecular mechanisms of TSD in reptiles: a search for the magic bullet. *J. Exp. Zool.* 270: 117–127.

Spratt, D. I., O'Dea, L. S., Schoenfeld, D., Butler, J., Rao, P. N. and Crowley, W. 1988. Neuroendocrine–gonadal axis in men: frequent sampling of LH, FSH and testosterone. *Am. J. Physiol.* 254: 658–666.

Stacey, N. and Kobayashi, M. 1996. Androgen induction of male sexual behaviors in female goldfish. *Horm. Behav.* 30: 434–445.

Staicer, C. A., Spector, D.A. and Horn, A. G. 1996. The dawn chorus and other diel patterns in acoustic signaling. In D. E. Kroodsma and E. H. Miller, eds., *Ecology and Evolution of Acoustic Communication in Birds*, 426–453. Ithaca, NY: Cornell University Press.

Stanislaw, H. and Rice, F. J. 1988. Correlation between sexual desire and menstrual cycle characteristics. *Archiv. Sex. Behav.* 17: 499–508.

Staub, N. L. and DeBeer, M. 1997. The role of androgens in female vertebrates. *Gen. Comp. Endocrinol.* 108: 1–24.

Stearns, S. C. 1989. Trade-offs in life history evolution. *Funct. Ecol.* 3: 259–268.

Stearns, S. C. 1992. *The Evolution of Life Histories*. New York: Oxford University Press.

Stearns, S. C. and Magwene, P. 2003. The naturalist in a world of genomics. *Am. Nat.* 161: 171–180.

Stern, J. M. 1990. Multisensory regulation of maternal behavior and masculine sexual behavior: a revised view. *Neurosci. Biobehav. Rev.* 14: 183–200.

Storey, A. E. 1994. Pre-implantation pregnancy disruption in female meadow voles *Microtus pennsylvanicus* (Rodentia: Muridae): male competition or female mate choice? *Ethology* 98: 89–100.

Streelman, J. T. and Danley, P. D. 2003. The stages of vertebrate evolutionary radiation. *Trends Ecol. Evol.* 18: 126–131.

Stribley, J. M. and Carter, C. S. 1999. Developmental exposure to vasopressin increases aggression in adult prairie voles. *Proc. Natl. Acad. Sci. U S A* 96: 12601–12604.

Striedter, G. F. 1994. The vocal control pathways in budgerigars differ from those in songbirds. *J. Comp. Neurol.* 343: 35–56.

Striedter, G. F. 1998. Progress in the study of brain evolution: from speculative theories to testable hypotheses. *Anat. Rec.* 253: 105–112.

Strier, K. B. and Ziegler, T. E. 2000. Lack of pubertal influences on female dispersal in muriqui monkeys, *Brachyteles arachnoides*. *Anim. Behav.* 59: 849–860.

Stutchbury, B.J.M. and E. S. Morton. 2001. *Behavioral Ecology of Tropical Birds*. Academic Press: London.

Sullivan, J. P., Fahrbach, S. E. and Robinson, G. E. 2000. Juvenile hormone paces behavioral development in the adult worker honey bee. *Horm. Behav.* 37: 1–14.

Suomi, S. J. 1991. Uptight and laid-back monkeys: individual differences in the response to social challenges. In S. E. Brauth, W. S. Hall and R. J. Dooling, eds., *Plasticity of Development*, 27–56. Cambridge, MA: MIT Press.

Suzuki, M., Kubokawa, K., Nagasawa, H. and Urbano, A. 1995. Sequence analysis of vasotocin c-DNAs of the lamprey, *Lampetra japonica*, and the hagfish, *Eptatretus burgeri*: evolution of cyclostome vasotocin precursors. *J. Mol. Endocrinol.* 14: 67–77.

Svare, B. B., ed. 1983. *Hormones and Aggressive Behavior.* New York: Plenum Press.

Swaab, D. F., Gooren, L.J.G. and Hofman, M. A. 1992. The human hypothalamus in relation to gender and sexual orientation. *Progr. Brain Res.* 93: 205–219.

Swaddle, J. P. and Reierson, G. W. 2002. Testosterone increases perceived dominance but not attractiveness in human males. *Proc. R. Soc. Lond. B* 269: 2285–2289.

Takahashi, L. K. 1990. Hormonal regulation of sociosexual behavior in female mammals. *Neurosci. Biobehav. Rev.* 14: 403–413.

Takase, M., Ukena, K., Yamazaki, T., Kominami, S. and Tsutsui, K. 1999. Pregnenolone, pregnenolone sulfate, and cytochrome P450 side-chain cleavage enzyme in the amphibian brain and their seasonal changes. *Endocrinology* 140: 1936–1944.

Taleisnik, S., Caligaris, L. and Astrada, J. J. 1966. Effect of copulation on the release of pituitary gonadotropins in male and female rats. *Endocrinology* 79: 49–54.

Tallamy, D. W. 2000. Sexual selection and the evolution of exclusive paternal care in arthropods. *Anim. Behav.* 60: 559–567.

Tanabe, Y., Nakamura, T., Fujioka, K. and Doi, O. 1979. Production and secretion of sex steroid hormones by the testes, the ovary, and the adrenal glands of embryonic and young chickens (*Gallus domesticus*). *Gen. Comp. Endocrinol.* 39: 26–33.

Tanabe, Y., Yano, T., and Nakamura, T. 1983. Steroid hormone synthesis and secretion by testes, ovary, and adrenals of embryonic and postembryonic ducks. *Gen. Comp. Endocrinol.* 49: 144–153.

Tanabe, Y., Saito, N. and Nakamura, T. 1986. Ontogenetic steroidogenesis by testes, ovary, and adrenals of embryonic and postembryonic chickens (*Gallus domesticus*). *Gen. Comp. Endocrinol.* 63: 456–463.

Tanaka, K., Mather, F. B., Wilson, W. O. and McFarland, L. Z. 1965. Effect of photoperiods on early growth of gonads and on potency of gonadotropins of the anterior pituitary in *Coturnix. Poultry Sci.* 44: 662–665.

Tarlow, E. M., Wikelski, M. and Anderson, D. J. 2001. Hormonal correlates of siblicide in Galápagos Nazca boobies. *Horm. Behav.* 40: 14–20.

Tarrant, A. M., Atkinson, S. and Atkinson, M. J. 1999. Estrone and estradiol-17beta concentration in tissue of the scleractinian coral, *Montipora verrucosa. Comp. Biochem. Physiol.* A 122: 85–92.

Tatar, M., Bartke, A. and Antebi, A. 2003. The endocrine regulation of aging by insulin-like signals. *Science* 299: 1346–1351.

Temple, J. L., Fugger, H. N., Li, X., Shetty, S. J., Gustafsson, J.-Å. and Rissman, E. F. 2001. Estrogen receptor β regulates sexually dimorphic neural responses to estradiol. *Endocrinology* 142: 510–513.

Temple, J. L., Millar, R. P. and Rissman, E. F. 2003. An evolutionarily conserved form of gonadotropin-releasing hormone coordinates energy and reproductive behavior. *Endocrinology* 144: 13–19.

Terkel, A. S., Moore, C. L. and Beer, C. G. 1976. The effects of testosterone and estrogen on the rate of long-calling vocalization in juvenile laughing gulls, *Larus atricilla. Horm. Behav.* 7: 49–57.

Terkel, J. 1988. Neuroendocrine processes in the establishment of pregnancy and pseudopregnancy in rats. *Psychoneuroendocrinology* 13: 5–28.

Thompson, C. W. and Moore, M. C. 1992. Behavioral and hormonal correlates of alternative reproductive strategies in a polygynous lizard: tests of the relative plasticity and challenge hypotheses. *Horm Behav.* 26: 568–585.

Thompson, R. R. and Moore, F. L. 2000. Vasotocin stimulates appetitive responses to the visual and pheromonal stimuli used by male roughskin newts during courtship. *Horm Behav.* 38: 75–85.

Thompson, R. R. and Moore, F. L. 2003. The effects of sex steroids and vasotocin on behavioral responses to visual and olfactory sexual stimuli in ovariectomized female roughskin newts. *Horm. Behav.* 44: 311–318.

Thompson, R. R., Goodson, J. L., Ruscio, M. G. and Adkins-Regan, E. 1998. Role of the archistriatal nucleus taeniae in the sexual behavior of male Japanese quail (*Coturnix japonica*): a comparison of function with the medial nucleus of the amygdala in mammals. *Brain Behav. Evol.* 51: 215–229.

Thornton, J. W. 2001. Evolution of vertebrate steroid receptors from an ancestral estrogen receptor by ligand exploitation and serial genome expansions. *Proc. Natl. Acad. Sci. U S A* 98: 5671–5676.

Thornton, J. W., Need, E. and Crews, D. 2003. Resurrecting the ancestral steroid receptor: ancient origin of estrogen signaling. *Science* 301: 1714–1717.

Tiersch, T. R., Mitchell, M. J. and Wachtel, S. S. 1991. Studies on the phylogenetic conservation of the SRY gene. *Hum. Genet.* 87: 571–573.

Tinbergen, N. 1963. On aims and methods of ethology. *Z. Tierpsychol.* 20: 410–433.

Tito, M. B., Hoover, M. A., Mingo, A. M. and Boyd, S. K. 1999. Vasotocin maintains multiple call types in the gray treefrog, *Hyla versicolor. Horm. Behav.* 36: 166–175.

Tobias, M. L., Barnard, C., O'Hagan, R., Horng, S. H., Rand, M. and Kelley, D. 2004. Vocal communication between male *Xenopus laevis. Anim. Behav.* 67: 353–365.

Todd, P. M. and Miller, G. F. 1999. From pride and prejudice to persuasion: satisficing in mate search. In G. Gigerenzer and P. Todd, eds., *Simple Heuristics That Make Us Smart*, 257–285. Oxford, UK: Oxford University Press.

Tong, S. K., Chiang, E. F., Hsiao, P. H. and Chung, B. 2001. Phylogeny, expression and enzyme activity of zebrafish cyp19 (P450 aromatase) genes. *J. Steroid Biochem. Mol. Biol.* 79: 299–303.

Townsend, D. S. and Moger, W. H. 1987. Plasma androgen levels during male parental care in a tropical frog (*Eleutherodactylus*). *Horm. Behav.* 21: 93–99.

Trainor, B. C. and Marler, C. A. 2001. Testosterone, paternal behavior, and aggression in the monogamous California mouse (*Peromyscus californicus*). *Horm. Behav.* 40: 32–42.

Trainor, B. C. and Marler, C A. 2002. Testosterone promotes paternal behaviour in a monogamous mammal via conversion to oestrogen. *Proc. R. Soc. Lond. B* 269: 823–829.

Trainor, B. C., Rouse, K. L. and Marler, C. A. 2003. Arginine vasotocin interacts with the social environment to regulate advertisement calling in the gray treefrog (*Hyla versicolor*). *Brain Behav. Evol.* 61: 165–171.

Trainor, B. C., Bird, I. M. and Marler, C. A. 2004. Opposing hormonal mechanisms of aggression revealed through short-lived testosterone manipulations and multiple winning experiences. *Horm. Behav.* 45: 115–121.

Tramontin, A. D. and Brenowitz, E. A. 2000. Seasonal plasticity in the adult brain. *Trends Neurosci.* 23: 251–271.

Trivers, R. L. 1972. Parental investment and sexual selection. In B. Campbell, ed., *Sexual Selection and the Descent of Man 1871–1971*, 136–179. Chicago: Aldine.

Trivers, R. L. and Willard, D. E. 1973. Natural selection of parental ability to vary the sex ratio of offspring. *Science* 179: 90–92.

Trobec, R. J. and Oring, L. W. 1972. Effects of testosterone propionate implantation on lek behavior of sharp-tailed grouse. *Am. Midland Naturalist* 117: 531–536.

Truman, J. W. 1994. The endocrine system and the organization and control of behavior. In R. J. Greenspan and C. P. Kyriacou, eds., *Flexibility and Constraint in Behavioral Systems*, 29–39. Chichester, UK: Wiley.

Truman, J. W. and Riddiford, L. M. 2002. In In D. W. Pfaff, A. P. Arnold, A. M. Etgen, S. E. Fahrbach and R. T. Rubin, eds., *Hormones, Brain and Behavior*, vol. 2, 841–873. Amsterdam: Academic Press (Elsevier).

Trumbo, S. T. 1996. Parental care in invertebrates. In J. S. Rosenblatt and C. T. Snowdon, eds., *Parental Care: Evolution, Mechanisms, and Adaptive Significance* (*Advances in the Study of Behavior*, vol. 25), 3–52. San Diego: Academic Press.

Trut, L. N. 1999. Early canid domestication: the farm-fox experiment. *Am. Scientist* 87: 160–169.

Tsai, L. W. and Sapolsky, R. M. 1996. Rapid stimulatory effects of testosterone upon myotubule metabolism and sugar transport, as assessed by silicon microphysiometry. *Aggr. Behav.* 22: 357–364.

Tsai, P. S., Maldonado, T. A. and Lunden, J. B. 2003. Localization of gonadotropin-releasing hormone in the central nervous system and a peripheral chemosensory organ of *Aplysia californica*. *Gen. Comp. Endocrinol.* 130: 20–28.

Tsutsui, K. and Schlinger, B. A. 2001. Steroidogenesis in the avian brain. In A. Dawson and C. M. Chaturvedi, eds., *Avian Endocrinology*, 59–77. New Delhi: Narosa Publishing House.

Tullberg, B. S., Ah-King, M. and Temrin, H. 2002. Phylogenetic reconstruction of parental-care systems in the ancestors of birds. *Proc. R. Soc. Lond. B* 357: 251–257.

Tuttle, E. M. 2003. Alternative reproductive strategies in the white-throated sparrow: behavioral and genetic evidence. *Behav. Ecol.* 14: 425–432.

Ulinski, P. S. and Margoliash, D. 1990. Neurobiology of the reptile-bird transition. In E. G. Jones and A. Peters, eds., *Cerebral Cortex*, 217–265. New York: Plenum Press.

Unwin, M. J., Kinnison, M. T. and Quinn, T. P. 1999. *Can. J. Fish. Aquat. Sci.* 56: 1172–1181.

Urano, A., Hyodo, S. and Suzuki, M. 1992. Molecular evolution of neurohypophysial hormone precursors. *Prog. Brain Res.* 92: 39–46.

Utami, S. S., Goossens, B., Bruford, M. W., de Ruiter, J. R. and van Hooff, J.A.R.A.M. 2002. Male bimaturism and reproductive success in Sumatran orang-utans. *Behav. Ecol.* 13: 643–652.

Vaiman, D. and Pailhoux, E. 2000. Mammalian sex reversal and intersexuality: deciphering the sex-determination cascade. *Trends Genet.* 16: 488–494.

Valenzuela, N., Adams, D. C. and Janzen, F. J. 2003. Pattern does not equal process: exactly when is sex environmentally determined? *Am. Nat.* 161: 676–683.

Valle, L. D., Ramina, A., Vianello, S., Belvedere, P. and Colombo, L. 2002. Cloning of two mRNA variants of brain aromatase cytochrome P450 in rainbow trout (*Oncorhynchus mykiss* Walbaum). *J. Steroid Biochem. Mol. Biol.* 82: 19–32.

Vandenbergh, J. G. 1994. Pheromones and mammalian reproduction. In E. Knobil and J. D. Neill, eds., *Physiology of Reproduction*, 2nd ed., vol. 2, 343–362. New York: Raven Press.

Vandenbergh, J. G. and Huggett, C. L. 1994. Mother's prior intrauterine position affects the sex ratio of her offspring in house mice. *Proc. Natl. Acad. Sci. U S A* 91: 11055–11059.

van de Poll, N. E. 1991. The physiological significance of acute activation of the gonadal hormonal axis. In T. Archer and S. Hansen, eds., *Behavioral Biology: Neuroendocrine Axis*, 69–74. Hillsdale, NJ: Lawrence Erlbaum Associates.

van der Westhuizen, L. A., Bennett, N. C. and Jarvis, J.U.M. 2002. Behavioural interactions, basal plasma luteinizing hormone concentrations and the differential pituitary responsiveness to exogenous gonadotrophin-releasing hormone in entire colonies of the naked mole-rat (*Heterocephalus glaber*). *J. Zool.* 256: 25–33.

Van Duyse, E., Pinxten, R. and Eens, M. 2002. Effects of testosterone on song, aggression, and nestling feeding behavior in male great tits, *Parus major. Horm. Behav.* 41: 178–186.

van Eerdenburg, F.J.C.M., et al. 1990. A vasopressin and oxytocin containing nucleus in the pig hypothalamus that shows neuronal changes during puberty. *J. Comp. Neurol.* 301: 138–146.

van Kesteren, R. E., Tensen, C. P., Smit, A. B., van Minnen, J., Kolakowski, L. F., Meyerhof, W., Richter, D., van Heerikhuizen, H., Vreugdenhil, E. and Geraerts, W. P. 1996. Co-evolution of ligand-receptor pairs in the vasopressin/oxytocin superfamily of bioactive peptides. *J. Biol. Chem.* 271: 3619–3626.

van Oortmerssen, G. A. and Bakker, T. C. 1981. Artificial selection for short and long attack latencies in wild *Mus musculus domesticus*. *Behav. Genet.* 11: 115–126.

Van Soest, P. F. and Kits, K. S. 2002. Role of lys-conopressin in the control of male sexual behavior in *Lymnaea stagnalis*. In D. W. Pfaff, A. P. Arnold, A. M. Etgen, S. E. Fahrbach and R. T. Rubin, eds., *Hormones, Brain and Behavior*, vol. 3, pp. 317–330. Amsterdam: Academic Press (Elsevier).

van Tienhoven, A. 1983. *Reproductive Physiology of Vertebrates*, 2nd ed. Ithaca, NY: Comstock (Cornell University Press).

Vasey, P. L. 1995. Homosexual behavior in primates: a review of evidence and theory. *Int. J. Primatol.* 16: 173–204.

Veenema, A. H., Meijer, O. C., de Kloet, E. R., Koolhaas, J. M., Bohus, B. G. 2003. Differences in basal and stress-induced HPA regulation of wild house mice selected for high and low aggression. *Horm. Behav.* 43: 197–204.

Vega Matuszczyk, J. V. and Larsson, K. 1995. Sexual preference and feminine and masculine sexual behavior of male rats prenatally exposed to antiandrogen or antiestrogen. *Horm. Behav.* 29: 191–206.

Vehrencamp, S. L. 1982. Testicular regression in relation to incubation effort in a tropical cuckoo. *Horm. Behav.* 16: 113–120.

Vehrencamp, S. L. 1983. Optimal degree of skew in cooperative societies. *Am. Zool.* 23: 327–335.

Vehrencamp, S. L., Bradbury, J. W. and Gibson, R. M. 1989. The energetic cost of display in male sage grouse. *Anim. Behav.* 38: 885–896.

Verhulst, S., Dieleman, S. J. and Parmentier, H. K. 1999. A tradeoff between immunocompetence and sexual ornamentation in domestic fowl. *Proc. Natl. Acad. Sci. U S A* 96: 4478–4481.

Vleck, C. M. and Brown, J. L. 1999. Testosterone and social and reproductive behaviour in *Aphelocoma* jays. *Anim. Behav.* 58: 943–951.

Vleck, C. M. and Dobrott, S. J. 1993. Testosterone, antiandrogen, and alloparental behavior in bobwhite quail foster fathers. *Horm. Behav.* 27: 92–107.

Vleck, C. M. and Patrick, D. J. 1999. Effects of vasoactive intestinal peptide on prolactin secretion in three species of passerine birds. *Gen. Comp. Endocrinol.* 113: 146–154.

Volny, V. P. and Gordon, D. M. 2002. Genetic basis for queen-worker dimorphism in a social insect. *Proc. Natl. Acad. Sci. U S A* 99: 6108–6111.

vom Saal, F. S. 1981. Variation in phenotype due to random intrauterine positioning of male and female fetuses in rodents. *J. Reprod. Fertil.* 62: 633–650.

vom Saal, F. S., Finch, C. E. and Nelson, J. F. 1994. Natural history and mechanisms of reproductive aging in humans, laboratory rodents, and other selected vertebrates. In E. Knobil and J. Neill et al., eds., *The Physiology of Reproduction*, 2nd ed., vol. 2, 1213–1314. New York: Raven Press.

von Engelhardt, N., Kappeler, P. M. and Heistermann, M. 2000. Androgen levels and female social dominance in *Lemur catta*. *Proc. R. Soc. Lond. B* 267: 1533–1539.

von Holst, D. 1998. The concept of stress and its relevance for animal behavior. *Adv. Study Behav.* 27: 1–131.

Wachtel, S. S., ed. 1994. *Molecular Genetics of Sex Determination*. San Diego: Academic Press.

Wade, G. N. 1976. Sex hormones, regulatory behaviors, and body weight. *Adv. Study Behav.* 6: 201–279.

Wade, G. N. and Schneider, J. E. 1992. Metabolic fuels and reproduction in female mammals. *Neurosci. Biobehav. Rev.* 16: 235–272.

Wade, J. 1999. Sexual dimorphisms in avian and reptilian courtship: two systems that do not play by the mammalian rules. *Brain Behav. Evol.* 54: 15–27.

Wade, J. and Arnold, A. P. 1994. Post-hatching inhibition of aromatase activity does not alter sexual differentiation of the zebra finch song system. *Brain Res.* 639: 347–350.

Wade, J. and Arnold, A. P. 1996. Functional testicular tissue does not masculinize development of the zebra finch song system. *Proc. Natl. Acad. Sci. U S A* 93: 5264–5268.

Wade, J., Huang, J.-M. and Crews, D. 1993. Hormonal control of sex differences in the brain, behavior and accessory sex structures of whiptail lizards (*Cnemidophorus* species). *J. Neuroendocrinol.* 5: 81–93.

Wade, J., Gong, A. and Arnold, A. P. 1997. Effects of embryonic estrogen on differentiation of the gonads and secondary sexual characteristics of male zebra finches. *J. Exp. Zool.* 278: 405–411.

Wade, J., Swender, D. A. and McElhinny, T. L. 1999. Sexual differentiation of the zebra finch song system parallels genetic, not gonadal, sex. *Horm. Behav.* 36: 141–152.

Wade, J., Buhlman, L. and Swender, D. 2002. Post-hatching hormonal modulation of a sexually dimorphic neuromuscular system controlling song in zebra finches. *Brain Res.* 929: 191–201.

Waddington, C. H. 1939. *An Introduction to Modern Genetics.* London: Allen & Unwin.

Wagner, H. and Luksch, H. 1998. Effect of ecological pressures on brains: examples from avian neuroethology and general meanings. *Z. Naturforsch.* 53: 560–581.

Wake, D. B. 1991. Homoplasy: the result of natural selection, or evidence of design limitations? *Am. Nat.* 138:543–567.

Walker, C.-D., Welberg, L.A.M. and Plotsky, P. M. 2002. Glucocorticoids, stress, and development. In D. W. Pfaff, A. P. Arnold, A. M. Etgen, S. E. Fahrbach and R. T. Rubin, eds., *Hormones, Brain and Behavior*, vol. 4, pp. 487–534. Amsterdam: Academic Press (Elsevier).

Wallen, K. 1990. Desire and ability: Hormones and the regulation of female sexual behavior. *Neurosci. Biobehav. Rev.* 14: 233–241.

Wallen, K. 1996. Nature needs nurture: the interaction of hormonal and social influences on the development of behavioral sex differences in rhesus monkeys. *Horm. Behav.* 30: 364–378.

Wallen, K. 1998. Ovarian influences on female development: revolutionary or evolutionary? *Brain Behav. Sci.* 21: 339–340.

Wallen, K. 2001. Sex and context: hormones and primate sexual motivation. *Horm. Behav.* 40: 339–357.

Wallen, K. 2000. Risky business: social context and hormonal modulation of primate sexual behavior. In K. Wallen and J. E. Schneider, eds., *Reproduction in Context: Social and Environmental Influences on Reproductive Physiology and Behavior*, 289–324. Cambridge, MA: MIT Press.

Wallen, K. and Baum, M. J. 2002. Masculinization and defeminization in altricial and precocial mammals: comparative aspects of steroid hormone action. In D. W. Pfaff,

A. P. Arnold, A. M. Etgen, S. E. Fahrbach and R. T. Rubin, eds., *Hormones, Brain and Behavior*, vol. 4, 385–424. Amsterdam: Academic Press (Elsevier).

Wang, Z.-Y., Seto, H., Fujioka, S., Yoshida, S. and Chory, J. 2001. BRI1 is a critical component of a plasma-membrane receptor for plant steroids. *Nature* 410: 380–383.

Ward, I. L., Ward, O. B., Winn, R. J. and Bielawski, D. 1994. Male and female sexual behavior potential of male rats prenatally exposed to the influence of alcohol, stress, or both factors. *Behav. Neurosci.* 108: 1188–1195.

Warner, R. R. 1988. Sex change in fishes: hypotheses, evidence, and objections. *Environ. Biol. Fishes* 22: 81–90.

Warner, R. R. 2002. Synthesis: Environment, mating systems, and life-history allocations in the bluehead wrasse. In L. Dugatkin, ed., *Model Systems in Behavioral Ecology*, 227–244. Princeton, NJ: Princeton University Press.

Warren, R. P. and Hinde, R. A. 1959. The effect of oestrogen and progesterone on the nest-building of domesticated canaries. *Anim. Behav.* 7: 209–213.

Wasser, S. K. and Barash, D. P. 1983. Reproductive suppression among female mammals: implications for biomedicine and sexual selection theory. *Q. Rev. Biol.* 58: 513–538.

Watson, J. T. and Adkins-Regan, E. 1989a. Activation of sexual behavior by implantation of testosterone propionate and estradiol benzoate into the preoptic area of the male Japanese quail *(Coturnix japonica)*. *Horm. Behav.* 23: 251–268.

Watson, J. T. and Adkins-Regan, E. 1989b. Testosterone implanted in the preoptic area of male Japanese quail must be aromatized to activate copulation. *Horm. Behav.* 23: 432–447.

Watson, J. T. and Kelley, D. B. 1992. Testicular masculinization of vocal behavior in juvenile female *Xenopus laevis* reveals sensitive periods for song duration, rate, and frequency spectra. *J. Comp. Physiol.* A 171: 343–350.

Watson, J. T., Robertson, J., Sachdev, U. and Kelley, D. B. 1993. Laryngeal muscle and motor neuron plasticity in *Xenopus laevis*: testicular masculinization of a developing neuromuscular system. *J. Neurobiol.* 24: 1615–1625.

Weeks, J. C. and Levine, R. B. 1995. Steroid hormone effects on neurons subserving behavior. *Curr. Opin. Neurobiol.* 5: 809–815.

Weiger, W. A. 1997. Serotonergic modulation of behaviour: a phylogenetic overview. *Biol. Rev.* 72: 61–95.

Weir, B. J. 1970. The management and breeding of some more hystricomorph rodents. *Lab. Anim.* 4: 83–97.

Wells, M. J. and Wells, J. 1972. Sexual displays and mating of *Octopus vulgaris* Cuvier and *O. cyanea* Gray and attempts to alter performance by manipulating the glandular condition of the animals. *Anim. Behav.* 20: 293–308.

Wennstrom, K. L. and Crews, D. 1998. Effect of long-term castration and long-term androgen treatment on sexually dimorphic estrogen-inducible progesterone receptor mRNA levels in the ventromedial hypothalamus of whiptail lizards. *Horm. Behav.* 34: 11–16.

Wennstrom, K. L., Reeves, B. J. and Brenowitz, E. A. 2001. Testosterone treatment increases the metabolic capacity of adult avian song control nuclei. *J. Neurobiol.* 48: 256–264.

Whalen, R. E. 1982. Current issues in the neurobiology of sexual differentiation. In A. Vernadakis and P. S. Timiras, eds., *Hormones in Development and Aging*, 273–304. New York: Spectrum.

Whaling, C. S., Nelson, D. A. and Marler, P. 1995. Testosterone-induced shortening of the storage phase of song development in birds interferes with vocal learning. *Dev. Psychobiol.* 28: 367–376

Whitacre, C. C., Reingold, S. C. and O'Looney, P. A. 1999. A gender gap in autoimmunity. *Science* 283: 1277–1278.

White, S. A., Nguyen, T. and Fernald, R. D. 2002. Social regulation of gonadotropin-releasing hormone. *J. Exp. Biol.* 205: 2567–2581.

Whittier, J. M. and Tokarz, R. R. 1992. Physiological regulation of sexual behavior in female reptiles. In C. Gans and D. Crews, eds., *Biology of the Reptilia*, vol. 18, *Physiology E: Hormones Brain and Behavior*, 24–69. Chicago: University of Chicago Press.

Whittingham, L. A. and Schwabl, H. 2002. Maternal testosterone in tree swallow eggs varies with female aggression. *Anim. Behav.* 63: 63–67.

Wibbels, T. and Crews, D. 1994. Putative aromatase inhibitor induces male sex determination in a female unisexual lizard and in a turtle with temperature-dependent sex determination. *J. Endocrinol.* 141: 295–299.

Wikelski, M. and Ricklefs, R. E. 2001. The physiology of life histories. *Trends Ecol. Evol.* 16: 479–481.

Wikelski, M., Hau, M. and Wingfield, J. C. 1999. Social instability increases plasma testosterone in a year-round territorial neotropical bird. *Proc. R. Soc. Lond. B* 266: 551–556.

Wiley, R. H., Steadman, L., Chadwick, L. and Wollerman, L. 1999. Social inertia in white-throated sparrows results from recognition of opponents. *Anim. Behav.* 57: 453–463.

Wilkins, A. S. 2002. *The Evolution of Developmental Pathways*. Sunderland, MA: Sinauer.

Wilkinson, G. 1992. Communal nursing in the evening bat, *Nycticeius humeralis*. *Behav. Ecol. Sociobiol.* 31: 225–235.

Willeke, H., Claus, R., Mueller, E., Pirchner, F. and Karg, H. 1987. Selection for high and low level of 5-alpha androst-16-en-3-one in boars I. Direct and correlated response of endocrinological traits. *J. Anim. Breeding Genet.* 104: 64–73.

Williams, C. L., Barnett, A. M. and Meck, W. H. 1990. Organizational effects of early gonadal secretions on sexual differentiation in spatial memory. *Behav. Neurosci.* 104: 84–97.

Williams, G. C. 1957. Pleiotropy, natural selection and the evolution of senescence. *Evolution* 11: 398–411.

Williams, G. C. 1966. *Adaptation and Natural Selection: A Critique of Some Current Evolutionary Thought*. Princeton, NJ: Princeton University Press.

Williams, J. R., Insel, T. R., Harbaugh, C. R. and Carter, C. S. 1994. Oxytocin administered centrally facilitates formation of a partner preference in female prairie voles (*Microtus ochrogaster*). *J. Neuroendocrinol.* 6: 247–250.

Williams, R. W. 2000. Mapping genes that modulate mouse brain development: a quantitative genetic approach. In A. F. Goffinet and P. Rakic, eds., *Mouse Brain Development*, 21–49. New York: Springer.

Williams, T. D., Kitaysky, A. S. and Vézina, F. 2004. Individual variation in plasma estradiol-17β and androgen levels during egg formation in the European starling *Sturnus vulgaris*: implications for regulation of yolk steroids. *Gen. Comp. Endocr.* 136: 346–352.

Wilson, J. A. and Glick, B. 1970. Ontogeny of mating behavior in the chicken. *Am. J. Physiol.* 218: 951.

Wilson, J. D., Leshin, M. and George, F. W. 1987. The Sebright bantam chicken and the genetic control of extraglandular aromatase. *Endocrine Rev.* 8: 363–376.

Wilson, J. D., George, F. W. and Renfree, M. B. 1995. The endocrine role in mammalian sexual differentiation. *Recent Progr. Horm. Res.* 50: 349–364.

Wingfield, J. C. 1984. Androgens and mating systems: testosterone-induced polygyny in normally monogamous birds. *Auk* 101: 665–671.

Wingfield, J. C. 1988. Changes in reproductive function of free-living birds in direct response to environmental perturbations. In M. H. Stetson, ed., *Processing of Environmental Information in Vertebrates*, 121–148. Berlin: Springer-Verlag.

Wingfield, J. C. 1994a. Hormone-behavior interactions and mating systems in male and female birds. In R. V. Short and E. Balaban, eds., *The Differences Between the Sexes*, 303–330. Cambridge, UK: Cambridge University Press.

Wingfield, J. C. 1994b. Control of territorial aggression in a changing environment. *Psychoneuroendocrinology* 19: 709–721.

Wingfield, J. C. and Farner, D. S. 1993. Endocrinology of reproduction in wild species. In D. S. Farner, J. R. King and K. C. Parkes, eds., *Avian Biology*, vol. IX, 164–328. London: Academic Press.

Wingfield, J. C. and Sapolsky, R. M. 2003. Reproduction and resistance to stress: when and how. *J. Neuroendocrinol.* 15: 711–724.

Wingfield, J. C. and Silverin, B. 1986. Effects of corticosterone on territorial behavior of free-living male song sparrows *Melospiza melodia*. *Horm. Behav.* 20: 405–417.

Wingfield, J. C. and Silverin, B. 2002. Ecophysiological studies of hormone-behavior relations in birds. In D. W. Pfaff, A. P. Arnold, A. M. Etgen, S. E. Fahrbach and R. T. Rubin, eds., *Hormones, Brain and Behavior*, vol. 2, 587–648. Amsterdam: Academic Press (Elsevier).

Wingfield, J. C., Hegner, R. E., Dufty, A. M. and Ball, G. F. 1990. The "challenge hypothesis": theoretical implications for patterns of testosterone secretion, mating systems, and breeding strategies. *Am. Nat.* 136: 829–846.

Wingfield, J. C., Hahn, T. P., Levin, R. and Honey, P. 1992. Environmental predictability and control of gonadal cycles in birds. *J. Exp. Zool.* 261: 214–231.

Wingfield, J. C., Whaling, C. S. and Marler, P. 1994. Communication in vertebrate aggression and reproduction: the role of hormones. In E. Knobil and J. D. Neill, eds., *The Physiology of Reproduction*, 2nd ed., vol. 2, 303–342. New York: Raven Press.

Wingfield, J. C., Maney, D. L., Breuner, C. W., Jacobs, J. D., Lynn, S., Ramenofsky, M. and Richardson, R. D. 1998. Ecological bases of hormone-behavior interactions: the "emergency life history stage." *Am. Zool.* 38: 191–206.

Wingfield, J. C., Lynn, S. E. and Soma, K. K. 2001. Avoiding the "costs" of testosterone: ecological bases of hormone-behavior interactions. *Brain Behav. Evol.* 57: 239–251.

Winkler, D. W. 1993. Testosterone in egg yolks: an ornithologist's perspective. *Proc. Natl. Acad. Sci. U S A* 90: 11439–11441.

Winkler, S. M. and Wade, J. 1998. Aromatase activity and regulation of sexual behaviors in the green anole lizard. *Physiol. Behav.* 64: 723–731.

Winslow, J. T., Hastings, N., Carter, C. S., Harbaugh, C. R. and Insel, T. R. 1993. A role for central vasopressin in pair bonding in monogamous prairie voles. *Nature* 365: 545–548.

Wise, T., Young, L. D. and Pond, W. G. 1993. Reproductive, endocrine, and organ weight differences of swine selected for high or low serum cholesterol. *J. Anim. Sci.* 71: 2732–2738.

Witschi, E. 1961. Sex and secondary sexual characters. In A. J. Marshall, ed., *Biology and Comparative Physiology of Birds*, vol. II, 115–168. New York: Academic Press.

Witt, D. M. 1997. Regulatory mechanisms of oxytocin-mediated sociosexual behavior. In C. S. Carter, I. I. Lederhendler and B. Kirkpatrick, eds., *The Integrative Neurobiology of Affiliation*, 287–301 (*Annals of the New York Academy of Sciences*, vol. 807). New York: New York Academy of Sciences.

Witte, K. and Sawka, N. 2003. Sexual imprinting on a novel trait in the dimorphic zebra finch: sexes differ. *Anim. Behav.* 65: 195–203.

Wolff, E. 1959. Endocrine function of the gonad in developing vertebrates. In A. Gorbman, ed., *Comparative Endocrinology*, 568–581. New York: Wiley.

Wood, D. E., Gleason, R. A. and Derby, C. D. 1995. Modulation of behavior by biogenic amines and peptides in the bluecrab, *Callinectes sapidus. J. Comp. Physiol.* 177A: 321–333.

Wood, R. I. 2004. Reinforcing aspects of androgens. *Physiol. Behav.* 83:279–289.

Wood, R. I. and Newman, S. W. 1995. Integration of chemosensory and hormonal cues is essential for mating in the male Syrian hamster. *J. Neurosci.* 15: 7261–7269.

Wood, R. I., Ebling, F. J., I'Anson, H., Bucholtz, D. C., Yellon, S. M. and Foster, D. L. 1991. Prenatal androgens time neuroendocrine sexual maturation. *Endocrinology* 128: 2457–2468.

Woodley, S. K. and Moore, M. C. 1999. Ovarian hormones influence territorial aggression in free-living female mountain spiny lizards. *Horm. Behav.* 35: 205–214.

Woods, J. E. 1987. Maturation of the hypothalamo-adenohypophyseal-gonadal (HAG) axes in the chick embryo. *J. Exp. Zool. Suppl.* 1: 265–271.

Woodson, J. C. 2002. Including "learned sexuality" in the organization of sexual behavior. *Neurosci. Biobehav. Rev.* 26: 69–80.

Woodson, J. C. and Gorski, R. A. 1999. Structural sex differences in the mammalian brain: reconsidering the male/female dichotomy. In A. Matsumoto, ed., *Sexual Differentiation of the Brain*, 229–256. Boca Raton, FL: CRC Press.

Woolley, C. S. and Cohen, R. S. 2002. Sex steroids and neuronal growth in adulthood. In In D. W. Pfaff, A. P. Arnold, A. M. Etgen, S. E. Fahrbach and R. T. Rubin, eds., *Hormones, Brain and Behavior*, vol. 4, 717–778. Amsterdam: Academic Press (Elsevier).

Wray, G. A. 2002. Do convergent developmental mechanisms underlie convergent phenotypes? *Brain Behav. Evol.* 59: 327–336.

Wu, K. H., Tobias, M. L., Thornton, J. W. and Kelley, D. B. 2003. Estrogen receptors in *Xenopus*: duplicate genes, splice variants, and tissue-specific expression. *Gen. Comp. Endocrinol.* 133: 38–49.

Wynne-Edwards, K. E. 2003. From dwarf hamster to daddy: the intersection of ecology, evolution, and physiology that produces paternal behavior. *Adv. Study Behav.* 32: 207–261.

Wynne-Edwards, K. E. and Reburn, C. J. 2000. Behavioral endocrinology of mammalian fatherhood. *Trends Ecol. Evol.* 15: 464–468.

Yahr, P. and Commins, D. 1983. The neuroendocrinology of scent marking. In Silverstein, R. M. and Muller-Schwarze, D., eds., *Chemical Signals in Vertebrates III.* Plenum, New York.

Yamamoto, T. 1969. Sex differentiation. In W. S. Hoar and D. J. Randall, eds., *Fish Physiology,* vol. 3, 117–175. New York: Academic Press.

Yang, N., Dunnington, E. A. and Siegel, P. B. 1998. Forty generations of bidirectional selection for mating frequency in male Japanese quail. *Poultry Sci.* 77: 1469–1477.

Yang, N., Dunnington, E. A. and Siegel, P. B. 1999. Heterosis following long-term bidirectional selection for mating frequency in male Japanese quail. *Poultry Sci.* 78: 1252–1256.

Yazaki, Y., Matsushima, T. and Aoki, K. 1999. Testosterone modulates stimulation-induced calling behavior in Japanese quails. *J. Comp. Physiol. A* 184: 13–19.

Young, K. M., Brown, J. L. and Goodrowe, K. L. 2001. Characterization of reproductive cycles and adrenal activity in the black-footed ferret (*Mustela nigripes*) by fecal hormone analysis. *Zoo Biol.* 20: 517–536.

Young, L. J. and Crews, D. 1995. Comparative neuroendocrinology of steroid receptor gene expression and regulation: relationship to physiology and behavior. *Trends Endocrinol. Metab.* 6: 317–323.

Young, L. J., Wang, Z. and Insel, T. R. 1998. Neuroendocrine bases of monogamy. *Trends Neurosci.* 21: 71–75.

Young, L. J., Nilsen, R., Waymire, K. G., MacGregor, G. R. and Insel, T. R. 1999. Increased affiliative response to vasopressin in mice expressing the V_{1a} receptor from a monogamous vole. *Nature* 400: 766–768.

Young, L. J., Lim, M. M., Gingrich, B. and Insel, T. R. 2001. Cellular mechanisms of social attachment. *Horm. Behav.* 40: 133–138.

Young, W. C. 1965. The organization of sexual behavior by hormonal action during the prenatal and larval periods in vertebrates. In F. A. Beach, ed., *Sex and Behavior,* 89–107. New York: Wiley.

Young, W. C., Goy, R. W. and Phoenix, C. H. 1964. Hormones and sexual behavior. *Science* 143: 212–218.

Zahavi, A. J. 1977. The cost of honesty (further remarks on the handicap principle). *Theor. Biol.* 67: 603–605.

Zakon, H. H. 1998. The effects of steroid hormones on electrical activity of excitable cells. *Trends Neurosci.* 21: 202–207.

Zann, R. A. 1996. *The Zebra Finch: A Synthesis of Field and Laboratory Studies.* Oxford, UK: Oxford University Press.

Zeh, J. A. and Zeh, D. W. 2003. Toward a new sexual selection paradigm: polyandry, conflict and incompatibility. *Ethology* 109: 929–950.

Zeifman, D. and Hazan, C. 1997. Attachment: The bond in pair-bonds. In J. A. Simpson and D. T. Kenrick, eds., *Evolutionary Social Psychology,* 237–263. Mahwah, NJ: Erlbaum.

Zhang, F.-P., Pakarainen, T., Poutanen, M., Toppari, J. and Huhtaniemi, I. 2003. The low gonadotropin-independent constitutive production of testicular testosterone is sufficient to maintain spermatogenesis. *Proc. Natl. Acad. Sci. U S A* 100: 13692–13697.

Zhang, L. I. and Poo, M.-M. 2001. Electrical activity and development of neural circuits. *Nature Neurosci.* 4 (suppl.): 1207–1214.

Ziegler, T. E. 2000. Hormones associated with non-maternal infant care: a review of mammalian and avian studies. *Folia Primatol.* 71: 6–21.

Zielinski, W. J., vom Saal, F. S. and Vandenbergh, J. G. 1992. The effect of intrauterine position on the survival, reproduction and home range size of female house mice *Mus musculus. Behav. Ecol. Sociobiol.* 30: 185–191.

Zikopoulos, B., Kentouri, M. and Dermon, C. R. 2000. Proliferation zones in the adult brain of a sequential hermaphrodite teleost species (*Sparus aurata*). *Brain Behav. Evol.* 56: 310–322.

Zucker, K. J. 1999. Intersexuality and gender identity differentiation. *Annu. Rev. Sex Res.* 10: 1–69.

Zuk, M., Johnsen, T. S. and Maclarty, T. 1995. Endocrine-immune interactions, ornaments and mate choice in red jungle fowl. *Proc. R. Soc. Lond. B* 260: 205–210.

Zumpe, D. and Michael, R. P. 1996. Social factors modulate the effects of hormones on the sexual and aggressive behavior of macaques. *Am. J. Primatol.* 38: 233–261.

Zupanc, G. K. 2001. A comparative approach towards the understanding of adult neurogenesis. *Brain Behav. Evol.* 58: 246–249.

Index

Milton Keynes UK
Ingram Content Group UK Ltd.
UKHW020242250924
448780UK00006B/106

9 780691 092478